BIOLOGY OF FISHES

BIOLOGY
OF
FISHES

By

Dr. D.R. Khanna

Reader

Reader in Zoology

Gurukul Kangri University

Haridwar (Uttaranchal)

&

Dr. P.R Yadav

Reader

Department of Zoology

D.A.V. College

Muzaffarnagar (U.P.)

D P H

DISCOVERY PUBLISHING HOUSE

NEW DELHI-110002

First Published-2004

ISBN 81-7141-908-9

Published by

DISCOVERY PUBLISHING HOUSE
4831/24, Ansari Road, Prahlad Street,
Darya Ganj, New Delhi-110002 (India)
Phone: 23279245 • Fax: 91-11-23253475
E-mail:dphtemp@indiatimes.com

Printed at:
Tarun Offset Printers,
Delhi-110 053

PREFACE

Fishes are the most widely distributed vertebrates in the vast water areas of the earth. The present title deals with the broad principles of Ichthyology including morphology, anatomy, physiology and development of fishes. The text is the coverage, commensurate, concise and comprehensive presentation. It is hoped that this publication will adequately meet the long felt need of the students for a simple and upto date book. The method of presentation is also very simple so as to impact to the students a clear and vivid understanding. Profusely illustrated the authoritative information contained in this text will prove of utmost use to the students, teachers and researchers.

To make the work more comprehensive and informative, we have consulted many authoritative books, research journals, abstracts, monographs etc. We are grateful to all those great scholars whose work are cited or substantially reproduced.

There can be no claim to originality except in the manner of treatment and much of the information has been obtained from the books and scientific journals available in the different libraries.

The authors express their thanks to their friends and colleagues whose continue inspirations have initiated them to bring out this book.

The authors are painfully aware of the shortcomings, errors and misprints that have crept in, and shall be greateful to receive suggestion for improvement of the next edition from all the readers.

The authors express their gratitute to Mr. Wasan and staff of M/s Discovery Publishing House for their whole hearted co-operation in the publication of this book.

Authors

Contents

1

FIRST APPEARANCE OF FISHES

The history of life, like human history, has been marked by certain great development rising above the general level of events that collectively make up the record. These outstanding evolutionary developments are in the nature of revolutions, affecting profoundly the phylogenetic trends that follow them, just as great historical revolutions, like the American Revolt against Britain or the French revolutions, have affected the subsequent histories of the peoples concerned with them, A better comparison might be with the peaceful revolution in human arts and industries, such as those brought about by the development of the printing press or the application of steam power to machinery.

Similarly animal life has also a significant history. The first vertebrates known in the fossil record are indicated, by scales that have been found in Ordovician freshwater sediments in Colorado. These remains are fragmentary and give no clue as to the nature of the animals represented by them; yet they are sufficiently well preserved to be studied. When placed under a microscope, they show a bony structure. Therefore, it is obvious from these scales, unsatisfactory and tantalizing though they may be as evidence of the first vertebrates in the geological record, that in Ordovician times there was armored vertebrates living in shallow rivers and lakes. The next evidence of vertebrate life comes from England in rocks of Middle Silurian ago. Two genera, *Jamoytius* and *Thelodus*, are represented by completed skeletons which, because of the manner of their preservation, are not easy to interpret. *Jamoytius* is of special interest because it appears to be a very primitive jawless vertebrate, perhaps, occupying a position close to the ancestry of the lamprey and its relatives.

The few, rather enigmatic fossil remains of *Jamoytius* show that this animal was small, elongated and tubular shaped. It had a terminal, sectorial mouth, and on each side of the head region, there was a row of circular gill openings, behind the eye. There was a propulsive tail fin with a long lower lobe and a shorter, deep upper lobe, and possibly there were lateral fin folds and a long dorsal fin, for maintaining balance. *Jamoytius* and *Thelodus* are found together in marine sediments. One of the great events or revolutions in the history of the vertebrates was the appearance of the jaws. The importance of this evolutionary development can hardly be overestimated, for it opened to the vertebrates new lines of adaptation and new possibilities of evolutionary advancement that expanded immeasurably the potentialities of these animals.

The jawless vertebrates were definitely restricted as to their adaptations for different modes of life, and it is possible that the ostracoderms of late Silurian and Devonian times had explored and virtually exhausted the evolutionary possibilities and the ecological niches open to animals of this type. Animal without jaws can evolve as bottom feeders, as did some of the ostracoderms, or they can develop movable plates around the mouth opening that serve after a fashion as weak "jaws," as did some other ostracoderms. Or they can become parasites, like the modern lampreys and hagfishes. Yet even under the most favourable conditions they are denied the possibilities of development that are open to animals with jaws; they simply do not have the structural mechanisms to exploit the opportunities that are available to jawed animals. So it is that the appearance of jaws marked

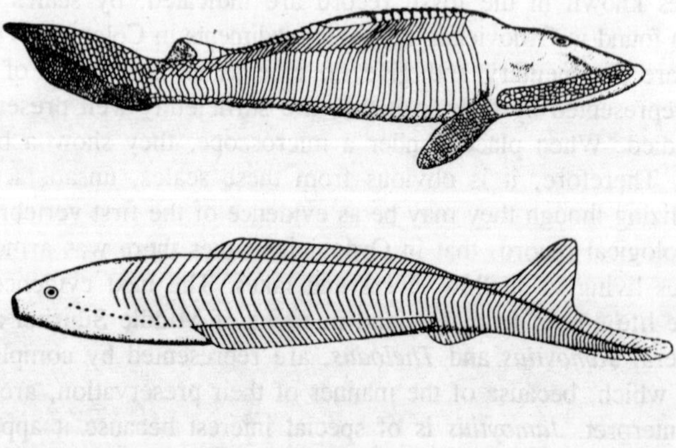

Fig. 1.1. Fossil ostracoderms. A—Hemicyclaspis, a member of the Osteostraci; B—Jamoytius, two members of the Anaspida.

a major turning point in vertebrate evolution. The appearance of the jaws in vertebrates was brought about by a transformation of anatomical elements that originally had performed a function quite different from the function of food gathering. Here we see the working of a process that has taken place innumerable times, in the course of animal evolution; indeed, much of the progress from earlier to later forms of animals has been brought about through the transformation of structures from one function to another.

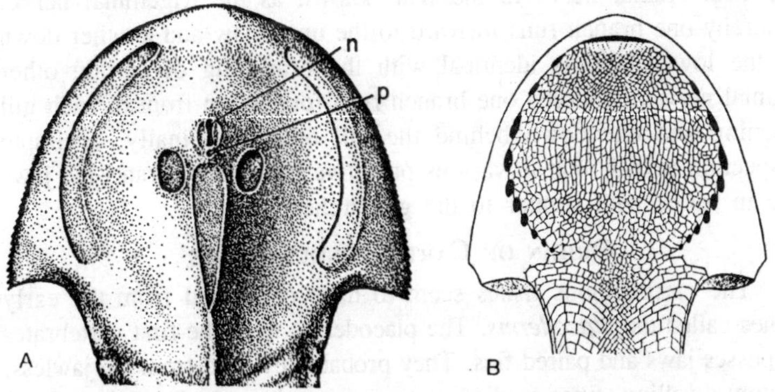

Fig. 1.2. A—Dorsal and B—ventral views of the head region of a fossil ostracoderm of the Cephalaspis type. Dorsally are seen openings for the paired eyes, median eye (p), and a median slit (n) for nostril and hypophyseal sac.

The origin and evolution of the jaws are an excellent example of this evolutionary principle. The jaws were derived originally from gill arched. It will be remembered that the ostracoderms had a large number of gills, as many as ten in *Cephalaspis*, and it seem probable that the presence of comparatively numerous gills with cartilaginous or bony supports was typical of the primitive vertebrates. At an early stage in the history of the vertebrates at least one and probably two of the original anterior gill 'arches" were eliminated, while another arch, probably the third one in the series, was changed from a gill support into a pair of jaws. This transformation actually was not as radical a shift as on first sight it may appear to have been. Each gill support or arch in the primitive vertebrate was formed by a series of several bones, arranged somewhat in the fashion of a V turned on its side, with the point directed posteriorly. Imagine several such lazy V's in series, thus : > > >. Imagine also the first of the these (morphologically the third of the original gill arches) supplied with teeth, and hinged at the point of the V, and you have the primitive vertebrate jaws, in alignment with the gill arches behind them.

It is immediately apparent that the transformation of a gill arch into jaws was a natural evolutionary development, perhaps the simplest possible to the basic problem of developing a pair of vertebrate jaws. There are various facts that support such an origin for the jaws in the vertebrates. For instance, a study of the embryological development of certain modern fishes indicates this origin very strongly. Moreover, the arrangement of the nerves in the head region of sharks shows that the jaws are in series with the gill apparatus. Thus the branching of the fifth cranial nerve in the fish, known as the trigeminal nerve, whereby one branch runs forward to the upper jaw and another down to the lower jaw, is identical with the branching of certain other cranial nerves, whereby one branch runs forward in front of each gill opening and one down behind the gill opening. Finally, a simple inspection reveals that in various primitive jawed vertebrates the jaws are in series with similar to the gill arches.

Origin of Cartilaginous Fishes

The cartilaginous fishes seem to have originated from the early fishes called the *placoderms*. The placoderms were the first vertebrates to posses jaws and paired fins. They probably evolved from the jawless, bottom-dwelling, filter feeding, ostracoderms in early Devonian period i.e., about 350 million yeas ago. However, no connections link between the jawless and the jawed fishes is available in the fossil record. The placoderms were small, mainly freshwater fishes. They retained the body armour of the ancestral ostracoderms. The armour consisted of large plates over the head and anterior part of the trunk and scales elsewhere. The endoskeleton had considerable ossification. The visceral arches were, however, cartilaginous. The notochord persisted throughout life. The first pair of gill-slits were not reduced. The second pair of gill arches were unaffected by jaws.

Paired fins were present and unpaired fins included one or two dorsal fins, a caudal fin and an anal fin. The tail was heterocercal. The jaws developed from the first pair of gill, arches in front of the first gill-slits. Their formation affected neither the first pair of gill-slits nor the second pair of gill-arches. In later fishes, both these structures changed due to posterior expansion of the jaws. Appearance of jaws caused a revolution in the vertebrate feeding as now the fishes could leave filter feeding and resort to other modes of feeding and spread into a variety of habitats. Many of them become predaceous and this led to the development of a more efficient muscular system for quick locomotion. Efficient locomotion in turn required and resulted

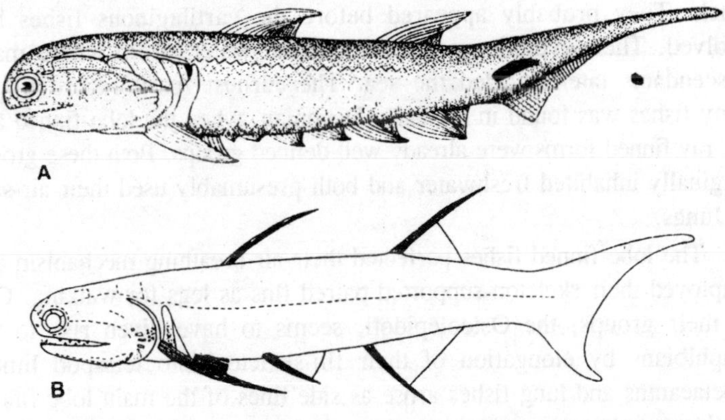

Fig. 1.3. Acanthodians, primitive jawed fishes probably related to the modern *Osteichthyes. A—Climatius, a form with heavy scales and very large fin spines; B—Ischnacanthus, one with greatly reduced armor and thin spines.*

in an improvement in the respiratory, circulatory, sensory and nervous system. The paired fins may have first appeared in the form of lateral folds, such as those found in the primitive ostracoderm *Jamoytius*, These folds may have broken up into a series of small paired fins such as are found in certain primitive placoderms e.g., *Climatius*.

The anterior and posterior pairs were larger than other, and ultimately developed into the pectoral and pelvic fins respectively. The paired fins between these appeared, some paleontology do not believe in the above "fin fold" theory of the origin of paired fins, and hold that these numerous paired fins may have originated as separate structures. Specific placoderm ancestors of higher fishes are not known. It is, however, possible that the cartilaginous fishes may have evolved from the stegoselachian stock and the bony fishes from the acanthodian stock. Evolution of both the groups occurred in freshwater, but many from later took to marine life. The cartilaginous fishes are known in the fossil records from the mid-Devonian period and the bony fishes from the early Devonian period. This indicates that the bony fishes are oeder than the cartilaginous fishes. This inference is further supported by the fact that the possession of bone is a primitive character and its absence in the cartilaginous fishes is a case degeneration.

Origin of Bony Fishes

Bony fishes, as mentioned earlier, arose from the early jawed fishes, the placoderms, which themselves originated from the jawless ostracoderms, specific placoderm ancestor of bony fishes is not known, but it seem possible that they may have evolved from the acanthodian stock. They probably appeared before the cartilaginous fishes had evolved. Their evolution occurred in fresh water, from where many descendants later invaded the sea. The earliest fossil record of the bony fishes was found in the early Devonian, when the lobe-finned and the ray-finned forms were already well-defined groups. Both these groups originally inhabited freshwater and both presumably used their air-sacs as Jungs.

The lobe-finned fishes perfected their air-breathing mechanism and employed their skeleton-supported paired fins as legs for walking. One of their groups, the Osteolepidoti, seems to have given rise to the amphibians by elongation of their fin-skeleton into tetrapod limbs. Coelacanths and lung-fishes arose as side lines of the main lobe-finned stock terminating in amphibians through the osteolepids. The ray-finned fishes started using air-sacs as hydrostatic organs and evolved through palaeoniscoid forms into Chondrostei, Holostei and Peleostei.

Evolution of Fishes

In a general way it is evident from the preceding summary of structure and classification that the Shark-like fishes or Elasmobranchs are the most primitive of existing fishes, and that the bony fishes can be derived from these by modifications of the various organs; also that the Ganoids are the more primitive of the bony fishes, and the Teleostei the most modified. Among the Ganoids, the "fringe-finned" forms or Crossopterygii are the most primitive, being nearest to the Elasmobranchs in the structure of the paired fins. The Dipnoi, or lung-fishes, on the other hand, might be supposed to be the latest stage in the evolution of fishes, especially on account of the advance stage of evolution shown by the lungs in adaptation to atmospheric respiration; but on the other hand in structure the Dipnoi show unmistakable affinities to the Crossopterygii.

In order, however, to discuss the evolution of fishes we must consider the evidence afforded by the chronological succession of fossil forms as well as the evidence of the structure of existing forms; we must study palaeontology as well as comparative anatomy, and include under the latter term the study of development or embryology. We will proceed therefore to give a brief outline of the most important facts concerning the palaeontology of fishes. There is no evidence of the derivation of fishes with jaws from those with sectorial mouths (Cyclostomes) resembling the lampreys and hags. Minute fossils known as Conodonts occur in the most ancient stratified rocks called by geologists palaeozoic, from the Lower Silurian to the Carboniferous

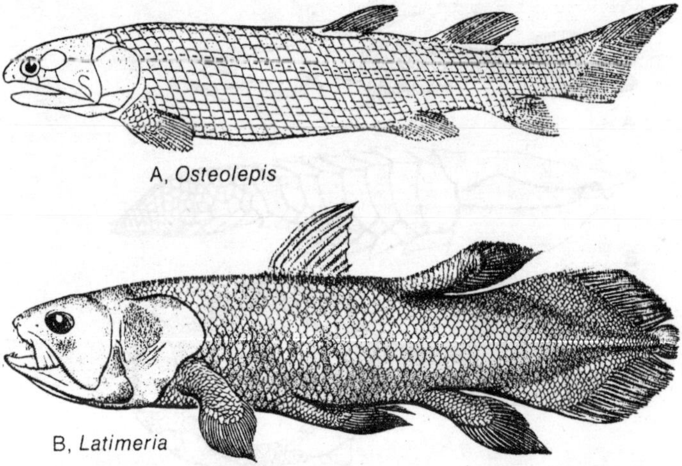

Fig. 1.4. Crossopterygians. A—Typical Devonian form; B—the only living coelacanth.

Limestone, and have been compared to the teeth of the Cyclostomes; but these fossils in microscopic structure do not resemble such teeth, and nothing can be said about them.

Small fossils about two inches long found by *Dr. Traquair* in the Caithness flag-stones near Thurso appear to have the structure of Cyclostomes, but they have a calcified internal skeleton, while the skeleton of existing Cyclostomes is soft and cartilaginous. Indeed, in deposits belonging to the Lower Silurian and Devonian, in Russia, England, and North America, minute, slender, pointed horny bodies, bent like a hook, with sharp opposite margins, have been found and described under the name of *Conodonts*. More frequently they possess an elongated basal portion, in which there is generally a larger tooth with rows of similar but smaller denticles on one or both sides of the larger tooth, according as this is central or at one end of the base. In other examples, there is no prominent central tooth, but a series or more or less similar teeth is implanted on a straight or curved base.

Modifications of these arrangements are very numerous, and many Palaeontologists entertain still doubts whether the origin of these remains is not rather from Annelids and Mollusks than from Fishes. The first undeniable evidence of a fish, or, indeed, of a vertebrate animal, occurs in the *Upper Silurian* Rocks, in a bone-bed of the Downton sandstone, near Ludlow. It consists of compressed, slightly curved, ribbed spines, of less than two inches in length (*Onchus*); of small shagreen-scales (*Thelodus*); the fragment of a jaw-like bar with

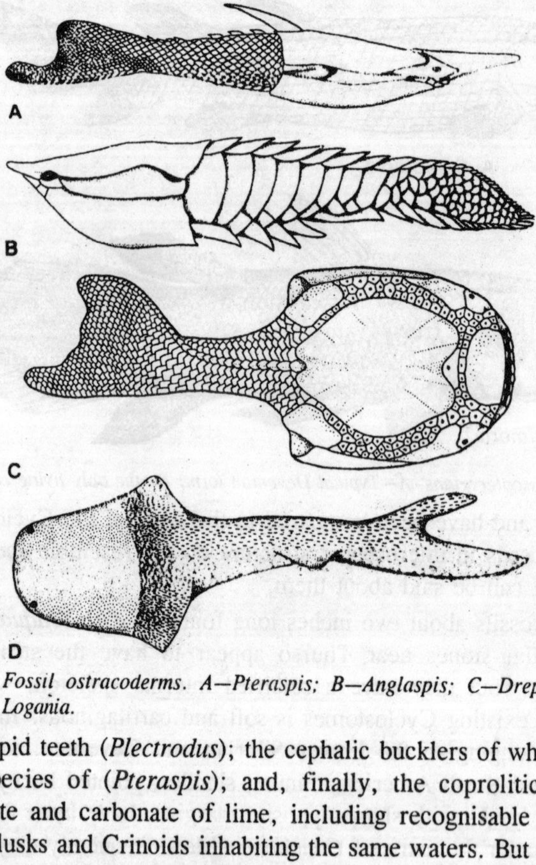

Fig. 1.5. Fossil ostracoderms. A—Pteraspis; B—Anglaspis; C—Drepanaspis; D—Logania.

pluricuspid teeth (*Plectrodus*); the cephalic bucklers of what seems to be a species of (*Pteraspis*); and, finally, the coprolitic bodies of phosphate and carbonate of lime, including recognisable remains of the Mollusks and Crinoids inhabiting the same waters. But no vertebra or other part of the skeleton has been found. The spines and scales seem to have belonged to the same kind of fish, which probably was a Plagiostome.

It is quite uncertain whether or not the jaw (if it be the jaw of a fish) belonged to the buckler-bearing I *teraspis*, the position of which among Ganoids, with which it is generally associate, is open to doubt. No detached undoubted tooth of a Plagiostome or Ganoid scale has been discovered in the Ludlow deposits; but so much is certain that those earliest remains in Palaeozoic rocks belonged to fishes closely allied to forms occurring in greater abundance in the succeeding formation, the Devonian, where they are associated with undoubted Palaeichtyes, Plagiostomes as well as Ganoids. These fish-remains of the *Devonian* or *Old Red Sandstone*, can be determined with greater certainty. They consist of spines or the so called *Ichthyodorulites*, which show sufficiently distinctive characters to be referred to several genera, one of them, *Onchus*, still surviving from the Silurian epoch.

All these spines are believed to be those of Chondropterygians, to which order some pluricuspid teeth (*Cladodus*) from the Old Sandstone in the vicinity of St. Petersburg have been referred likewise. The remains of the Ganoid fishes are in a much more perfect state of preservation, so that it is even possible to obtain a tolerably certain idea of the general appearance and habits of some of them, especially of such as were provided with hard carapaces, solid scales, and ordinary or bony fin-rays. A certain proportion of them, as might have been expected, remind us, with regard to external form, of Teleosteous fishes rather than of any of the few still existing Ganoid types; but it is contrary to all analogy and to all palaeontological evidence to suppose that those fishes were, with regard to their internal structure, more nearly allied to Teleosteans than to Ganoids. If they were not true Ganoids, they may be justly supposed to have the essential characters of Palaeichtyes.

Other forms exhibit even that remote geological epoch so unmistakably the characteristics of existing Ganoids, that no one can entertain any doubt with regard to their place in the system. In none of these fishes is there any trace of vertebral segmentation. The Palaeichtyes of the Old Red Sandstone, the systematic position of which is still obscure, are the *Cephalaspidae* from the Lower Old Red Sandstone of Great Britain and Eastern Canada; *Pterichthys*, *Coccosteus*, and *Dinichthyas*: genera which have been combined in one group—*Placodermi*; and *Acanthodes* and allied genera, which combined numerous branchiostegals with chondropterygian spines and a shagreen-like dermal covering. Among the other Devonian fishes (and they formed the majority) two types may be recognised, both of which are unmistakably Ganoids.

The first approaches the still living *Polypterus*, with which some of the genera like *Diplopterus* singularly agree in the form and armature of the head, the lepidosis of the body, the lobate pectoral fins, and the termination of the vertebral column. Other genera, as *Holoptychius*, have cycloid scales; many have two dorsal fins (*Holoptychius*), and, instead of branchiostegals, jugular scutes; other one long dorsal confluent with the caudal (*Phaneropleuron*). In the second type the principal characters of the *Dipnoi* are manifest, and some of them, for example *Dipterus*, *Paloedaphus*, *Holodus*, approach so closely the Dipnoi which still survive, that the differences existing between them warrant a separation into families only.

Devonian fishes are frequently found under peculiar circumstances, enclosed in the so-called *nodules*. These bodies are elliptical fattened pebbles, which have resisted the action of water in consequence of their greater hardness, whilst the surrounding rock has been reduced to detritus by that agency. Their greater density is due to the dispersion in their substance of the fat of the animal which decomposed in them. Frequently, on cleaving one of these nodules with the stroke of the hammer, a fish is found embedded in the centre. At certain localities of the Devonian, fossil fishes are so abundant that the whole of the stratum is affected by the decomposing remains emitting a peculiar small when newly opened, and acquiring a density and durability not possessed by surata without fishes. The flagstones of Caithness are a remarkable instance of this.

The fish-remains of the *Carboniferous* formation show a great similarity to those of the preceding. They occur throughout the series, but are very irregularly distributed, being extremely scarce in some countries, whilst in others entire beds (the so-called bone-beds) are composed of ichthyolites. In the ironstones they frequently form the nuclei of nodules, as in the Devonian. Of Chondropterygians the spines of *Onchus* and others still occur, with the additions of teeth indicative of the existence of fishes allied to the Cestracion-type (*Cochliodus*, *Psammodus*): a type which henceforth plays an important part in the composition of the extinct marine fish faunae. Another extinct Selachina family, that of Hobodontes, make its appearance, but is known from the teeth only. Of the Ganoid fishes, the family *Palaeoniscidae* (Traquair) is numerously represented; others are Coelacanths (*Coelacanthus*, *Rhizodus*), and *Saurodipteridue* (*Megalichthys*). None of these fishes have an ossified vertebral column, but in some (*Megalichthys*) the outer surface of the vertebrae is ossified into a ring; the termination of their tail is heterocercal.

The carboniferous *Uronemus* and the Devonian *Phaneropleurone* are probably generically the same; and the Devonian *Dipnoi* are continued as, and well represented by, *Ctenodus*. In the Carboniferous and Permian strata some remarkable forms of the shark type have been obtained in recent year in a very well-preserved condition. The most primitive and earliest of these is *Cladoselache*, found in Lower Carboniferous strata in Ohio. Its mouth is at the end of the snout. It is from two to six feet long. The paired fins are attached horizontally

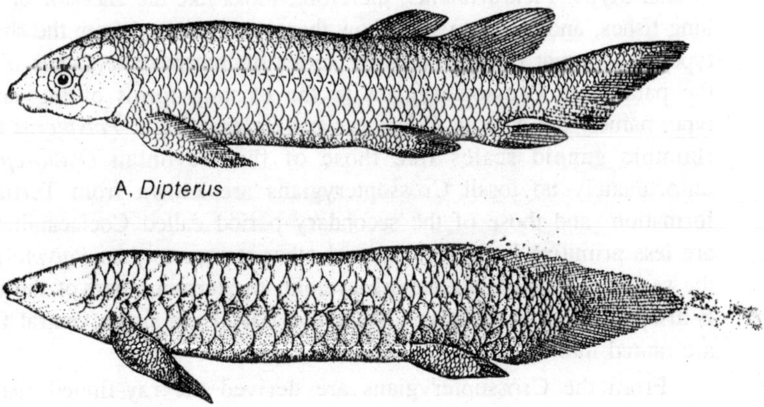

Fig. 1.6. Lungfishes. A—An ancient Devonian fossil type; B—Neoceratodus of Australia.

by a wide base, and have the appearance of enlarged portions of an originally continuous fold. The tail is short and high vertically, with an extremely heterocercal structure; that is to say, the termination of the trunk is bent upwards almost at right angles and the ventral fin-rays are very long.

The Acanthodians already mentioned extend from the Silurian to the end of the Permian and therefore go back to an earlier period than *Cladoselache*; but *Professor Bashford Dean* considers nevertheless that the latter represents the ancestral form of the Acanthodians. *Pleuracanthus*, which occurs in the Carboniferous and Permian both in Europe and North America, on the other hand, differs in many respects from sharks, and in these same features approximates to the Dipnoi. It was a shark in the following features: the gillslits opened separately on the surface; the jaws consisted of the same primitive cartilages as in sharks, claspers occurred in the male. On the other hand, the paired fins show a transition to the featherlike or pinnate structure which is in reality not primitive but secondary, the basal cartilages being separated from the body at the posterior end, and the radial cartilages extending round to the posterior, originally the internal, side, so that a structure is produced which has a central axis and rays along each side.

But the feature most similar to the lung-fish was that the dermal denticles had disappeared from the skin of the trunk, and on the head had developed into dermal bones arranged like those of the lung-fish. The tail retained the primitive symmetry and was not of the heterocercal of shark type. *Pleuracanthus*, therefore, looks like the ancestor of the lung-fishes, and seems to show how these were derived from the shark type. We cannot trace a complete series of intermediate stages from the palaeozoic Crossopterygians to the existing forms of the same type, namely the African *Polypterus* and *Calamochthys*. *Polypterus* has rhombic ganoid scales like those of the Devonian *Osteolepis*; unfortunately no fossil Crossopterygians are known from Tertiary formation, and those of the secondary period called Coelacanthidae are less primitive in their scales and other structures than *Polypterus*; the scales of Coelacanthidae are cycloid, the terminal part of the tail is disappearing and the basals of the two dorsal and single ventral fins are united into a single, forked bone.

From the Crossopterygians are derived the ray-finned fishes (Actinopterygians), in which the lungs gradually lose their respiratory function and paired character, and become more and more completely adapted to serve as an air-bladder which diminishes the specific gravity of the fish. In the evolution of these forms we can trace a gradual return to the sea, for among existing fishes the more primitive forms live more or less completely in fresh water, and the most recent and most modified are typically marine. The fins of Actinopterygians are characterized by the reduction of the basal lobe and its internal skeleton, the dermal rays arising from the surface of the body.

The oldest form is *Chirolepis* which occurs in the Lower Old Red Sandstone together with the earliest Crossopterygians. This genus belongs

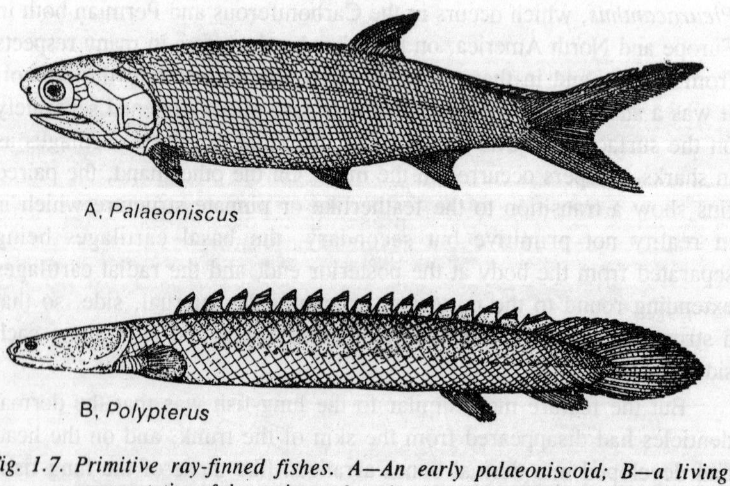

Fig. 1.7. Primitive ray-finned fishes. A—An early palaeoniscoid; B—a living representative of the ancient palaeoniscoids, with modified fin structure.

to the group or branch of the Chondrostei and is the first of the great family Palaeoniscidae which developed an abundance of species and individuals in the Corboniferous and Permian formations and gradually diminished in the Secondary period till the end of the Jurassic. These fishes had a complete armour of ganoid scales, a large heterocercal tail, and a single dorsal fin in the middle of the back. The dermal bones of the head were external and were simply the ganoid plates of that region. As the Palaeoniscidae are disappearing the family Chondrosteidae comes into existence, distinguished by the degeneration of the scales on the body; specimens occur in the Lias or Lower Jurassic, and these forms must be regarded as directly descended from the Palaeoniscidae.

The Chondrosteidae, on the other hand, are the ancestors of the existing Sturgeons, which are not known before the Tertiary. A branch from the Palaeoniscidae gave rise to the Platysomidae which existed in the Carboniferous and Permian periods; they are deep-bodied compressed fishes completely covered with ganoid scales, but with short jaws carrying blunt crushing teeth; they do not appear to have give rise to any later forms. The passage from the Palaeozoic into the *Mesozoic* era is not indicated by any marked changed as far as fishes are concerned. The more remarkable forms of the Trias are Sharklike fishes represented by ichthyodorulithes like *Nemacanthus*, *Liacanthus*, and *Hybodus*; and Cestracionsts represented by species of *Acrodus* and *Strophodus*. Of the Ganoid genera *Coelacanthus*, *Ambiypterus* (*Palaeoniscidae*, *Saurichthys*) persists from the Carboniferous epoch. *Ceratodus* appears for the first time.

The Hybodonts and Cestracionts continue in their fullest development Holocephales (*Ischyodus*), true Sharks (*Paloeoscellium*), Rays (*Squaloraja*, *Arthropterus*), and Sturgeons (*Chondrosteus*) make their first appearance; but they are sufficiently distinct from living types to be classed in separate genera, or even families. The Ganoids, especially Lepidosteoids, predominate over all the other fishes: *Lepidotus*, *Semionotus*, *Pholidophorus*, *Pachycormus*, *Eugnathus*, *Tetragonolepis*, are represented by numerous species; other remarkable genera are *Aspidorhynchus*, *Belonostomus*, *Saurostomus*, *Sauropsis*, *Thrissonotus*, *Conodus*, *Ptycholepis*, *Endactis*, *Centrolepis*, *Legnonotus*, *Oxygnathus*, *Heterolepidotus*, *Isocolum*, *Osteorhachis*, *Mesodon*. These genera offer evidence of great change since the preceding period, the majority not being represented in older strata, whilst, on the other hand, many are continued into the succeeding oolithic formations.

The homocercal termination of the vertebral column commences to supersede the heterocercal, and many of the genera have well ossified and distinctly segmented spinal columns. Also the cycloid form of scales becomes more common: one genus (*Leptolepis*) being, with regard to the preserved hard portions of its organisation, so similar to the Teleosteous type that some Palaeontologists refer it (with much reason) to that sub-class. The Cretaceous group offers clear evidence of the further advance towards the existing fauna. Teeth of Sharks of existing genera *Carcharias* (*Corax*), *Scyllium Notidanus*, and *Galeocerdo*, are common is some of the marine strata, whilst Hybodonts and Cestracionts are represented by a small number of species only; of the latter one new genus, *Ptychodus*, appears and disappears.

A very characteristic Ganoid genus, *Macropoma*, comprises homocercal fishes with rounded ganoid scales sculptured externally and pierced by prominent mucous tubes. *Caturus* becomes extinct. Teeth and scales of *Lepidotus* (with *Sphoerodus* as sub-genus), clearly a fresh water fish, are widely distributed in the Wealden, and finally disappear in the chalk; its body was covered with large rhomboidal ganoid scales. *Gyrodus* and *Aspidorhynchus* occur in the beds of Variance, *Coelodus* and *Amiopsis* (allied to Amra), in those of Comen, in Istria. But the Palaeichthyes are now in the minority; undoubted Teleosteans have appeared, for the first time, on the stage of life in numerous genera, many of which are identical with still existing fishes. Assuming that the ancestral fishes were Elasmobranchs, as we are bound to do from the fact that the flat bones of the head and of the pectoral girdle of bony fishes are derived from the dermal denticles of the Elasmobranch, the earliest bony fishes must have had a Crossopterygian or fringed structure in their lateral fins.

From the Crossopterygian fringe finned type was derived the ray-finned type by the reduction of the internal radials, and the lung-fish (Dipnoan) in another direction by the development of an elongated axis to form the biserial or pinnate type of fin. The Crossopterygian is therefore a central type and its fossil forms should be among the oldest. The more primitive and earlier forms retain the ganoid scaling, while some of the later have cycloid scales and perfectly homocercal tails. They first appear in the Permian, but were only abundant in the Trias and still more so in the Jurassic, while in the Cretaceous the Teleostei become the dominant type.

The family Eugnathidae, fishes much like a tarpon of large herring in shape and fins, appear first in the Trias and are common in the

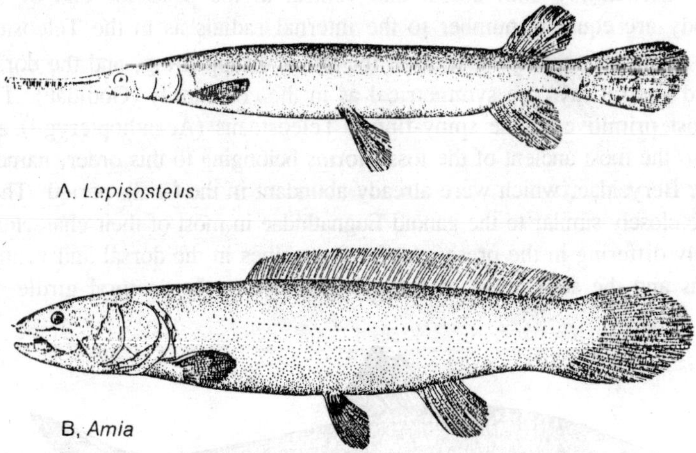

Fig. 1.8. Holosteans. A—The gar pike; B—the bowfin.

Jurassic; some of them have the Ganoid scaling, while in others the scales are thin, cycloid, and over lapping. The American Bowfin, *Amia*, is a surviving member of this group, retaining the incomplete vertebrae in the tail, and a bony plate under the throat, which are primitive features. *Caturus* and *Exrycormus* of the Jurassic lead directly to the soft-finned Teleostei (Malacopterygii) such as Tarpons (Elopidae) and Herrings (Clupeidae). The latter family were already abundant in Cretaceous times and are sometimes fossilized in dense shoals. The Pycnodonts, deep-bodied ganoids with crushing teeth, seem to have became extinct in the Lower Eocene without leaving any descendants.

Lepidosteus, the bony Pike of North America, is a pike-like, predaceous modification of the earlier Holostei, but in the fossil condition it is only known as far back as the Eocene, when it lived in Europe as well as in America. The Ancestry of the several divisions of Teleostei has not yet been worked out, but it is probable that they have descended separately from the fossil ganoids, and not all from the most primitive Malacopterygii above mentioned. It has been suggested that the Cat-fishes (Siluridae) were descended directly from the Chondrostei, and are therefore more allied to the Surgeons than to the Malacopterygii, but it is a question whether the equality of dermal rays and radials could have been evolved twice independently. There seems greater probability that the cod-like fishes (Anacanthini) may have been derived directly from the extinct Coelacanthidae; although the latter are placed among the Crossopterygians it is remarkable that the dermal ray both dorsal and ventral at the posterior end of the body are equal in number to the internal radials as in the Teleostei.

The original termination of the tail is disappearing, and the dorsal and ventral rays are symmetrical as in the cod family (Godidae). The most primitive of the spiny-finned Teleosteans (Acanthopterygii) are also the most ancient of the fossil forms belonging to this order, namely the Berycidae, which were already abundant in the Chalk period. They are closely similar to the ganoid Eugnathidae in most of their characters, only differing in the presence of a few spines in the dorsal and ventral fins and the attachment of the pelvic bones to the pectoral girdle.

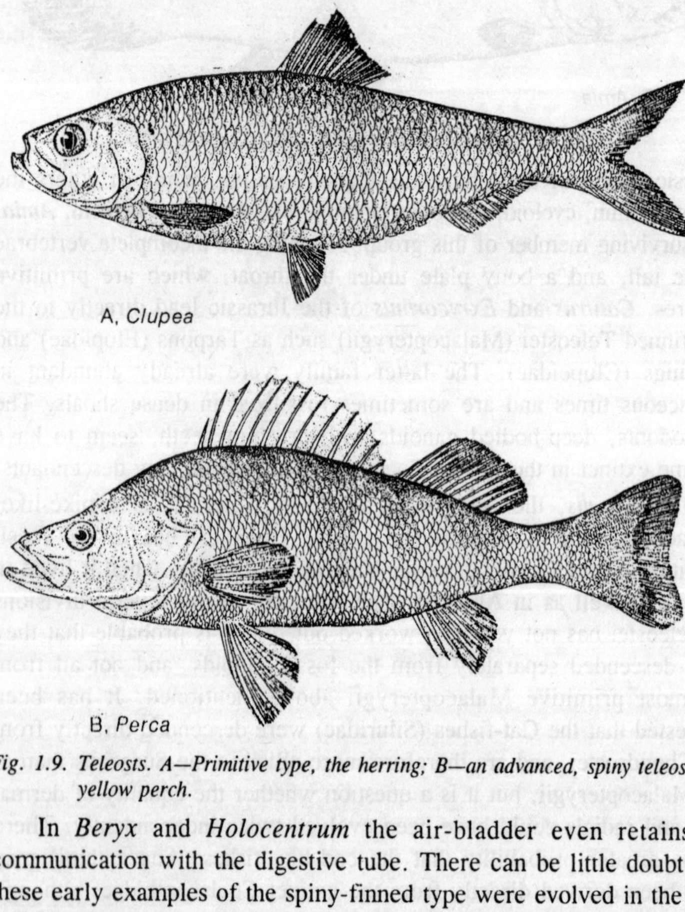

Fig. 1.9. Teleosts. A—Primitive type, the herring; B—an advanced, spiny teleost, the yellow perch.

In *Beryx* and *Holocentrum* the air-bladder even retains its communication with the digestive tube. There can be little doubt that these early examples of the spiny-finned type were evolved in the sea, and that the comparatively few existing fresh-water forms, such as the

Perch, have ascended from the sea at some later period. The above may seem a somewhat technical and condensed discussion, beyond the easy comprehension of the average reader. It is however impossible to state the most important facts concerning fossil fishes, and the relative age of the geological strata in which they occur, without the use of some technical terms. We will endeavours to give in simple language the chief conclusions which our present knowledge enables us to draw concerning the evolution of fishes.

In the broadest sense it may be said that there are two principal types of structure in fishes, the shark type and the bony type. At the present time fishes of these two types exist in large numbers and in great variety side by side in the sea. We cannot maintain that the differences of structure between sharks and bony fishes correspond to equally marked differences mode of life: the shark and the funny are both voracious monsters living as far as we can see much in the same way, both powerful swimmers both feeding on other fishes, both ranging through the surface water of the open ocean. We cannot perceive that the bony fish was derived from the shark-like fish by adaptation to a different mode of life in the sea, as the Amphibian was derived from the fish by adaptation to a life on dry land in its adult condition. What then is the explanation offered by the doctrine of evolution for the existence of these two types? When we turn to the study of fossil fishes, we find that in the most ancient stratified rocks both were represented, but the fossil forms of the bony belonged exclusively to certain groups of which at the present time very few species exist, and these only in fresh water, namely the Fringe-finned Ganoids, represented now by the African *Polypterus*, and the Lung fishes which occur now in tropical rivers or swamps. Both these kinds of fishes have lungs, or open air-bladders actually used for breathing air.

The Lung-fishes evidently resemble the ancestors of the Amphibia, while from the Fringefinned Ganoids we have a series of diverging forms leading to the great variety of existing bony fishes. The conclusion is that the bony fishes of the sea at the present day are descended from fresh-water fishes of ancient times, which were adapted to breathe air in order to supplement the original respiration by gill. In this way we can understand the origin of the airbladder which is wanting in the shark type. The air or gas-bladder is evolved from lungs not from air-bladder. At the same time the skeletal structures both within the body and in the skin underwent a change to the bony type with lungs the change of habitat from the sea to fresh water.

From the Fringe-finned Ganoids descended numerous diverging forms which populated the various fresh waters of the globe, and ultimately reached the sea again and established themselves there along with the fishes of the shark type. The first important change in the transition from the fringe-finned type to the modern fishes was the reduction of the internal skeleton of the paired fins, so that the basal fleshy lobe disappeared and the fin became fan-like. It is difficult to say with certainty, which of the extinct fan-finned Ganoids were first to reach the sea: it is probable enough that many of the groups lived along the sea coast, but we know that, excepting the Sturgeons, the only surviving forms *Lepidosteus* and *Amia* of N. America, live in fresh water, and the sea is inhabited by the latest results of evolution in fishes, the Teleostei, together with Elasmobranchs or fishes of the shark type.

2

The Skin

The integument or skin forms the external covering of the body and performs a number of important functions in fishes. Besides protecting the body against injury and infections, the integument has respiratory, excretory and osmoregulatory functions. The derivatives of the integument play an important part in the metabolic activities of the body. Special structures of some species like the electric organs, poisonous glands, phosphorescent organs are also the integumentary derivatives. The integument of a fish, like that of any other vertebrate, is composed of two layers, a thin outer *epidermis* and an inner thick *dermis*.

Epidermis

The epidermis is ectodermal in origin and consists of several layers of flattened cells forming the *stratified epithelium*, of which the deepest layers are made up of columnar cells forming the *straum germinativum*. The cells of this layer always undergo mitotic divisions to replace the warn out cells. A superficial layer of dead horny cells, forming the *stratum corneum* in terrestinal vertebrates, is not present in fishes. The cells at the surface are not fully keratinized as in cyclostomes.

Dermis

The lower thick layer of integument is called *dermis*. It is mesodermal in origin. It is formed from embyonic dermatomes of mesoderm. It is composed of mesenchyme cells, connective tissues, blood vessels, and nerves. In some fishes the dermis has three layers: (a) an outer layer around the neck of scales, which is nearly free of fibres, is called *stratum laxum*. (b) a laminated fibrous layer to which the base of the scale is tied by sharpey's fibres, is called *stratum compactum*, and (c) *subcutaneous* consisting of a network of fine fibres.

In the Cyclostomata, the epidermis is particularly rich in goblet-shaped, mucous-secreting, gland-cells. The Myxinoids also possess numerous pockets of so-called "*thread-cells*". In each of these cells the protoplasm secretes a long spirally-coiled thread, and under the influence of appropriate stimuli the thread is shot out and unwound to a great length. The threads and the mucous are so abundant that one of these animals will convert a bucket of water into a thick mass of jelly. No scales or other hard exoskeletal structures are present in any of the Cyclostomata. In Fishes muscous glands are also abundant in the epidermis and to their activity is due to the slimy mucous which lubricates the surface of the body. They are specially numerous in the Dipnoi (e.g., *Protopterus*), where, in addition, there are many simple multicellular glands which secrete the "cocoon" or capsule is which the fish is enclosed during the dry season.

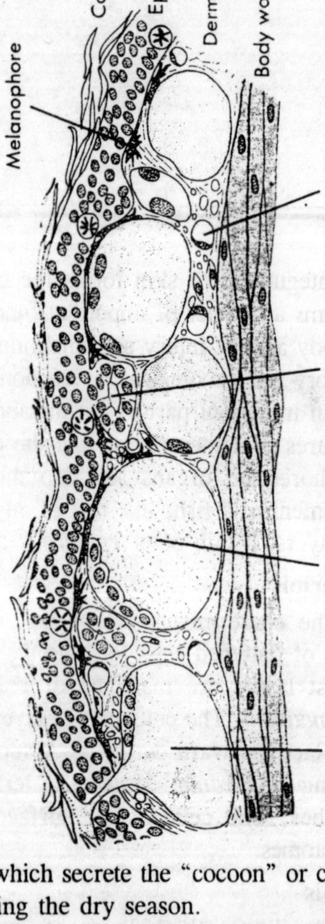

Fig. 2.1. Skin of the red eft Notophthalmus.

From the epidermis are derived the *poison-glands* in some Teleosts, and also the "glandula pterygopodia" in relation with the claspers of the male Elasmobranchs. The glandular structures in connection with the phosphorescent organs of the deep-sea fishes will no doubt be traced to the same source. In the great majority of fishes the skin becomes the seat of calcareous deposit, and gives rise to such diverse exoskeletal

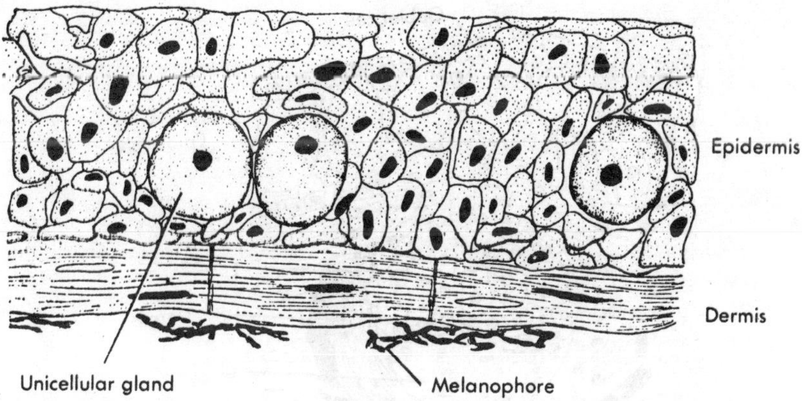

Fig. 2.2. Skin of a larval lamprey. The dermis contains very dense collagenous fibers.

structures as the varied forms of spines and scales with which the surface of a fish is invested. These structures, probably the most ancient form of vertebrate skeleton owing its existence to the presence of lime slats in the tissues of the body, present highly characteristic modifications in the different groups.

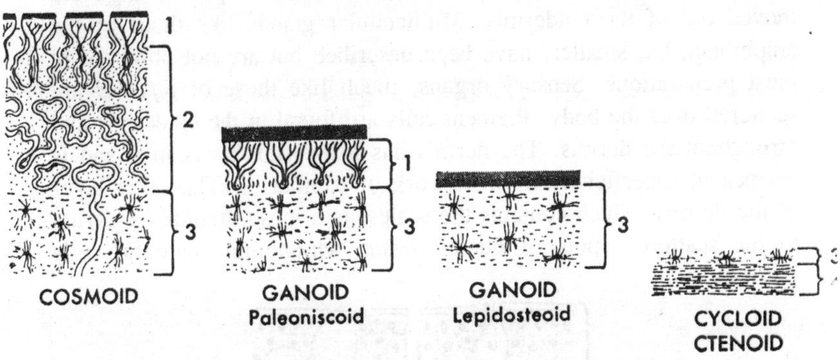

Fig. 2.3. Dermal scale structure, diagrammatical. 1-Dentine-type bone (cosmine); 2-spongy bone; 3-lamellar bone.

Integument in Certain Fishes

Choanate Fishes

The skin of *Latimeria* is described as simple in structure. The epidermis is stratified and has a thick basement membrane. There are mucous cells and chromatocytes in the epidermis. The dermis is thick, fibrous and has many fat cells. In *Protopterus*, a dipnoan, the epidermis is stratified with a basal, low columnar stratum germinativum and four or five layers of cells. There is a well-marked external squamous

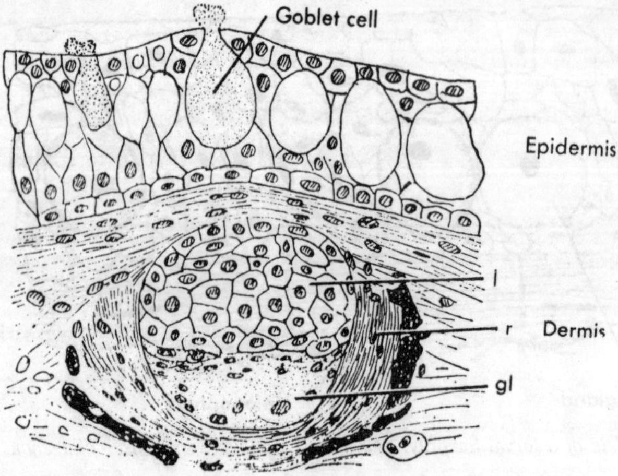

Fig. 2.4. Skin and light organ (photophore) of a luminous fish. gl—Luminous cells; l—lens cells; r—reflector cells surrounded by pigment cells.

layer and a thick basement membrane. In the epidermis there are large *mucous cells* (Becker's cells) which move outward as they develop. They shed their contents from a narrow surface tip, and then are moved out of the epidermis. Multicellular glands like those of the amphibian, but smaller, have been described but are not observed in most preparations. Sensory organs, much like those of *Squalus* are scattered over the body. Pigment cells are found in the epidermis and throughout the dermis. The dermis has many fibroblasts in it and is permeated superficially by a network of capillaries. The outer layer of the dermis, which contains the scales, is a loose *areolar* (referring to the scattered spots or nuclei) *tissue* which may have fat cells

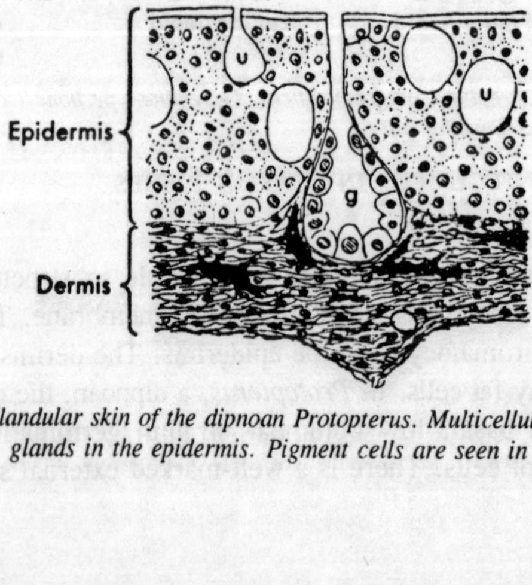

Fig. 2.5. Glandular skin of the dipnoan Protopterus. Multicellular, g, and unicellular, u, glands in the epidermis. Pigment cells are seen in the dermis.

scattered through it. A dense layer or collagenous fibres lies next to the body muscles. In cross sections of the dermis, one sees parts of several scales, which lie in dermal pockets.

Actinopterygian Fishes

The skin of the antinopterygian fish is like that of the choanate. The epidermis is dotted by *pearl* (or sensory) organs and overlies the scales. The pearl organs are perhaps related to the prototrichs of higher forms. In addition to the mucous cells of the surface, two types of secretory cells are observed. These are the *granular secretory cells*, or serous bladders, and the *club-cells*, The former begins as a large *Leydig's cell*, then becomes filled with a granular secretion which is expelled to the surface by a tubular extension of the cell. The club cells produce the *intercellular matrix* of the epidermis. Tubular, multicellular glands are observed on the snout and with decreasing frequency further back on the body. *Poison glands*, formed by clumps of holocrine cells, are associated with opercular or dorsal spines in some fishes.

The epidermis is occasionally invaded by blood capillaries as in the upper lip of *Periophthalmus*. As in the choanates, the dermis encloses the scales lying in sacs. *Sharpey's fibres* penetrate the ganoid type of scales, tying them together as well as into the fibrous network of the dermis. Sharpey's fibres are evident only in teleosts with the thick bony scales. The dermis contains several colours of *chromatocytes* and also guanin containing *irridocytes*, which are responsible for the silver colour of many fishes. The innermost layer of the dermis is densely fibrous.

Chondrichthyes

In the sharks, the points of the placoid scales extend out through the epidermis. The dermis has three layers: an outer layer around the necks of the scales, which is nearly free of fibre; the *stratum laxum*, a laminated fibrous layer to which the base of the scale is tied by Sharpey's fibres; the *stratum compactum*; and a *subcutaneous layer* of variable thickness, in which a reticulum of fine fibres is observed. There is a fairly sharp line of demarcation between the last two. The young shark has ciliated cells intermixed with large mucoid cells in the epidermis. As the animal increases in size the cilia are lost and the stratified condition is achieved. Sensory organs presumed to be related to prototrichs, are scattered over the body. The scales do not appear till quite late.

Cyclostomes

In the lamprery, the epidermis is many cells thick. The main *epidermal cells* are very small, and among them are *granular* and club-shaped *secretory cells*. The outer layer of this epithelium is modified into "calcareous cells", which have a modified, impregnated, outer surface. These cells are overlaid by a thin *cuticle* and the entire epidermis is supported by *perpendicular fibres* from the thin *basement membrane*. The *dermis* is stratified into an outer, thick, laminated, *fibrous layer* below the epidermis; next is a thin *pigment layer*, and below this a reticular *subcutaneous layer*. There are no scales. *Myxine* is similar in having a thick epidermis covered by mucous. The outer cell layer is made up of columnar *mucous cells*, while deeper in the epidermis are large *bladder cells* and somewhat smaller *thread* or *filament cells*. Along the ventrolateral margins of the body, the epidermis forms saccular *slime glands*, which utilize the same kinds of secretory cells as the general epidermis. The dermis is stratified into an outer layer of *fibrous laminae* enclosing *pigment cells* and a deeper *subcutaneous layer* of fatty areloar tissue.

Derivatives of Skin

Excoskeleton in the form of scale in the most important derivative of the skin, which provides procactin to the body.

Scales

In most of the fishes the skin is covered with scales, and only a few are naked having a scales on the body, as in cat fishes and lamprays.

Types of Scales

Structurally there are two types of fish scales:

(*a*) Placoid scale

(*b*) Non-placoid scale.

Placoid Scale

The scales which are developed from secretory activity of the both dermis and epidermis like the scales of the "Elasmobranches". Placoid scale also called *dermal denticles* which is the characteristics feature of the marks and their relatives (chondrichthys). Each has a disc like basal plate embedded in the dermis and a spine projecting out through the epidermis. In structure, a placoid scale resembles a tooth. The spine has an external covering of enamel like, hard, transparent material called '*Vitrodentine*'. This is followed by a layer

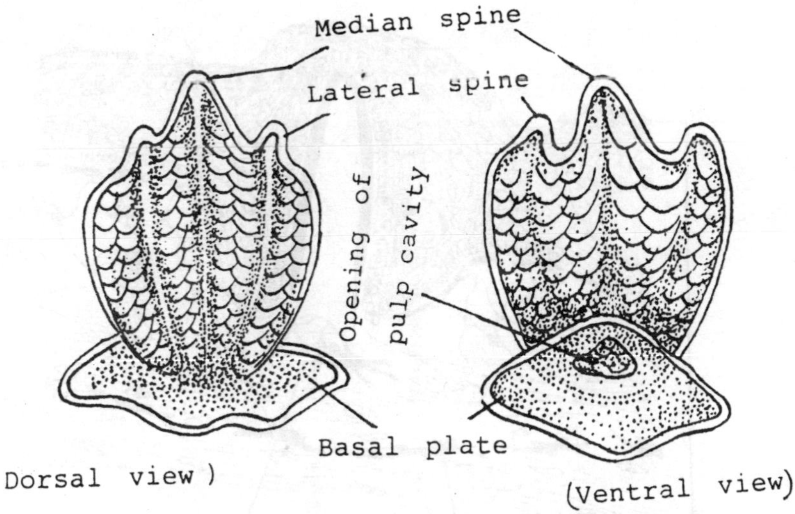

Fig. 2.6. Scales in fishes.

of dentine enclosing a pulp cavity from where branching dentinal lubules radiate in different directions. The centre of the basal plates is perforated by an aperture to provide entrance to the blood vessels and nerves from the dermis. The scales are closely set but do not overlap to each other.

Development of a placoid scale

The development of a placoid scale begins with the multiplication of dermal cells which accumulate at different points below the stratum germinativum (Malpighian layer). These dermal cells form a papilla which grows upwards pushing the Malpighian layer of this region and the cells become columnar and glandular forming the enamel organ. The airnal papii assumes the shape of a basal plate and spine. The cells at the surface of the papilla form a distinct layer and called *odontoblasts*. They secret dentine towards their outer surface to cover the papilla. The central cells of the papilla do not calcify and form the pulp. The spine of the scale is gradually pushed towards the surface.

Before eruption, the cells of the enamel organ, called the *ameloblasts*, secrete a layer of *Vitrodentine* to form a cap over the spine. The mesenchyme cells of the dermis, surrounding the dermal papilla, secrete a substance like the cement of teeth to form the basal plate of the scale. Thus, the placoid scale is partly dermal and partly epidermal in origin. The spine projects above the skin while the basal

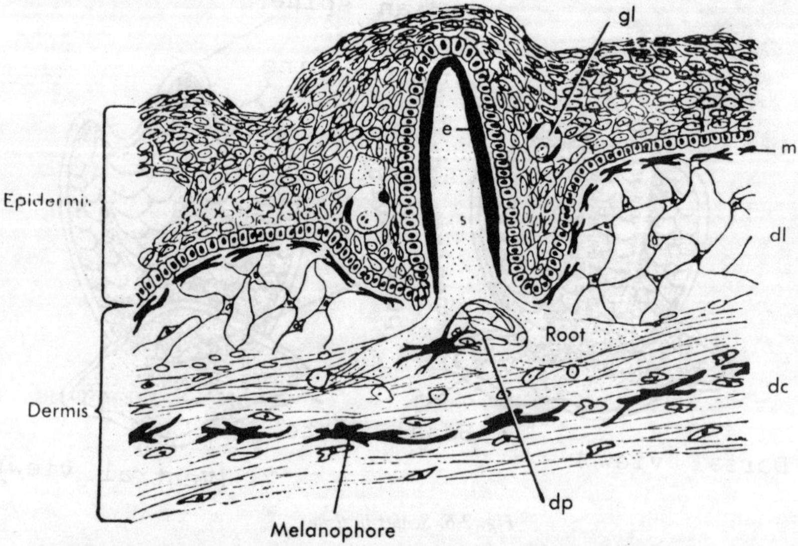

Fig. 2.7. Developing placoid scale in skin of embryonic shark. Dermis consists of compact layer, dc, and loose layer, dl; dp dermal pulp within root and spine; e, enamel; gl, unicellular epidermal gland; m, melanophore; s, spine, composed of dentin, enamel, and a core of pulp.

plate remains embedded in the dermis. The teeth of shark appear to be enlarged and modified placoid scales. Vertebrate teeth also develop similarly and have a comparable structure, hence placoid scales and vertebrates teeth are considered to be homologus structure. However, the hypothesis that the placoid scales are modified into teeth of the shark is subject to many objections and is said that both placoid scales and teeth are modified remmants of the dermal plates of ancestral fishes and are therefore homologus structure.

Non-placoid Scales

They are derived from dermis only as a scale of bellost. Non-placoid scales are basically of three types:

A. Cosmoid scale.

B. Ganoid scale.

C. Bony ridge scale.

Cosmoid scale

The cosmoid scales are found in the extinct "Crossopterygean and Depnoi". The external layer of the scale is thin and enamel like and is called the *vitrodentine*. The middle layer is made up of hard, non-

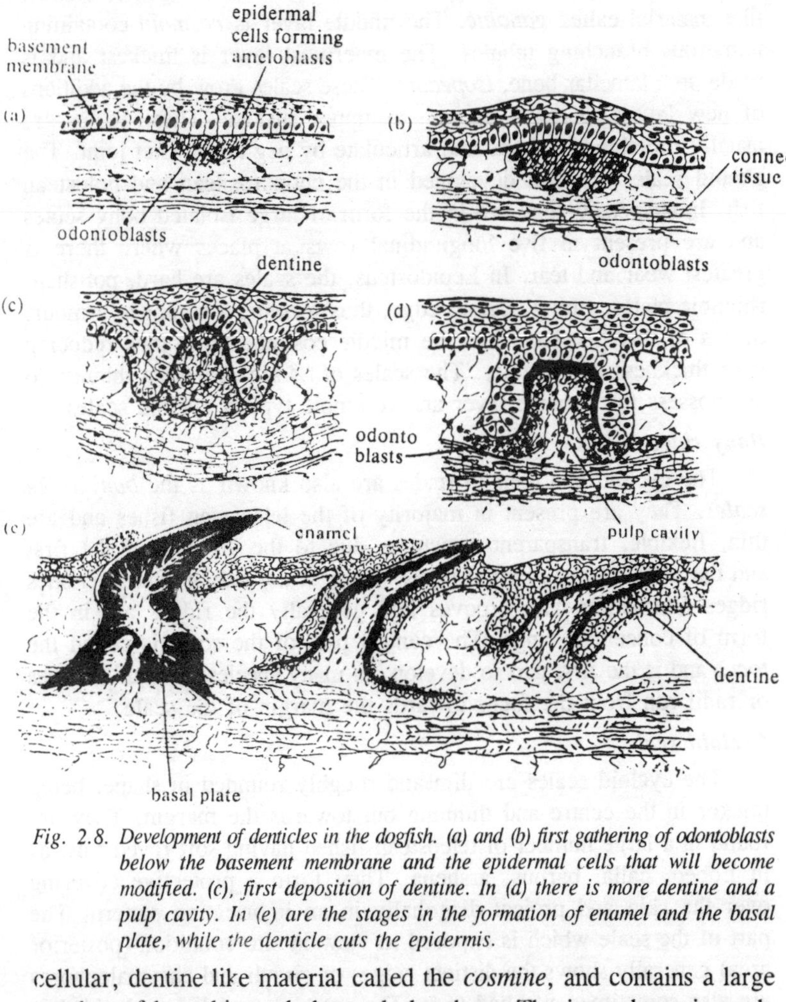

Fig. 2.8. Development of denticles in the dogfish. (a) and (b) first gathering of odontoblasts below the basement membrane and the epidermal cells that will become modified. (c), first deposition of dentine. In (d) there is more dentine and a pulp cavity. In (e) are the stages in the formation of enamel and the basal plate, while the denticle cuts the epidermis.

cellular, dentine like material called the *cosmine*, and contains a large number of branching tubules and chambers. The inner layer is made up of vascularised bony substance, *isopedine*. These scales grew by the addition of new isopedine material from bells along the edges. Cosmoid scales are not found in the living fishes. The living Dipnoi possess thin cycloid scales and lost the cosin layer. The scales of "*Latimeria*" resembles the ctenoid scales but have a denticular outer surface.

Ganoid scale

These are characteristics of the primitive Actinoptergyians called the ganoid fishes and are of the various forms and structures. These scales are heavy and have a outer layer of hard inorganic, enamel like material called *ganoine*. The middle layer is *cosmoin* containing numerous branching tubules. The innermost layer is thickest and is made up a lamellar bone, *isopedine*. These scales grow by the additions of new layers to lower as well as upper surface. These scales are usually rhomboid in shape and articulate by peg and socket joint. The ganoid scales are best developed in the chondrosilean and holostean fish. In *Acipenser* they are in the form of large isolated bony scutes and are present in five longitudinal rows at places where there is greatest wear and tear. In Lepidosteus, the scales are hard, polished, rhombic plates, fitting edge to edge, thus forming a complete armour. Scales of this fish have lost the middle cosmine layer thus reducing their thickness and rigidity. The scales of '*Amia*' are even thinner do not possess the ganoine layer and resemble typical cycloid scales.

Bony ridge scales

The cycloid and ctenoid scales are also known as the *bony ridge scales*. They are present in majority of the teleostean fishes and are thin, flexible, transparent structures due to the absence of the first and the middle layer of other types. These scales exhibit characteristics ridges alternating with grooves and generally the ridges are in the form of concentric rings. The central part of the scale is called the *focus* and is the first part to develop. In many species, oblique grooves or radii run from the focus towards the margin of the scale.

Cycloid scale

The cycloid scales are thin and roughly rounded in shape, being thicker in the centre and thinning out towards the margin. They are found in a large number of teleostean fished having soft rayed fins, as in Lobeo, calla, barbus, arshena. They form a protective covering over the skin and project diagonally in an imbricating pattern. The part of the scale which is exposed to view in situ condition (posterior area) generally shows the distinct ridges or circuli and chromalophores are also sometimes attached to it. The anterior area lies embedded in the skin.

Ctenoid scales

Ctenoid scales are also circular and can be distinguished from the cycloid by having a more or less serrated free edge. Moreover, several spines are present on the surface of the posterior area of the scale. These scales are found in a large number of fishes with spiny rayed fins. All intermediate type between the cycloid and ctenoid scales are found on different parts of the body.

Development of cycloid and ctenoid scales

The cycloid and ctenoid scales develop in the form of a dermal papilla formed by the multiplication of cells. The epidermis does not take part in the formation of these scales, which appear of an average body length between 12-50 mm. The scales are generally formed first at the candle peduacle and then spread to the rest of the body. The papilla first secrets the central part of the scale or focus and then the ridges are departed on the growing scale.

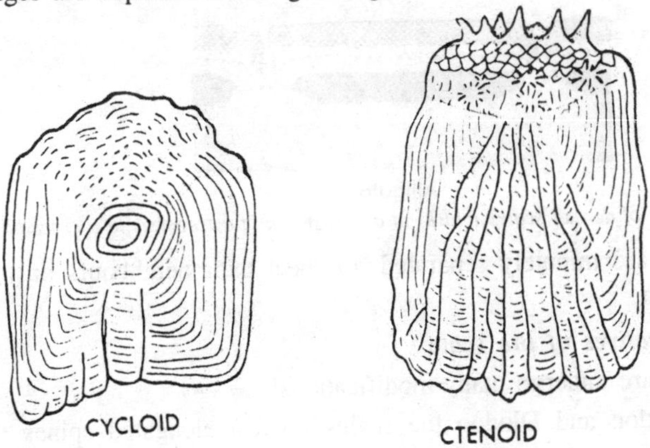

Fig. 2.9. Modern flexible fish scales. The upper edges are caudal free borders.

Squamation

Some kinds of fishes are "naked" in the technical sense of having no scales. Examples include the lampreys (Petromyzontidae) and North American freshwater cat fishes (Idaluridae). An intermediate structure category comprises fishes that are nearly naked and have no scales only on a few places of the body, or have scales reduced to a few prickles. Some of these prickled are in very localized positions. Examples of partly scales fishes include the paddlefish (*Polyodon*). Scales patterns are fundamentally associated with body segmentation as manifested initially in embyonic development by the vertebrae and the myomeres.

In arrangement, scales are most often imbricated and thus overlap like the shringees on a roof with the free margin directed towards the tail in a manner that minimizes friction with the water. Rarely, aberrent individuals, exhibit total or partial reversal of this arrangement. In some fishes, such as the burbot (Coca) and the freshwater eels (Anguilla) the pattern is mosaic rather than overlapping one another.

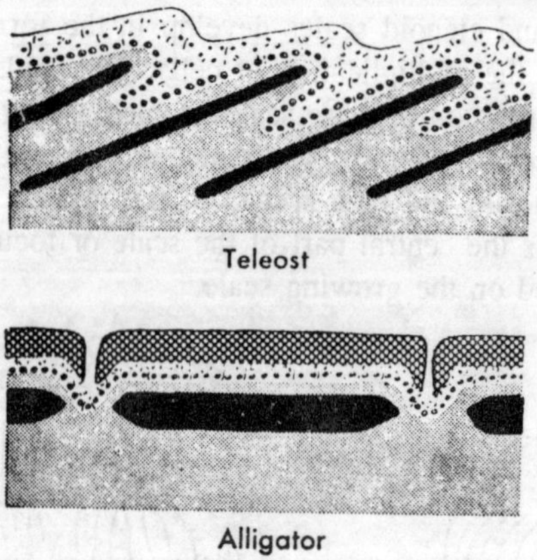

Fig. 2.10. Site of dermal scales (black) in a teleost and of osteoderms in a crocodilian.

The scales are minutely separated, or meet their neighbours only at their margin.

Other derivatives of the scale

Scales are undergo many modifications:

1. In Tetrdon and Diodon the scales form a elongated spines for protection.
2. Glandular cell of epidermis are modified to form a poisonous gland in scarpecue and toad fishes and are used as a organ for the offense and defense.
3. Light producing phosphorescent organ of marine fishes are also the derivatives of the epidermis.
4. Jaw and teeth in the sharks are modification of placoid scale.
5. Jaw teeth of Pristis and string of string rays are also modified scale.

Uses of Scale in Classification and Life History

Presence of the scale on the body distinguished the fish for various groups where the scales covering the body denote the group of fishes. For ex-placoid scales are present in cartilagenous fishes which belongs the group Ecasmobranch. Similarly Teleostein fishes possess a scale which are bony ridge scale known as ctenoid and cycloid scale. These scales are covering the body surface. Thereby distinguished fishes which are called as cypronian fishes. Cosmoid scales are present in the fishes which belong the group of Sarcopterygii, thereby indicating the use of scale in classification.

Functions of the Skin

1. The skin is also an important respiratory organ in certain species like *Anguilla*. They arc ablc to survive for considerable time out of water due to highly vascular dermis of their skin.
2. Skin has also the power to absorb dissolved nutrients substances from the surrounding water.
3. The mucous secreted by mucous gland 'which performs several important functions. The mucous covering the body of the fish. It lubricates the fish so as to reduce the body friction in water while swimming this enabling the fish to move with greater speed. It protects the fish from fungus, bacteria and microorganism. Certain species like the Gastrosteres, *Belta utelise* the stricky mucous for preparing nests for egg laying.
4. The skin performs and important function in repairing the surface wounds. Immediately after an injury the wound is closed by the mucous containing a large number of lymphocytes. The marginal cells of the wound then multiply rapidly to form a thin protective layer and gradually healing of wound is completed.

3

ORIGIN OF FINS

Fins are the chief organs of locomotion in fishes and are of two kinds. Median or *unpaired* and *paired fins*. The Median fins include a *dorsal* on the back, an *anal* on the ventral side behind the anal and *caudal* at the end of the tail. The paired fins are the *pectorals* and *pelvics* corresponding to the fore and hind limbs of the terrestial vertebrates. Both the median and paired fins are supported by skeletal rods called the *radials* and *dermal fins rays*. The appendages of fishes comprise the fins and cirste (flaps of flesh) which attain extreme development in the sargossum (*Pterophryne*) and the leafy sea dragan (*Phyllopteryse*).

Median Fins

The median fins are provided with supporting structure. Those of the dorsal and anal fins consists of cartilagenous or bony radials, supporting suitably a series of horny or bony rays. The radials are usually trisegmented. Although they are simple cartilageous rods in holocephali, and equally simple bony rods in polypterus. In all bony fishes the proximal segment is the largest, middle short and distal further reduced. In Elasmobranch the segments are of equal length. In all fishes the proximal segment is more persistent and the reduction is at the expense of middle and distal segments. The reduction is often due to the fin rays taking the form of spines. There is usually one radial to each body segment, but two may be present:

1. Opposite to the neural and heamal spine.
2. Between the adjace spines as in the case of certain cat fishes.

The fin rays are soluble, disposed on both sides of the fin. The horny unjointed fin rays of cartilageous fishes are called '*ceratotrichia*'. The bony segmented and ranched rays of bony fishes are called '*lepidotrichia*'. Delicate horny rays on the embryonic fins and on the tips of adult fins of bony fishes are called 'Actinotrichia'. The bony jointed rays formed by the combination of Lepidotrichia and Actinotrichia are known as '*compotrichia*'. It is found in Dipnoi. The proportional share taken by radial and fin rays in the support of the fin varies. In cartilageous fishes the radials are the main support and may extend up to the margin of the fin. In most bony fishes the radial lies at the base in the musculature of the body and the fin rays support the fin.

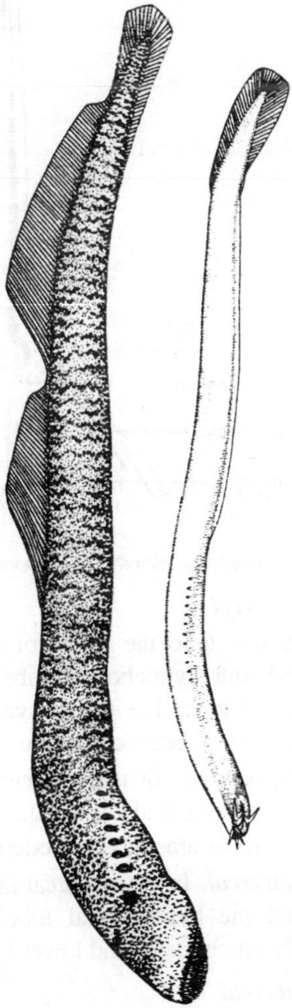

Fig. 3.1. Lamprey Petromyzon above; hagfish Bdellostoma below.

The segmental value of the fin rays and radials is also variable. The fin rays are much more numerous than the radials in 'Elasmobranchia'. The ray become gradually reduced in number. In bony fishes till in higher bony fishes. They correspond in number with supporting radials. Numericles correspondence between the neural and haemal spines, and the radials and fin rays is vary rare and has been seen in the caudal fin of certain crossopterygian fishes.

Caudal Fin

The caudal fin is supported by the neural spines and radials in the dossal lobe and greatly expanded haemal spines in the ventral lobe. A number of types of tail are distinguished on the basis of the skeleton:

Protocercal (first tail)

It is primitively symmetrical caudal fin, the symmetry is both externally and internally and the continues dorsal epichordal and ventral hypochordal lobes are equally developed such fins are found in larva fishes and cyclostomes.

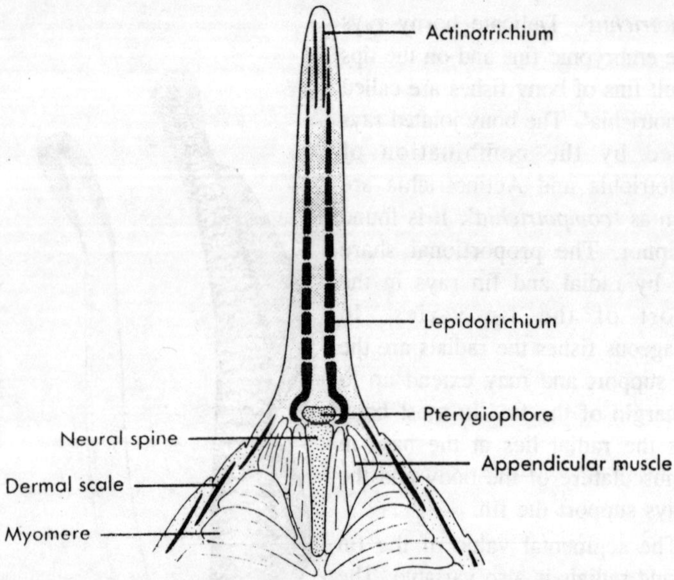

Fig. 3.2. Supporting elements of the dorsal fin of a teleost in diagrammatic cross section.

Heterocercal

In this type the posterior end of the vertebral column is bent upward and hypochordal lobe is much more developed then the epichordal lobe. The fin is asymetrical both externally and internally. Two grades of heterocercal caudal fin are distinguished, *i.e.* as follows:

(i) *Hypocercal.* In the hypocercal tail the vertebral column is bent downwards and the epichordal lobe is well developed, *e.g.*, are Anaspida among Ostracoderms.

(ii) *Epicercal.* In the epicercal tail the vertebral column is bent upward and the hypochordal lobe is well developed, *e.g.*, Dog fish (Elasmobranchs and lower bony fish).

Dephycercal

This is heterocercal fin which has become secondary symmetrical both. Externally and internally either by the reduction of larger lobe or growth of the smaller lobe so that both the lobes of the tail become of the same size. *e.g.* Dipnoi and Chimera.

Homocercal

It is modified heterocercal fin in which the fin is externally symmetrical but internally asymmetrical. The notochordal axis is

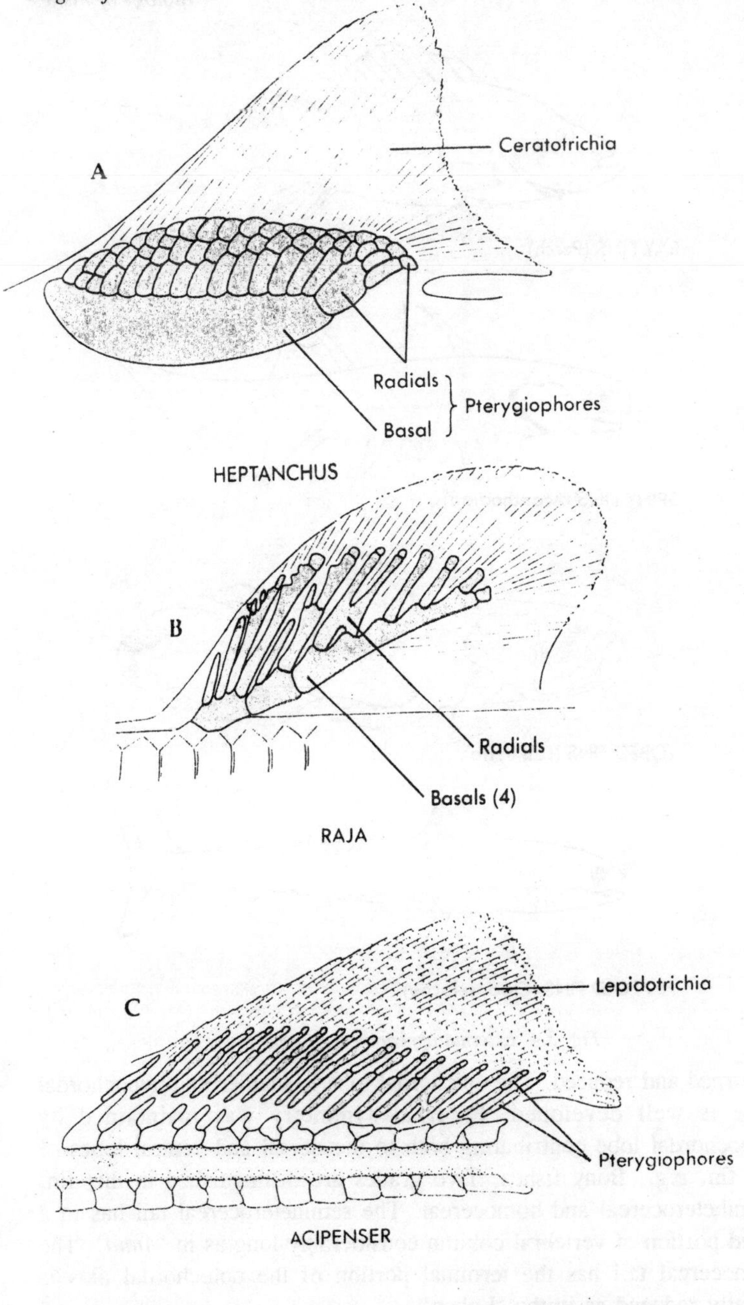

Fig. 3.3. Median dorsal fin skeletons of a generalized shark, a ray, and a sturgeon.

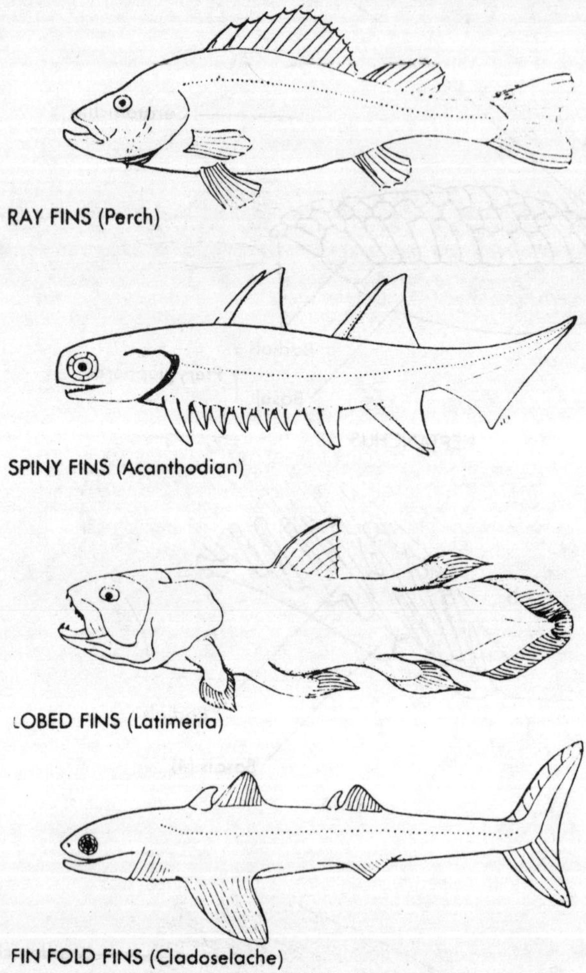

Fig. 3.4. External appearances of some fins.

upturned and reduced. The epichordal lobe is smaller and hypochordal lobe is well developed. External symmetry was maintained by hypochordal lobe contributing both to the dorsal and ventral lobes of the fin. *e.g.*, Bony fishes. Two grades are distinguished in this fin. Semiheterocercal and homocercal. The semiheterocercal tail has in a lined portion of vertebral column considerably long as in '*Amia*'. The homocercal tail has the terminal portion of the notochordal axis is greatly reduced as in the 'Lobeo'.

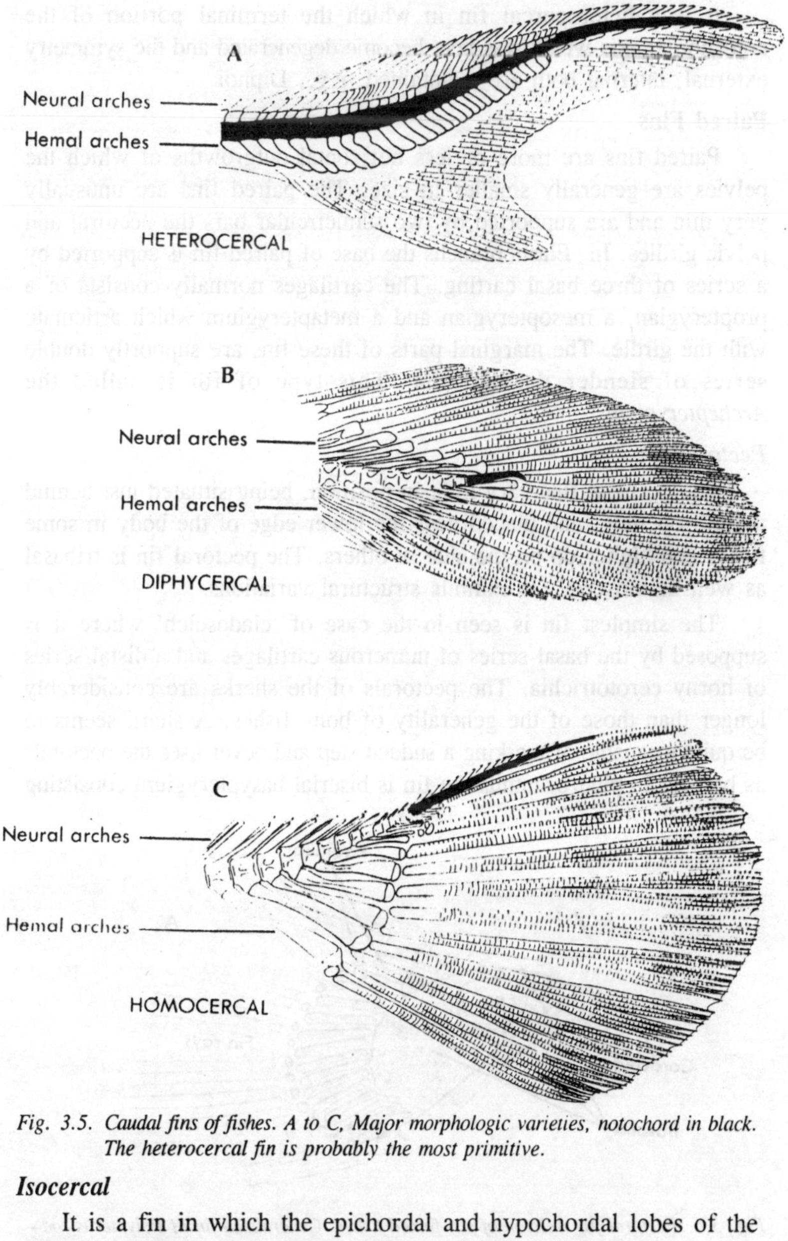

Fig. 3.5. Caudal fins of fishes. A to C, Major morphologic varieties, notochord in black. The heterocercal fin is probably the most primitive.

Isocercal

It is a fin in which the epichordal and hypochordal lobes of the fin get reduced and the dorsal and anal fin fold is established. e.g., *Anguilla* and *Notopterus*.

Gyphyrocercal

It is a heterocercal fin in which the terminal portion of the notochordal axis which was bent become degenerated and the symmetry external, internal both are maintained. *e.g.*, Dipnoi.

Paired Fins

Paired fins are more or less horizontal outgrowths of which the pelvics are generally smaller in size. The paired fins are unusually very thin and are supported by two semicircular bars the pectoral and pelvic girdles. In Elasnobranchs the base of paired fin is supported by a series of three basal carting. The cartilages normally consists of a propterygian, a mesopterygian and a metapterygium which articulate with the girdle. The marginal parts of these fins are supportly double series of slender dermatricha. This type of fin is called the *Archepterygium*.

Pectoral fin

The pectoral fins vary little in position, being situated just behind the gill opening, and placed near the lower edge of the body in some fishes and higher up on the side in others. The pectoral fin is tribasal as well as uniserial. It exhibits structural variations.

The simplest fin is seen in the case of 'cladoselch' where it is supposed by the basal series of numerous cartilages and a distal series of horny cerototrichia. The pectorals of the sharks are considerably longer than those of the generality of bony fishes. A shark seems to be quite incapable of marking a sudden step and never uses the pectorals as brakes. In pleurocanthus the fin is biserial basypterygium consisting

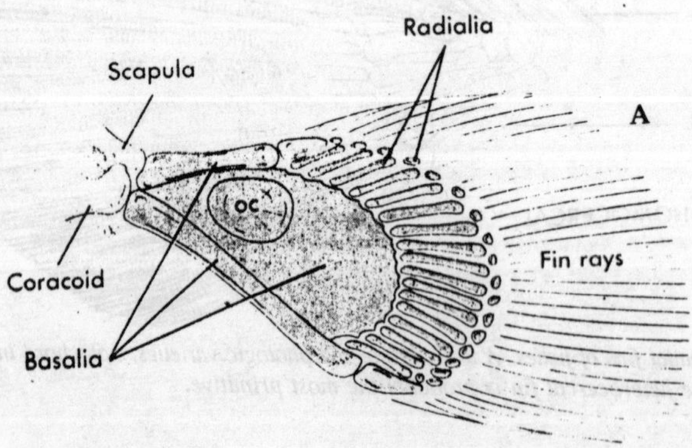

Fig. 3.6. Pectoral fin skeleton of ray-finned fishes. Generalized fin of Polypterus. oc—Ossification center in central basal cartilage (mesopterygium).

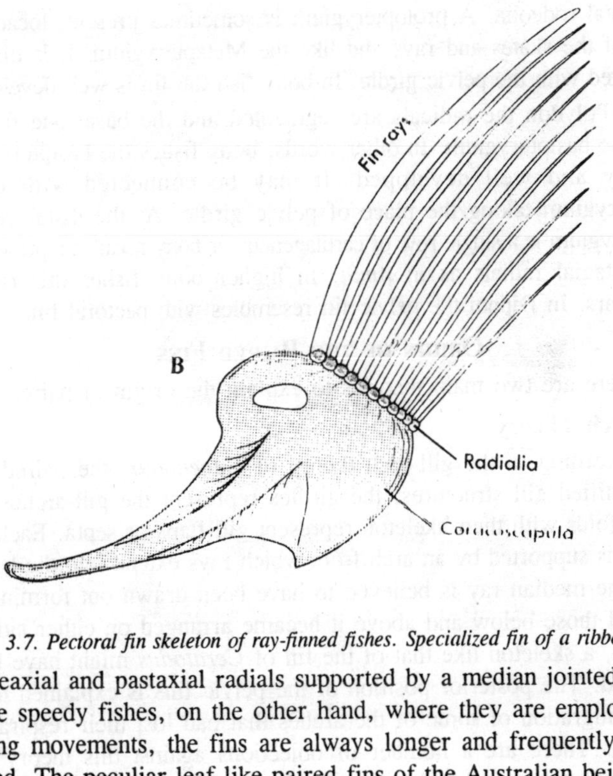

Fig. 3.7. Pectoral fin skeleton of ray-finned fishes. Specialized fin of a ribbonfish.

of preaxial and pastaxial radials supported by a median jointed axial. In the speedy fishes, on the other hand, where they are employed in turning movements, the fins are always longer and frequently sickle shaped. The peculiar leaf like paired fins of the Australian being fish *Meoceratody* have been described.

In the *Protopterus* and *Lepidosiren* the central lobe is long and narrow fins take the form of long tupering filaments. In some of the pipe fishes however and in certain of the eels, they are absent. In the remaining Teleostomi the endoskeletal elements become gradually reduced in the number and importance.

Pelvic fin

In the simplicity of their endoskeletal supports the pelvic fins of 'Clacloselachae' are the most primitive type of paired fins at present known. In general structure they resemble the pectorals but the radials are few in number, less modified by concentration and exhibit little of any, trace of basal fusion. In '*Pleurocalthus*' the pelvic fins differ from the corresponding pectorals in being uniserial instead of beserial. All other Elasmobranchs, including the Holocephali have uniserial fins, which consist of a large metapterygium, supporting a preaxial fluange of segmental rodeolia. A protopterygium is sometimes present, locably in some of the skates and rays and like the Metapterygium. It is directly connected with the pelvic girdle. In bony fish the fin is well developed.

In Polydon the radicals are segmented and the basal one fuse to form the basipterygium. In other words, bony fishes the basipterygium is bony and well developed. It may be connected with other basipterygium taking the place of pelvic girdle. At the distal end of basipterygium is a single row of cartilagenous or bony nodules representing the vestagial radials as in *Amia*. In higher bony fishes this radials disappears. In Dipnoi the pelvic fin resembles with pectoral fin.

Origin of the Paired Fins

There are two main theories to explain the origin of paired fins:

Gill Arch Theory

According to the gill-arch theory of *Gegenbaur*, the paired fins are modified gill structures, the girdles represent the gill-arches and the fin folds with their skeleton represent gill-flaps or septa. Each gill septum is supported by an arch from which rays extend into the lateral part. The median ray is believed to have been drawn out forming an axis and those below and above it became arranged on either side of it. Thus, a skeleton like that of the fin of *Ceratodus* might have been produced. The posterior position of the pelvic fins is explained to be due to migration of some of the arches that had lost their respiratory function. There are a number of objections against this theory and several morphological and embryological facts do not lend support to it as mentioned below:

1. The presence of some vestigial muscle buds in front of the pelvic fins is believed to indicate their backward migration. But the muscle buds are found the pelvic fins also, as well as on both sides of the pectoral fins.
2. During development, the branchial arches arise from the splanchnic wall of the coelom but the girdles of the paired appendages are somatic in origin. Thus, the gill arches and the limb girdles differ basically from each other regarding their position relative to the coelom, nerves, blood vessels etc. Morphologically, the gill arches are internal while the girdles are external to these structures.
3. If the paired fins are derived from the gill septum, they would be transverse to the long axis of the body and would be an obstruction to swimming movements. During development also, a paired fin does not appear as a dorso-ventral fold and is always in the form of a longitudinal fold.

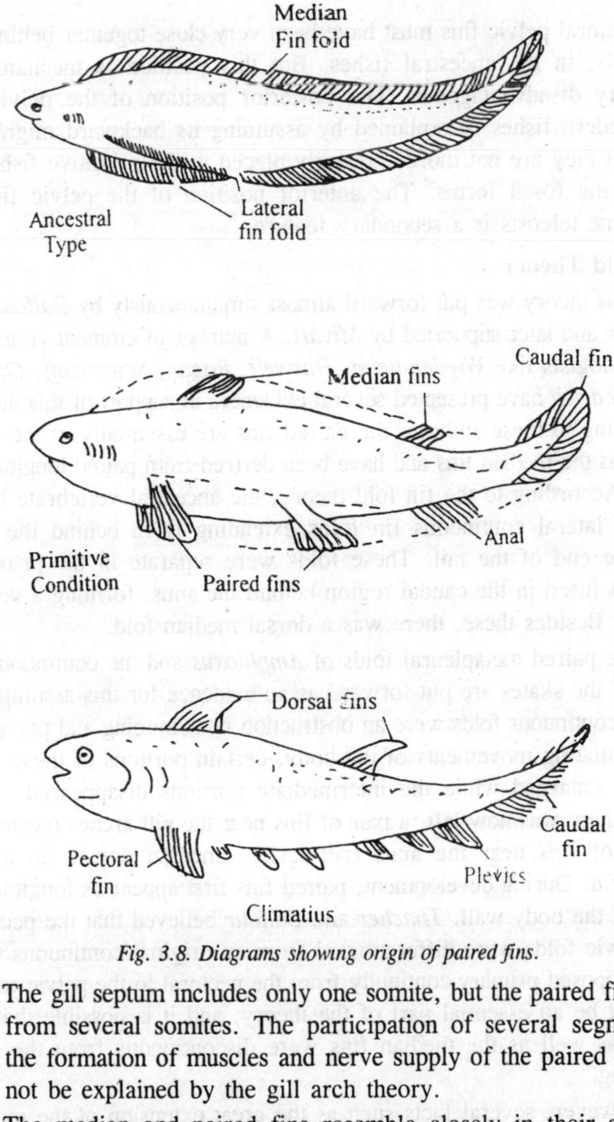

Fig. 3.8. Diagrams showng origin of paired fins.

4. The gill septum includes only one somite, but the paired fins arise from several somites. The participation of several segments in the formation of muscles and nerve supply of the paired fins can not be explained by the gill arch theory.
5. The median and paired fins resemble closely in their structure and development. This can not be explained by the gill arch theory.
6. The posterior position of the pelvic fins may be due to the ancestral Gnathostome having a large number of gill arches. But there is no evidence to support this supposition.

7. Pectoral pelvic fins must have been very close together behind the gills, in the ancestral fishes. But this position is mechanically very disadvantageous. The posterior position of the pelvics in modern fishes is explained by assuming its backward migration. But they are not more anteriorly placed in the primitive fishes or in the fossil forms. The anterior position of the pelvic fins in some teleosts is a secondary feature.

Fin Fold Theory

This theory was put forward almost simultaneously by *Balfour* and *Thacher* and later supported by *Mivart*. A number of eminent vertebrate morphologists like *Wiedersheim*, *Haswell*, *Ragan*, *Sewertzoff*, *Osburn* and *Goodrich* have presented several evidences in support of this theory. According to these authors, the paired fins are essentially of the same nature as the median fins and have been derived from paired longitudinal folds. According to the fin fold theory, the ancestral vertebrate had a pair of lateral continuous fin folds extending from behind the gills upto the end of the tail. These folds were separate in the proximal part but fused in the caudal region behind the anus, forming a ventral finfold. Besides these, there was a dorsal median fold.

The paired metapleural folds of *Amphioxus* and the continuous fin folds of the skates are put forward as an evidence for this assumption. As the continuous folds were an obstruction in swimming and prevented the undulating movements of the body, certain portions of these folds became enlarged while the intermediate portions disappeared. As a result, there was now left, a pair of fins near the gill arches (pectorals) a pair of fins near the anus (pelvics), a median dorsal fin and a caudal fin. During development, paired fins first appear as longitudinal folds of the body wall. *Thacher* and *Balfour* believed that the pectoral and pelvic folds were differentiated from an original continuous fold. This supposed primary continuity from the pectoral to the pelvic region may not be an essential part of the theory, and it is possible that the paired as well as the median fins were discontinuous from the very beginning.

However, several facts such as the great extension of the muscle bud areas and the nerves, both in front and behind the paired fins, the concentration of the fin elements, shortening of the fin-base, and the frequent presence of a larger number of muscle buds in early stages, may be considered as evidences in favour of the presence of continuous fin folds in the ancestral animals. This theory receives further support from a number of facts from embryology and palaeontology: (i) The mode of development and early structures of the paired and median fins of fishes are identical. (ii) In the elasmobranch embryos and in the extinct shark *Cladoselache*, the paired fins are in the form of longitudinal folds of the body wall and have parallel arrangements of their fin rays. This suggests their origin from a continuous fin fold. (iii) In *Climatius*, there was a series of smaller spines and finlets between the pectoral and pelvic fins. This suggests that they were originally joined by a continuous fold. The absence of paired appendages in Cyclostomes may be mentioned the view that the ancestral craniates had continuous paired fin folds. But as the Cyclostomes are probably the degenerate of Ostracoderms, it appears that the lampreys and hagfishes have probably lost their paired fin folds which were possibly present in the Ostracoderms.

LOCOMOTION

Many invertebrates spend the greater part of their lives attached to or crawling sluggishly over a small area of the sea bottom; others float more or less passively at the surface or in the middle layers of the water, their movements largely dependent on the action of the tides and currents. Squids and cuttle-fishes alone approach the fishes in speed and grace of form, but lack their agility in the water, and are generally far inferior to them in their mastery of the medium. Of the many are varied forms of animal life found in the sea and in fresh water few are more perfectly adapted for life in a liquid medium than are the fishes. Fishes are variously adopted for locomotion in water. The body in spindle shaped and is perfectly streamlined for movement through water. The body is covered with a layer of mucous which reduces drag on it.

Typically, the shape of the body is fusiform but variations occur when the fish become adapted to specialized mode of life. In fishes, body form is the result of interaction of the skeleton of the muscle, and in evolutionary adoption of both these systems to the lives of individual kinds of fishes. The adaptation are for various kinds of swimming although some fishes have also become adopted to limited locomotion on land. The generalized body shape is fusiform for most aquatic animals of a free swimming kind, be they fish, mammals or invertebrates. The ideal fusiform shape are retained by many pelagic fishes such as tunas and mackerels (scombridae) and several of the require sharks (carcharhinidae). Fish forms deviate from the ideal in three directions—compression, depression and elongation. Dorsoventral flatening is characteristic in varying degree of many stream fishes, ostensibly as a means for place holding in stream curras as well as of those which take shelter benefit stones or other objects on the bottom. Common examples of such flattening, especially in the head region are among the sculpins (cottidal) and widely distributed catfishes (cypriniform sub-order siluroides) of all the major continents.

Fins and Locomotion

The fins are important in fish locomotion. The caudal fin is of greatest importance for swimming in fishes with a more or less round

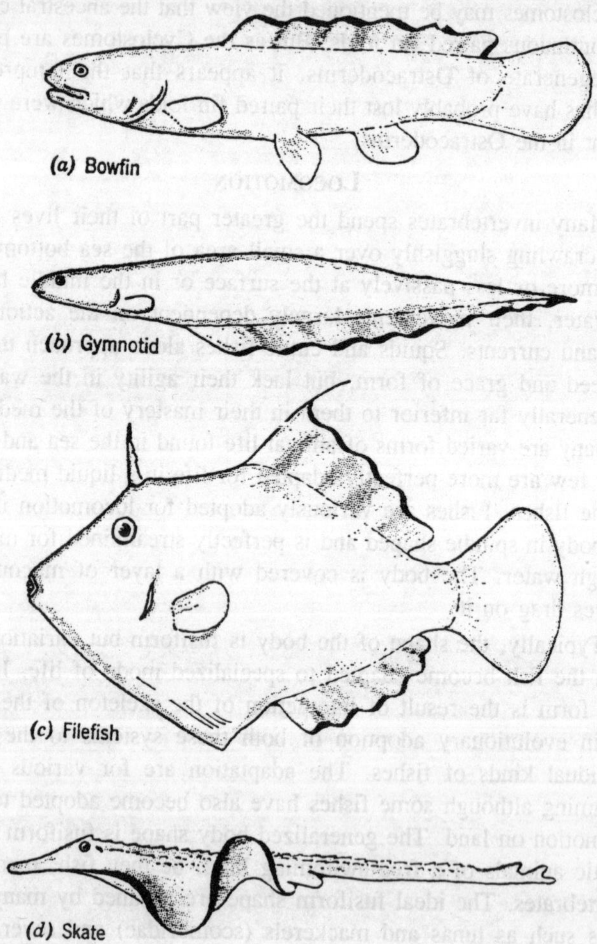

Fig. 3.9. Locomotion through undulation of longitudinal structures in: (a) the bowfin (Amia calva); (b) a gymnotid "eel" (Gymnotus); (c) a filefish (Monacanthus); and (d) a skate (Raja).

cross section such as several of the sharks, whereas in laterally compressed fishes the caudal peduncle is of greatest importance. A few highly specialized kinds of fishes move exclusively by means of their fins, as for example, the seaborns (*Hippocampus*), and the trunkfishes (*Ostraciidae*). All of the various fins, excepting, the pelvic, are used in some species as primary organs of propulsion.

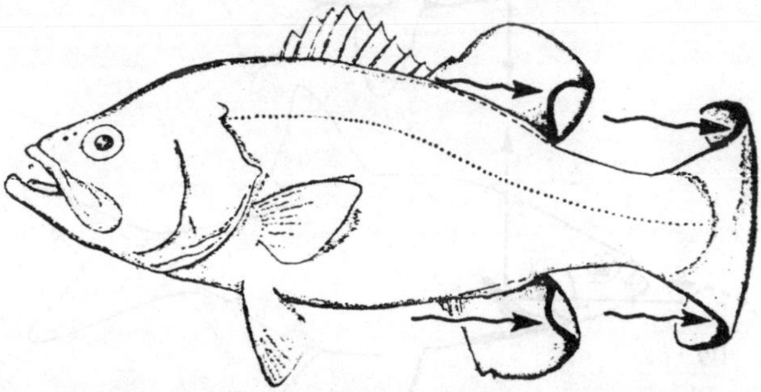

Fig. 3.10. Median fins curved to opposite sides to act as a sea anchor in the largemouth bass (Micropterus salmoides).

The long based dorsal of the bowfin (*Amia*) can be undulated to give forward or backward movement to the fish. The anal is similar in elongation and function in the featherbacks (*Notopterus*) and in the electric eel (*Electrophorus*), in both of which is actually extends candal for enough to fuse with the tail fin. Both the dorsal and the anal fins are moderately long based in the triggerfishes and filefishes (*Balistidae*) and undulate to give slow forward or backward progress to the fish, even when the tail fin is at rest. The pectorals, by undulation are the basic locomotor structure in the skates, stringray, eagarays and mantas (*Rajiformes*).

The pelvics and pectorals too can serve as beilge keels to retard rolling although the principle functions seems to be for such manoeurers as climbing and diving, banking, turning and stopping in the water. For these action, the pectorals are the primary organs and the pelvics the secondary organs, as learned from experimental removal of fins from living fishes. In the sharks (squaliformes), the paired fins tend to over, elevated the body apparently in part of offset the lack of a gas bladder.

TYPES OF LOCOMOTION

Locomotion is fish may be two types: (1) passive; and (2) active.

Passive Locomotion

The pelagic eggs and young of many fishes are often passively transported by currents. This happens in oceans, lakes and streams

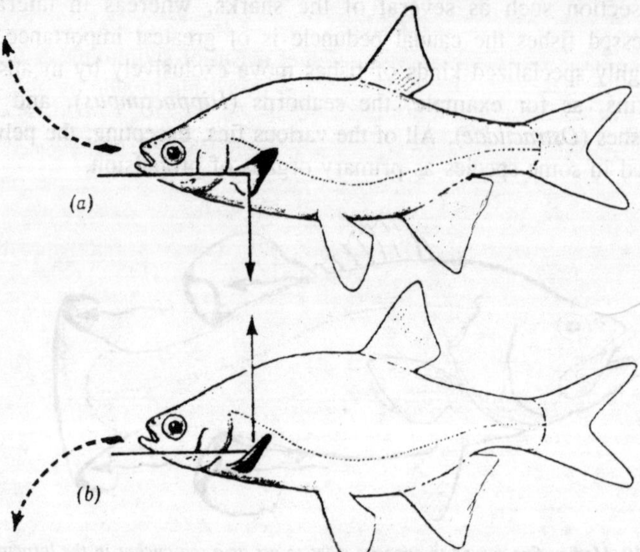

Fig. 3.11. Use of the pectorals to direct climbing (a) or diving (b) in the water by the golden shiner (Notemigonus crysoleucas). The curved arrows indicate direction of travel and the angled arrows, vectors of force.

where such eggs and youngs are induced a part of the plankton and often appear in thousand samples. Unbearing fishes afford passive conveyarto their young inside the mother. After birth or hatching there is also some protective transport of the young among fishes by temporary protective engulfment in the mouth cavity of a parent, as in some tilapicus (such as *Tilapia mossambica*) catfishes (*Cyprinefornes* suborder silluroidea) sticklebacks (Wastesosteidae) and other species which actively guard the young. A brood pouch is also a sort of tonnean in which the young ride about, as in the pipefish (*Syngnathus*) and seahorses (*Hippocampus*). Passive travel may also be by hitchhiking on another organism or object.

The most renowned hitchhikers among fishes are the marine suckerfish the remoras (*Echeneidae*), which attack harmlessly to sharks and other fishes, sea turtles and even to whales. The organ of attachment is on top of the head is a modification of the dorsal fin. A most unusual kind of passive transport of fishes occurs when water masses containing fish are picked up into the air by strong wind currents resulting in water spouts. Later the fishes may come down again in a rain. Man accounts for considerable passive transport of fishes by land, sea and air.

Active Locomotion or Swimming

The tail and caudal fin are the chief organ of swimming in fishes. During the swimming, the tail is laished from side to side by alternate contraction and relaxation of muscles on the opposite sides of the vertebral column. During such movements, the tail is first bent to one side. This is called the *non-effective* or *backstroke*. By a stroke in the reverse direction, the tail is extended and straightened. This is the *forward* or *effective stroke*. Swimming movements divided into three broad categories which are as follows:

1. Carangiform or Jack like.
2. Anguilliform or eel like.
3. Ostraciform or trunkfish like.

Carangiform or jack like

The most common type of locomotion is carangiform locomotion when a fish wants to move forwards, the first action is the contraction of first few myomeres of the anterior and of the body or one side only, so that the head is bent sharply to one side. This is followed by alternate contraction and relaxation of the successive myomeres, first on one side of the body and then of the other, from head towards the tail and the curve of the body passes backwards. The fish thus shows undulating movements and its body during swimming is thrown into a 'S' shaped curve. The actual forward thrust is affected by the pressure of the tail of the fish against the surrounding water.

Anguilliform or eel like

The anguilliform type of locomotion is serpentine in nature, as seen in a crawling snake, vaving flag, or swimming eel. In this the movement is initiated by the contraction of first few myomeres on one side, so that the anterior part of the body converted is seen into a curve, which passes backwards in a series of waves by alternate contraction and relation of the muscle segments, on each side of the body. The movement is of serpentine nature and the first looks like a crawling snake. In this case the forwards thrust is obtained entirely by the pressure of the fish body against the water and the caudal fin has little function in anguilliform type of movement. Hence, the caudal fin is very reduced or absent in eel like fishes. Instead, the hinder part of the body is laterally compressed and the blade like structure thus formed, serves to provide a large amount of thrust of fish. Fishes that have long, laterally compressed body as the Ribbon fishes (*Trichiurus*) also swim very actively by undulating movement of their body.

Ostraciform or trunk like

This is commonly seen in trunk fish. In this case the head and body is enclosed in hard and rigid bony case and the tail with a fan like caudal fin projects behind. In such fishes, undulating movements of the body are not possible and the dorsal and anal fins form the chief propelling agents for slow movements. Rapid swimming takes place by lashing movements of the tail.

Myomeres and swimming

When the muscle fibres of myomere contract they shorten. If the myomere on one side of a fish are shortened, the head and tail bend towards that side and the myomeres on the opposite side of the body stretch. The active of a muscle segment is affected by two forces. One of these is muscular, curving the body, the other is resistance which straightens the body. Thus alternative serial contractions of the myomeres on the two sides flex the fish, or successive parts of the fish into bows. The number of contraction waves passing down the body range from about 5 per minute (in sharks, *i.e.*, Squaliformes) to 170 in Mackerels, *i.e.*, Scombridae). Slow contraction waves are of high amplitude, whereas rapid waves are of low amplitude. The wave resulting from the serial contraction passes smoothly along the body, from head towards the tail, because of the overlapping of the muscle segments as in the gnathosiomes.

Non-swimming Locomotion

Although the swimming is the typical form of locomotion for fishes, beside these they are also jumping burrow, climbing, leap, soar and propel themselves by jet. Some fishes jump out of water with a considerable force either escape from an enemy or to clear an obstacle. Thus salmon, mugil, cyprinus, megalops, (Tarpon), sail fish etc. are able to take a leap out of water. Some species like the Exocoetus, have enlarged pectoral fin and glide like a parachute for some distance, after jumping out the water. The pelvics are also enlarged along with the pectoral in *Cypselurus* for gliding in air.

Some species burrow into the mud, making use to their swimming movements for the purpose. Fishes like the eel (Muraenidae) crawl on the bottom making use of body muscles and fins. Some like the toad fish, lophius, frog fish (*i.e.,* Antennarins) etc. Crawl on the bottom with the help of pectoral fins that are used like arms. The climbing perch (*Anabas*) can even climb up the trees with the help of its pectorals and operculline. The mud skepper (*Periopthalmus*) uses its pectoral fins which are bent at an angle like an elbow joint, for jumping over the land.

Speed of Travel

Speed of travel in fishes may be thought of as cruising speed, marcunea sustainable speed and top speed. Cruising speed and those involved in ordinary travel. Maximum sustainable speeds are those which a fish can maintain for lengthy inter svales. They are higher than cruising speeds but lower than top speed. Top speed involves maximum thrust for short duration. Actual travel speeds vary by species, size of the individuals and temperature of water. In general they approximate about ten times the length of the body, in feet per second, thus for any species the larger the fish greater the speed.

4

DIGESTIVE SYSTEM

Like the epidermis, cells that the line gastrointestinal tract have direct encounter with the external world. A major function of the epidermis of most vertebrates (amphibians excepted) is to limit strictly the entry of chemical substances and parasitic organisms from the outside, while permitting egress from the body of specialized secretions (slime, oil, sweat, and so on). In striking contrast, the major function of the epithelium of the digestive tract is to facilitate the entry of a wide range of chemical substances into the body. This is not an easy task. In involves the formation of enzyme-containing secretions needed for the chemical breakdown of food materials. Often it must be supported by the mechanical breakdown of food objects, and transfer processes are needed to bring nutrients into suitable relationship to the surfaces where entry takes place.

Various self-protective devices are present that, in general, limit to a certain extent the ingress of harmful substances and invading organisms. The digestive system also has other, somewhat less conspicuous roles. It removes undesirable substances from the body through the excretory processes of the liver and the mechanical egestion of food materials taken in but undigested. Relatively large masses of bacteria and of dead cells lost from the walls of the digestive tract itself also are discarded. As will be noted later, the tract contains structures important for a wide range of additional important bodily functions—search for the selection of food, self-defense respiration, reproduction, and kidney excretion—through activities of some of its specialized regions (*buccal cavity*, *pharynx*, *aboral intestine*, *cloaca*). Fishes have become adapted to a wide variety of food. Some of them feed exclusively on plants, others feed on animals, while a large number of species are omnivorous, deriving their requirements from both animals and plants.

A few fishes depend for their nutrition entirely on zoo- and phyto-plankton, and are called *plankton feeders*. A wide range of organisms are used as food by the fishes Zooplankton includes various kinds of protozoa, microcrustaceans, microscopic invertebrates, larvae and eggs of various animals. Among the larger animals consumed by fishes, are the annelid worms, snails, mussels, crustaceans, insects, their larvae, smaller fishes, and even frogs and tadpoles. The plant material includes unicellular algae, filamentous algae and portions of higher aquatic plants. Sand or mud is also taken in small quantities by several species of fishes along with other items of food.

FEEDING HABITS

Fresh water fishes can be divided into the following feeding types:

Herbivorous Fishes

A number of fresh water fishes feed mainly on unicellular algae, filamentous algae, portions of higher aquatic plants along with some sand or mud. As the plant material in their food is 75% or more of the total gut contents they are considered herbivorous in habit. Animal food usually varies from 1–10% in their diet. These species are *Labeo rohita*, *L. gonius*, *L. bata*, *L. fimbriatus*, *Schizothorax sinuatus*, *S. plagiostomus*, *Ctenopharyngodon idella*, *Amblypharyngodon mola*, *Tilapia mossambica*, *Osphronemus goramy* etc.

Omnivorous Fishes

Food of a large number of fishes consists of unicellular and filamentous algae, portions of higher aquatic plants, rotifers, insects and their larvae, crustaceans, Bryozoans, mud and sand. They are regarded *omnivorous* species and include *Cyprinus carpio*, *Tor putitora*, *Tor tor*, *Cirrhina mrigala*, *Puntius sophere*, *P. tioto*, *P. sarana*, *Heteropneustes fossilis*, *Clarias batrachus*, *Gadusia chapra*, *Osteobrama cotio*, *Esomus danricus*, *Eutropichthys vacha*, *Colisa fasciatus* etc. The food of these fishes consists of a varying percentage of plant and animal material, and they form a sort of link between the herbivorous and carnivorous fishes. Some of these feed on much larger amount of plant material like *Gadusia chapra*, *P. sarana*, *P. sophore*. A few like, *Cirrhina mrigala*, *Colisa fasciatus* feed on nearly equal amount of plant and animal material, while others like *O. cotio* and *Entropichthys vacha* have a greater amount of animal food in their diet.

Carnivorous Fishes

Food of a number of fishes consists of a very high percentage of animals has crustaceans (Copepods and Daphinds), insects (water-bugs, beetles, dragon flies, damsel flies), larvae, molluscs, smaller fishes and tadpoles. These species are *Wallago attu*, *Mystus seenghala*, *M. cavassius*, *M. vittatus*, *Channa striatus*, *C. marulius*, *C. gachua*, *C. Punctatus*, *Ompok pabda*, *Notopterus chitala*, *Rita rita*, *Chanda nama*, *C. ranga*, *Glossogobiun giuris*, etc. Some of these are active predators such as *W. attu*, *C. striatus*, *C. marutius*, *N. chitala* and *M. seenghala*.

Plankton Feeders

Some species feed on phyto and zoo-plankton, which they obtain by filtering water through their gill rakers. These are *Catla catla*, *Hilsa ilisha*, *Gadusia chapra*, *Cirrhina reba*, and *Hypopthalmichthys molitrix*. They can be both omnivorous and carnivorous.

Surface, Mid and Bottom Feeders

Das and *Moitra* (1955, 1963) have divided fresh water fishes into three groups according to the niche they occupy in the water. These are surface feeders, mid- or column-feeders, and bottom feeders. The surface feeders are *Catla catla*, *Hypopthalmichthys molitrix*, *Puntius ticto*, *Oxygaster bacaila*, *Chanda nama*, *C. ranga*, *Glossogobius giuris*, *Gadusia chapra*, *Hilsa ilisha*. Some species are neither true surface feeders, nor true bottom feeders and mostly depend on the organisms of the mid water. Such species are also called *column feeders* and include *Labeo rohita*, *Puntius sophore*, *Wallago attu*, *Mystus cavassius* and *M. vittatus*. The bottom feeders depend mainly on bottom organisms. Fishes included in this group are *Labeo calbasu*, *L. bata*, *L. gonius*, *Cirrhina mrigala*, *C. reba*, *Puntius sarana*, *Amblypharyngodon mola*, *Clarias batrachus*, *Heteropneustes fossilis*, *Channa marulius*, *C. striatus* etc. Of these the surface feeders are either omnivorous or carnivorous, while the mid and bottom feeders may be herbivorous, omnivorous or carnivorous.

ALIMENTARY CANAL

The alimentary canal is a muscular tube with an epithelial lining, formed for the reception and the digestion of the food. It begins with a mouth, and from thence it extends backwards through the coelom, finally communicating with the exterior either by a cloacal or by an anal orifice. The oral or buccal cavity into which the mouth leads is a stomodaeum, and is lined by inpushed epidermis, while the hinder portion of the cloaca and the anus are lined by a somewhat similar inpushing of the epidermis which forms the proctodaeum. The rest of

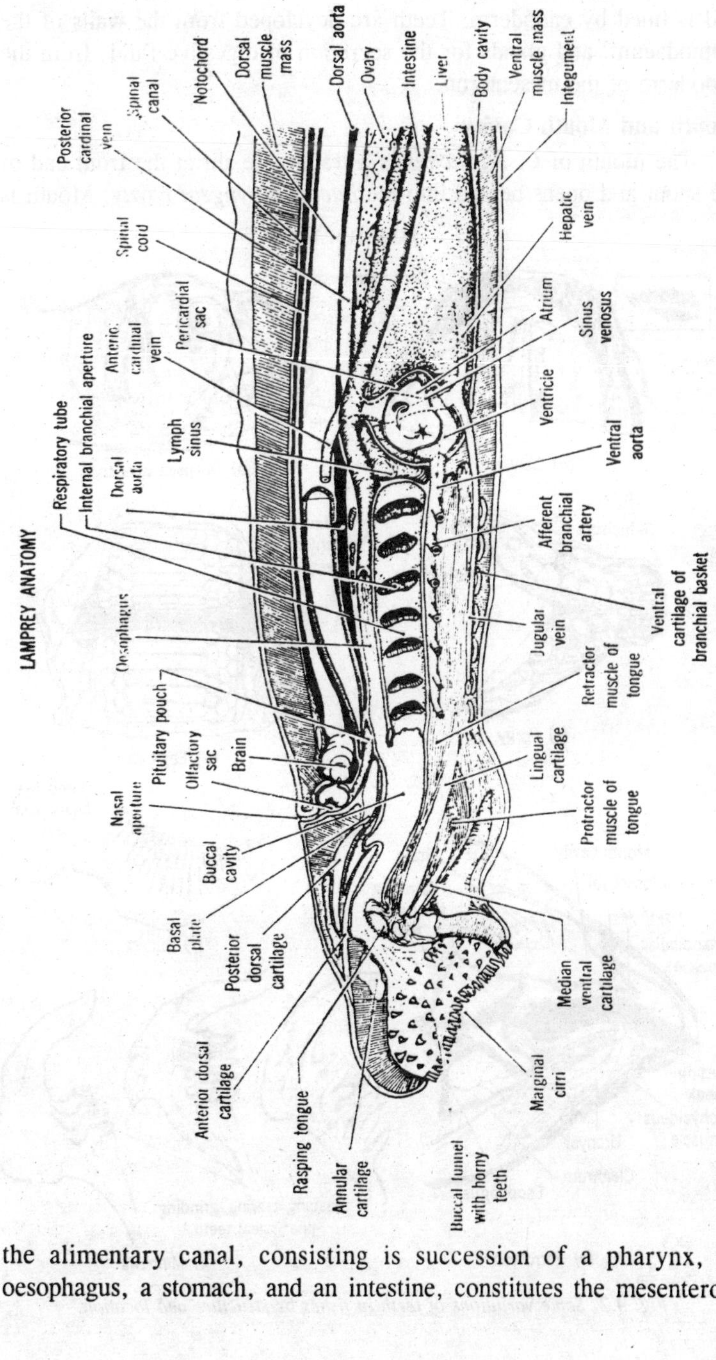

Fig. 4.1. Jawless mouth, foregut, and related structures of an adult sea lamprey (Petromyzon marinus) as seen in sagital section.

the alimentary canal, consisting is succession of a pharynx, an oesophagus, a stomach, and an intestine, constitutes the mesenteron, and is lined by endoderm. Teeth are developed from the walls of the stomodaeum, and glands for the secretion of digestive fluids from the endoderm of the mesenteron.

Mouth and Mouth Cavity

The mouth of *C. batrachus* is a transverse slit at the front end of the snout and opens behind into the *buccopharyngeal cavity*. Mouth is

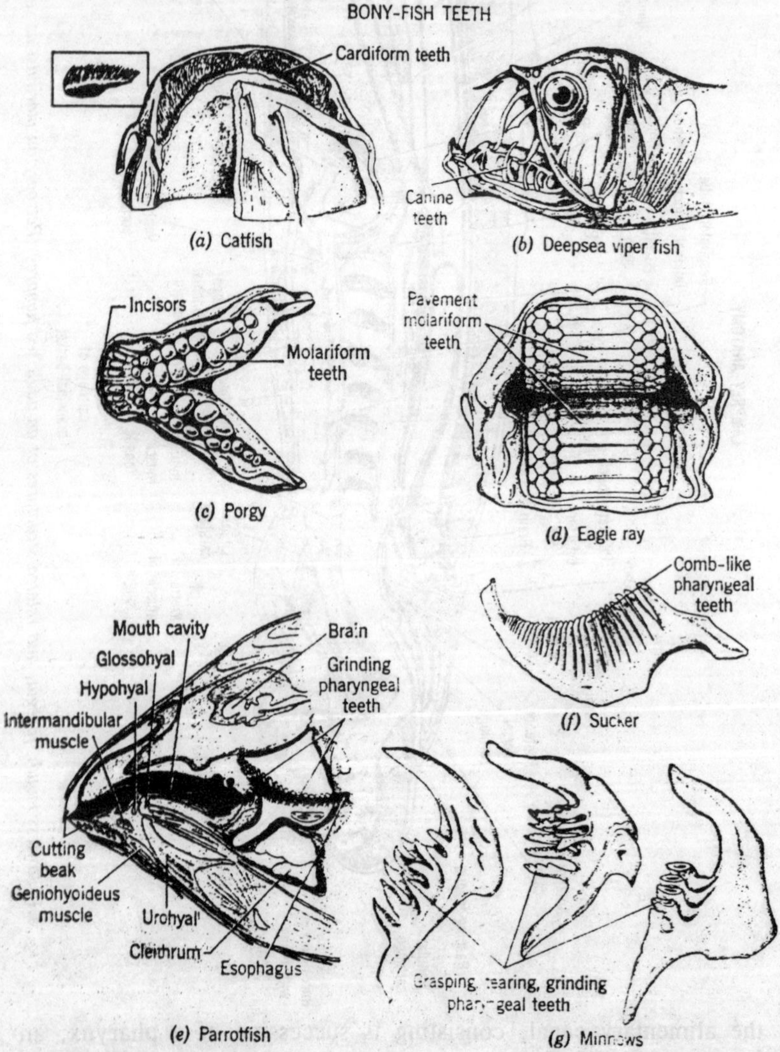

Fig. 4.2. Some variations of teeth in fishes by structure and location.

surrounded by four pairs of barbels that are used for searching food. Numerous find pointed teeth are present in two groups, in the roof of the buccal cavity, behind the upper lip. These are the *maxillary* and *vomerine teeth*. *Mandibular teeth* are present behind the lower jaw. The rest of the buccal cavity is lined with soft mucous membrane that may be folded slightly. The ventro-lateral wall of the pharynx is perforated by oblique gill slits, separated by means of four pairs of *gills*. Each gill has a longer lower limb and a short upper limb and bears gill lamellae towards the outer side and gill rakers internally.

The gill rakers are fairly long, hard and pointed structures projecting into the pharyngeal cavity and are arranged either in one or two rows on each arch. The gill rakers of the first arch are the longest while those on the fifth arch are the smallest in size. They serve to prevent the food from escaping out through the respiratory current of water. Behind the gill arches, at the hind end of the pharyngeal cavity, is a large circular aperture, the *gullet*, leading into the oesophagus.

Oesophagus

In fishes the oesophagus is very short and its junction, with the stomach is almost imperceptible. In certain elasmobranchs, as in *Squalus acanthias*, numerous backward-projecting papillae line the oesophagus as well as the first part of the stomach. The oesophagus commonly bears longitudinal folds which permit a considerable degree of distention. Some fishes are actually capable of swallowing others larger than themselves. Histologically, oesophageal mucosa is composed of stratified epithelium in the anterior region and columnar epithelium in the hinder part or a mixed epithelium is present.

A large number of mucous-secreting cells are scattered in the mucosa and taste buds are also present in some species (*Labea rohita*, *Schizothorax plagiostomus*, *Tor tor*, *Catla catla*, *Mystus gulio*) but may be absent from others (*Trichiurus*, *Caranx*, *Rita rita*). The muscularis consists of a thick layer of a circular muscle fibres. Longitudinal muscle fibres are scattered bundles in the inner side.

Stomach

The stomach is basically a dilatation of the digestive tract for the temporary storage of food. Only when its lining epithelium contains gastric glands (with certain special exceptions) it is properly called a *true stomach*. Its digestive function is apparently a secondary acquisition. The end of the stomach which connects with the oesophagus is nearer the heart and, therefore, is called the *cardiac end*. The main portion

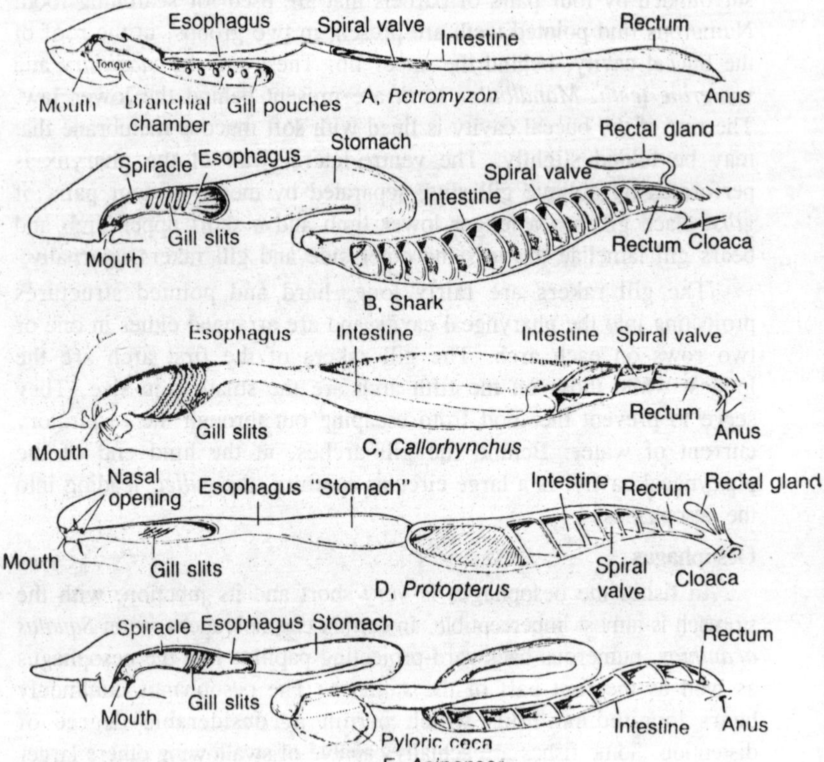

Fig. 4.3. Digestive tracts of A—a lamprey; B—a shark; C—a chimaera; D—a lungfish; E—a sturgeon; F—a teleost (perch). The "stomach" of the lungfish is nonglandular and is simply a somewhat enlarged section of the oesophagus.

is called the *body*. The *pyloric end* connects with the intestine and terminates at a valve, the *pylorus* or *pyloric valve*. This consists of a fold of the lining mucous membrane surrounded by a thick, involuntary, sphincter muscle which regulates the passage of the contents of the stomach into the intestine. There is practically no distinction between oesophagus and stomach in fishes and the longitudinal folds of the former may extend for some distance into the stomach.

A considerable variety of stomach shapes may be observed. Some are simple, straight tubes without any digestive function, as a Dipnoi, Chimaeras, and a number of teleosts. In others, as in *Polypterus*, the cardiac and pyloric limbs have fused along the line of the lesser curvature so that the stomach appears much as a blind pouch. All the fishes do not posses a true stomach and it is absent in a number of species.

In such species, the anterior part of the intestine is swollen to form a sac behind the oesophagus. This structure serves for the storage of food and is known as *intestinal bulb* or intestinal swelling as in *Lobeo rohita*, *L. gonius*, *C. mrigala*, *Tor tor*, *Catla ctala*, *Puntius sophore*, *B. sarana*. This is a special feature of the cyprinids but the presence or the absence of a true stomach is not related to the feeding habits of the fish. Stomach is absent in Holocephali and Dipnoi also. Besides, the cyprinids, some other fishes, *Scomberesox*, *Hippocampus* and *Xennetodon* are without stomach. In Elasmobranchs are J-shaped stomach is typical. The pyloric limb is smaller than the cardiac portion.

Many other fishes have similarly shaped stomachs. Among teleosts there is greater variety of stomach shapes than in the other groups of fishes. In some teteosts a ciliated lining is present. *Gastric glands* are not present in the intestinal bulb and its mucosa resembles closely with that of the intestine. Absorptive cells and the mucous-secreting cells are the two main types of cells in the mucosal epithelium of the intestinal bulb.

Intestine

The length of the intestine varies in different species. Generally, intestine is short and nearly straight in carnivorous fishes, but long, thin walled and highly coiled in herbivorous species, while the omnivorous forms show and intermediate condition. *Das* and *Moitra* (1956) have shown that the ratio between the gut length and body length is fairly constant in a large number of species. However, some carnivorous and voracious fishes as *Pterois*, have a long looped intestine. In Elasmobranchs the small intestine in shorter than the stomach. It is wide, but rather straight, and contains a well-developed *spiral valve* which greatly increases the surface for digestion and absorption. In *Squalus acanthias* the spiral valve takes 14-1/2 turns.

In some elasmobranchs the number of turns is far in excess of this. Spiral valves are also found in the small intestines of the Chimaeras, Dipnoi, *Latimeria*, members of the superorder *Chondrostei*, *Amia*, *Lepisosteus*, and others. Traces are to be found in certain teleosts. The presence of a spiral valve in some fossil placoderms is an indication of its primitive character. The spiral valve in certain elasmobranchs, as in *Charcharias*, has been modified to form a scroll-like structure extending the length of the small intestine. The elasmobranch large intestine is a short passageway extending from the small intestine to the anus. It bends slightly before opening into the cloaca. A long, slender, fingerlike *rectal gland* connects to the intestine by means of a duct near the point where small and large intestines join.

Some biologists in the past have homologized the rectal gland with the colic caecum of higher forms. This is highly unlikely. It function was long in doubt. It is now known that the rectal gland secrets a highly concentrated solution of NaCl, thus ridding the blood of excess salt. In the Dipnoi a *cloacal caecum* is present. Many fishes have *pyloric caeca* coming off the part of the intestine which immediately follows the pylorus. The *Polypterus* there is but one. In some teleosts, as in the mackerel, for example, approximately 200 separate pyloric caeca may be present. Sometimes the pyloric caeca are bound together in a compact mass by connective tissue.

Rectum

The rectum is not generally distinguishable externally but an *ileo-rectal* is present in some species to demarcate it from the intestine (*Sciana*, *Tetrodon*, *Muraenosox*). The mucosal folds of the rectum differ from those of the intestine in being shorter and broader, possess a large number of mucous-secreting cells and a thicker muscularis. Large number of mucous-secreting cells produce copious mucous to lubricate food and help in easy defecation.

Histology of Alimentary Canal

The whole length of the alimentary canal from the oesophagus to the rectum is invested externally by the visceral layer of the peritoneum which histologically consists of a stratum of connective tissue, supporting on its free surface an epithelial stratum (coelomic epithelium). Primarily, the investing peritoneum is continued both dorsally and ventrally into bilaminar suspensory folds, the dorsal and ventral mesenteries which extend to the mid-dorsal or mid-ventral line of the abdominal cavity. The two layers then separate and become continuous with the parietal layer of the peritoneum lining the whole of the inner surface of the body-wall. The lining epithelium differs considerably in character in different portions of the alimentary canal.

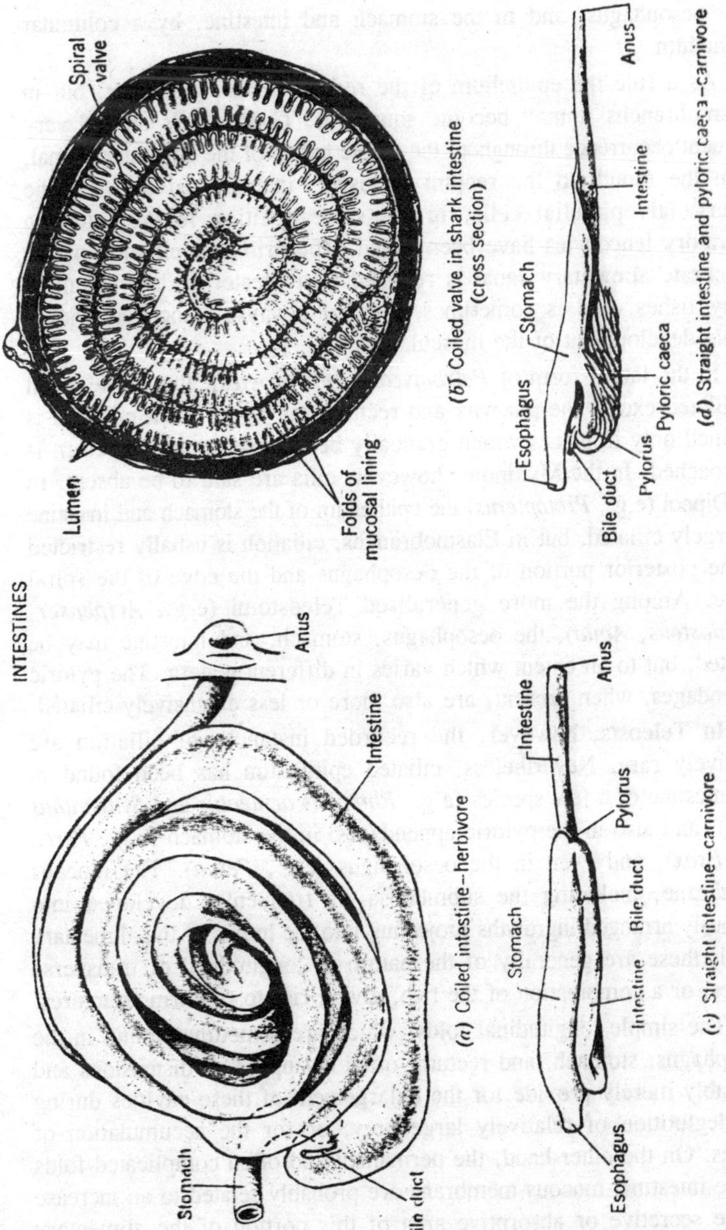

Fig. 4.4. Variation in intestinal length and other features among carnivorous and herbivorous fishes.

The epithelium of the mouth, pharynx, and anterior section of the oesophagus is often squamous and is succeeded in the hinder part of the oesophagus, and in the stomach and intestine, by a columnar epithelium.

As a rule the epithelium of the rectum is also columnar, but in Elasmobranchs it may become squamous. *Goblet cells* are of very frequent occurrence throughout the whole length of the alimentary canal, from the mouth to the rectum inclusive, interspersed between the superficial epithelial cells; in the same position in the intestine migratory leucocytes have been found. The primitive ciliation of the vertebrate alimentary canal is retained to a greater or less extent in many fishes, and is sometimes, but not always, associated with a feeble development of the musculature.

In the larval form of *Petromyzon* (*Ammocoetes*), the whole canal is ciliated except the pharynx and rectum; but in the adult ciliation is retained only in places which gradually become fewer as the return is approached. In the Myxinoids however, cilia are said to be absent. In the Dipnoi (e.g., *Protopterus*) the epithelium of the stomach and intestine is largely ciliated, but in Elasmobranchs, ciliation is usually restricted to the posterior portion of the oesophagus and the edge of the spiral valve. Among the more generalised Teleostomi (e.g., *Acipenser*, *Lepidosteus*, *Amia*), the oesophagus, stomach, and intestine may be ciliated, but to an extent which varies in different genera. The pyloric appendages, when present, are also more or less extensively ciliated.

In Teleosts, however, the recorded instances of ciliation are relatively rare. Nevertheless, ciliated epithelium has been found in the intestine of a few species (e.g., *Rhombus aculeatus* and *Syngnathus acus*), and also in the pyloric appendages: in the stomach (e.g., *Perca* and *Esox*), and even in the oesophagus (e.g., *Perca*). The mucous membrane, including the submucosa, is frequently developed into variously arranged ingrowths projecting into the lumen of the alimentary canal; these are generally of the nature of longitudinal or transverse ridges, or a combination of the two, giving rise to retiform structures.

The simple longitudinal folds, which are sometimes found in the oesophagus, stomach, and rectum, often disappear on distensions and probably merely provide for the enlargement of these cavities during the deglutition of relatively large prey, or for the accumulation of faeces. On the other hand, the permanent and often complicated folds of the intestinal mucous membrane are probably related to an increase in the secretive or absorptive area of this portion of the alimentary canal. In the stomach, the mucous membrane is usually smooth, rarely, as in the "Electric eel" (*Gymnotus*), reticulate. In the intestine the folds assume a highly characteristic and often complicated disposition. In the Cyclostomata the folds are simple and longitudinally arranged.

In Elasmobranchs obliquely transverse folds are present in addition, and, uniting with longitudinal ridges, bound linear depressions. In the Dipnoi (e.g., *Protopterus*) the mucous membrane of the imperfect truncated cones with their apices directed backwards, the successive cones adhering so closely to one another that they combine to form a central conical chamber with a spirally disposed cavity winding round it. In D, on the contrary, the free edge of the valve is deflected forwards, so that, as in C, a nest of cones is formed, but the apices of the successive cones are directed forwards instead of backwards. Not with standing these variations in the structure of the valve as a whole, the first coil or half coil nearly always resembles that described in A. It is obvious that the structure of the valve considerably within the limits of the genus, and it may be added that various intermediate types of structure occur between A and B, A and C, and A and D.

The individual variations are perhaps even more remarkable, and appear to be quite independent of age and sex. By way of example it may be mentioned that valves approximating to one or other of those represented by C and D occur in different individuals of *Raia maculata* of the same sex and similar in size, even in young specimens not more than three inches in length. As regards other Elasmobranchs, the common Dog-Fish (*Scyllium canicula*) has a well-developed spiral valve disposed in twelve coils, which structurally represents a more highly developed example of the type D. The existence of considerable individual variation in nevertheless indicated by the fact that in one specimen examined the valve was intermediate between C and D, five of the eight cones projecting forwards and three backwards. In a specimen of *Notidanus* sp. there were as many as twenty coils, which in disposition were intermediate between B and C, approximating however, more nearly to B.

In a specimen of the Port Jackson Shark (*Heterodonlus*) the valve had eight coils, and in structure was also intermediate between B and C, but approached more nearly to C. Some of the Hammer-headed sharks (e.g., *Sphyrna malleus*) possess a type of spiral valve which differs considerably from any of those hitherto described, and is termed a *scroll valve*. The attached edge of the valve pursues a straight longitudinal course, or at any rate only describes a half turn and back again in passing from the pyloric to the cloacal extremity of the gut. In the middle of its course the width of the valve is about equal to

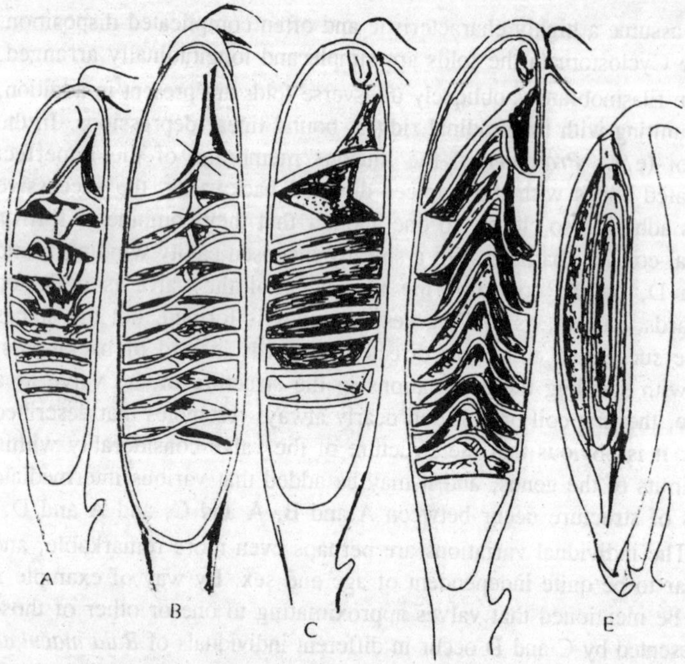

Fig. 4.5. Various types of the spiral valve in Elasmobranchs. A, B, C, and D in specimens of Raia spp.; E, in Sphyrna malleus.

two-thirds of its length, but towards either extremity it gradually diminishes until the free and attached margins meet.

The valve thus constituted is rolled upon itself from left to right, the successive coils being comparable to a series of cylinders placed one inside the other, and becoming gradually larger both in length and diameter from within outwards. A similar valve is present in some of the Carchariidae. In the Holocephali (e.g., *Chimaera monstrosa*) the valve describes only three and a half coils, and is further remarkable in that the attached margin, for a considerable portion of its extent, does not form a regular spiral but describes only a slightly sinuous course. Posteriorly, the valve is more normal, and consists of about two cones with their apices directed forwards.

In the Dipnoi the spiral valve is well developed, and in *Neoceratodus* describes nine coils, and in *Protopterus* six or seven. The structure of the valve in the latter Dipnoid resembles that of *Scyllium canicula*, except for the smaller number of cones. In the more generalised Teleostomi the valve is best developed in the Sturgeon

(*Acipenser*) and in *Polypterus*. In the former the valve is restricted to the posterior half of the total length of the intestine, often extending to within an inch of the anal aperture, and describing in its backward course about seven or eight coils. The width of the valve is about equal to the semi-diameter of the intestine, and the thickened free margin forms a well-marked axial columella, round which the cavity of the gut winds, as in the type B, except that the spiral is a more open one. In *Polypterus* the valve begins close to the solitary pyloric caecum, and for some distance pursues a straight, longitudinal course, but eventually forms a few spiral coils, ceasing, however, at a considerable distance from the anus. The evidence afforded by petrified faeces or "coprolites" proves that certain extinct Crossopterygii (e.g., *Macropoma*, *Megalichthys*), like their living representative, *Polypterus*, possessed a spiral valve.

In *Amia* and *Lepidosteus* the valve is almost vestigial, being restricted to the terminal portion of the intestine, and is somewhat variable as to the precise number of its coils. In *Amia* there are nearly four coils, extending over 3 cm. that is less than a tenth of the total length of the intestine, but in some specimens the coils do not exceed two and a half or three in number. *Lepidosteus* has a still shorter valve which, in specimens of 7-10 cm in length, may not consist of more than three and half coils, and in much larger specimens may be reduced to less than two coils, a variation which suggests that a reduction takes place in the number of coils as the fish increases in age and size.

The structure of the valve in the three last-mentioned genera resembles that described in *Acipenser*, and in none of them does the width of the valve so far exceed the semi-diameter of intestine as, by forward or backward deflection, to give rise to the highly characteristic cones of Elasmobranchs and Dipnoi. In the more specialised Teleostomi (Teleostei) The spiral valve is wholly wanting, except perhaps as a vestigial structure in certain Clupeoids, as, example, *Chirocentrus*, and possibly also in some Salmonidae.

GLANDS

The glands associated with the alimentary canal is different fishes (1) the gastric glands, (2) the liver, (3) the pancreas, (4) the pyloric appendages, and (5) the "rectal" gland. Oral salivary are wanting in all fishes, the only secretoary structures in the mouth being numerous mucous-secreting goblet cells, which here, as elsewhere throughout the alimentary canal, are intermixed with the ordinary epithelial cells.

Gastric Glands

The Cyclostomata and Dipnoi do not possess any specially differentiated gastric, and it is probable that in these fishes the secretion of the digestive fluids is effected by the ordinary lining epithelium of the stomach or intestine, or both. In the remaining groups gastric glands are generally present in the form of simple caecal structures embedded in the submucosa and opening on the surface of the mucous membrane into the cavity of the stomach. The glands differ in different fishes in the character of their lining epithelium and in the extent to which their component cells are differentiated from the epithelium of the stomach. There does not appear, however, to be any distinction into "*central*" (pepsinforming) and "*parietal*" (acid-secreting) cells, as is the case in the higher vertebrata. Towards the pyloric end of the stomach the true gastric glands are often replaced by mucous glands.

There are, nevertheless, not a few Teleosts in which special gastric glands are absent, as for example, *Syngnathus acus*, and several species of Cyprinidae, Labridae, and Blenniidae, etc. In at least two genera (*Gastrosteus* and *Cobitis*), belonging to widely different families, gastric glands are present in certain species but absent in others. As suggested by *Edinger*, the absence of these glands may possibly be due to degeneration. It may be remarked that the formation of such digestive ferments as *pepsin* and *trypsin*, which are associated with the stomach and pancreas respectively, in the higher vertebrates, is not nearly so strictly localised in Cyclostomes and Fishes.

So far from peptic digestion being limited to the stomach, it may take place in the pharynx, stomach and intestine of *Ammocoetes*, and in some Elasmobranchs (e.g., *Scyltium*), and in such Teleosts as the Pike, Eel, and Carp, the peptic region extends from stomach for some distance along the intestine, while trypsin has been obtained from the mucous membrane of the stomach intestine and pyloric caeca, as well as from the pancreas. *Intestinal glands* analogous to the glands of *Lieberkuhn* in higher vertebrates seem to be entirely wanting in fishes, unless presented by the sac-like or tubular crypts which are so generally present in the Teleostomi.

Liver

Phylogenetically the oldest gland in connection with the vertebrate alimentary canal, and in size by far the largest, the liver arises as a caecal outgrowth from the embryonic mesenteron, and in this primitive stage recapitulates a condition which is retained throughout life in *Amphioxus*. By the subsequent division and branching of this outgrowth the massive compound gland of the adult fish is eventually formed. The *liver* of fishes is very variable in size, shape, colour, and degree of lobulation. Anteriorly, it is usually moulded to the posterior face of the transverse septum between the pericardial and abdominal portions of the coelom, and from thence extends backwards in the abdominal cavity to a varying distance, in some sharks as far as the cloaca. Externally, the gland is invested by the peritoneum, which extends on to it from the pericardial septum and forms a suspensory fold, and also from the oesophagus and stomach. The shape of the liver usually bears some relation to that of body, being, for example, longest in the Eels and broadest in the Rays.

In the great majority of fishes the liver is bilobed, consisting of two sub-equal lateral lobes, disposed longitudinally and confluent anteriorly for a portion of their extent. From this normal type there are a few minor variations. In *Petromyzon*, *Lepidosteus* and a few Teleosts (*e.g.*, the Gymnodontes, Lophobranchii, and some Salmonidae) the liver is unilobed. In the Myxinoids and in the Dipnoi (e.g., *Protopterus*), the organ is bilobed, but the small anterior lobe lies immediately in front of the much larger posterior lobe, with the gall-bladder between the two. In some Teleosts (e.g., *Scomber*), the liver is trilobed. A gall-bladder is invariably present in either the larval or adult Cyclostomata, in the Chrondrostei, Holostei, Crossopterygii and Dipnoi and generally also in Elasmobranchs and Teleosts.

In the Elasmobranchs it is rarely entirely wanting, as in *Sphyrna* and *Pristis*, and in the Teleosts in some of the Gurnards (*Trigla*). The gall-bladder and bile-duct of *Petromyzon fluviatilis* atrophy after the metamorphosis which follows the larval *Ammocoetes* stage, but in *Petromyzon marinus* in duct, although usually absent, is sometimes retained. In the *Ammocoetes* the epithelium lining the gall-bladder is ciliated. In some fishes, as, for example, in many Elasmobranchs, the gall-bladder is more or less completely embedded in the substance of the liver; in others, as in most Teleostomi, the organ is quite distinct from the gland. A simple arrangement of the ducts from the liver and gall-bladder is that found in the common Dog-Fish (*Scyllium canicula*). In this Elasmobranch a *cystic duct* leaves the gall-bladder, and, after receiving several hepatic ducts from the lobes of the liver, becomes the bile-duct and open into the commencement of the intestine.

In the Myxinoids and in the Dipnoi (e.g., *Protopterus*), there are but two hepatic ducts, one from each lobe of the liver; these unite and then met the cystic duct to form the *bile duct*. The number of hepatic ducts may, however, be considerally increased, as, for example, in the Siluroid *Amiurue*, where 8-10 separate ducts join the cystic duct. In a fcw instances one of the hepatic ducts opens directly into the intestine, independently of that which unites with the cystic duct in forming the bile-duct. In the Dipnoi (e.g., *Protopterus*), and in some Teleostomi (e.g., *Lepidosteus*), the bile-duct receives the duct from the pancreas before opening into the intestine.

Pancreas

In the Cyclostomes (e.g., *Petromyzon*, *Bdellostoma*, *Myxine*) a rudimentary pancreas is apparently present, but the evidence as to its identity is not wholly conclusive. A well- developed pancreas occurs in Elasmobranchs, in at least one of the Dipnoi, and probably in most Teleostomi. In Elasmobranchs the pancreas is a compact structure, uni or bi-lobed, and entirely distinct from the liver. In *Scyllium canicula*, the bilobed gland lies in the angle between the distal limb of the stomach and the adjacent portion of the intestine, and from the smaller of its two lobes the duct issues to pass to its intestinal aperture near the commencement of the spiral valve. In most of the Teleostomi in which its existence has hitherto been recorded, the pancreas in a singularly diffuse gland; and usually a considerable portion, or even the whole of it, is embedded in the substance of the liver, its lobules accompanying the ramifications of the hepatic artery and duct, and the portal vein. The pancreatic duct usually opens into the intestine near the aperture of the bile duct (e.g., *Amiurus*); sometimes the two ducts open on the apex of a common papilla (e.g., *Acipenser* and *Amia*), or by their union form a common duct (e.g., *Lepidosteus*).

Among the Dipnoi a well-developed pancreas is present in *Protopterus*, embedded in the wall of the stomach and intestine, internal to the peritoneal investment of these organs, and extending even into the first fold of the spiral valve. The gland is traversed by fine ductules which unite together and open into the bile-duct just before the latter enters the intestine. In the remaining Dipnoi the existence of a pancreas has yet to be ascertained. Developmentally, the pancreas resembles the liver, and, histologically, is very similar to that of the higher vertebrates, consisting of terminal glandular alveoli continuous with intermediary tubular portions, and eventually with the finer ductules, which, by their union, form the main efferent duct.

Pyloric Caeca

These structures are caecal outgrowth from the intestine, and are situated close to the pyloric extremity of the stomach and the intestinal apertures of the bile and pancreatic ducts. Wholly wanting in the Cyclostomata and Dipnoi, and unless represented by a pair of caeca opening into the long, tubular, non-valvate anterior portion of the intestine of the Greenland Shark (*Laemargus borealis*), in the Elasmobranchs also, they are very generally present in the Teleostomi although extremely variably both in number and arrangement in different families. An *Amia* there is no trace of pyloric caeca. *Polypterus* has a single short caecum with a thick muscular wall. In *Acipenser*, *Polyodon*, and *Lepidosteus*, on the contrary, pyloric caeca are unusually well developed. In *Acipenser* the caeca are not only numerous, but are so connected together by connective tissue and blood-vessels and so invested dexternally by the peritoneum as to form a large compact, gland-like mass, communicating with the intestine by a single wide duct. In *Polyodon* the organ is essentially similar but is lobed externally. In *Lepidosteus*, the caeca are also very numerous, but relatively short, and, although united into a compact, mass, open by four pit-like orfices into the intestinal cavity.

In Teleosts and caeca are subject to extraordinary variations in number, size, and arrangement. In some families, and even in groups of higher taxonomic value, they are entirely absent, as is the case with the Siluridae, Esocidae, Cyprinodontidae, Labridae, Plectognathi, and Lophobranchii. The "Sand-eel" (*Ammodytes*) has but a single caecum; the Turbot (*Rhombus maximus*) two, and other Pleuronectide three to five; and the Perch (*Perca*), three. In other Teleosts, on the contrary, these structures are much more numerous. In *Labrus labrax* there are about 60, in the Whiting (*Gadus merlangus*) 120, while in the Mackerel (*Scomber scombrus*) there are no fewer than 191. If few in number the caeca open separately into the intestine, but when numerous, more or fewer of them may unite to form a smaller number of efferent ducts, as in the Whiting, where four such ducts are formed.

In some instances, as in the Tunny (*Thunnus*), the union of the caeca by connective tissue leads to the formation of a compact mass. As regards their arrangement, the caeca may either be disposed in a whorl round the intestine, as in the Whiting, or in a linear series, as in the Salmon (*Salmo*) and in some of the Clupeidae. The mucous membrane lining the anterior pyloric caeca is often developed into a network of ridges, limiting crypt-like or tubular depressions; and not infrequently the epithelium is ciliated. The precise function of these organs, whether digestive or absorptive is still uncertain. That they may be digestive is suggested by the presence of certain amylolytic and proteolytic enzymes, but this obvious conclusion is to some extent vitiated by the close proximity of these organs to the stomach, and more especially to the intestinal orifice of the pancreatic duct.

It is by no means improbable, however, that the caeca are both digestive and absorptive organs. An attempt has been made to show that the pyloric caeca and the spiral valve vary inversely as regards the extent of their development in different groups of fishes. To some extent the reciprocal variation of these structures supports this view, but it is also evident that there are obvious objections to its unqualified acceptance. Thus, in some Teleostomi (e.g., *Acipenser*, *Polyodon*), exceptionally well-developed and numerous caeca and a spiral valve are both present. *Amia* with an almost vestagial spiral valve has no trace of pyloric caeca, and in Teleosts the absence of a spiral valve is associated with the complete suppression of the caeca in many large and important groups.

Rectal Gland

The *rectal gland*, or *appendix digitiformis*, is a small organ of unknown function with complex glandular walls, and a central duct opening dorsally into the terminal portion of the intestine. The organ is generally present in Elasmobranchs, in which group the intestinal orifice of its duct may either be close to the termination of the spiral valve, or, as in *Chlamydoselachus*, near the cloacal outlet of the gut. An apparent representative of the gland, the "caecum cloacae", is also present in the Dipnoi, but communicates directly with the cloaca. The *rectal gland* is perhaps homologous with the intestinal caecum which is to be found in some Teleosts (e.g., *Box vulgaris*), and possibly also with the "caecum" (caecum cell), and its vermiform appendix in the higher vertebrata. The caecum cloacae, on the contrary, is morphologically a urogenital sinus, formed as a dilation of the fused hinder portions of the mesonephric ducts, and probably comparable with the sperm sacs of male Elasmobranchs, and also with the urinary bladder of Teleostomes.

DIGESTION

Very little is known about the physiology of digestion in the Indian fishes. *Agarwal* and *Singh* (1961, 1964) have shown that the *pH* of different parts of the gut in *Colisa fasciatus* and *Notopterus notopterus* is either neutral or weakly acidic. It is about 5.6 in the stomach and 6.7 in the intestine of *C. fasciatus*. In *Notopterus*, the *pH* was found to be 6.8 in the stomach and 6.4 in the intestine and phyloric caeca. In their opinion most of the enzymes are secreted in the hepatopancreas of the fish. Several workers have shown that an acidic gastric juice is secreted in the stomach of teleosts (average *pH* 5.6). The *pH* of the stomach content was found to be about 4.6. Pepsin is secreted in the stomach of fishes by the granular cells of the gastric glands and hydrochloric acid also appears to be secreted by the same cells.

There is some evidence of a weak *amylase* and weak *lipase* in the stomach of teleosts. In the stomachless fishes as *Rutilus* pepsin and hydrochloric acid is absent. The *pH* of the intestinal contents varies from 6.12 to 7.72 in *Rutilus*, *Gobio* and *Cyprinus* (*Al Hussaini*, 1949). The bile juice has been found to be slightly acidic (*pH* 5.5 to 6.8) in *Fundulus*, *Cyprinus* and *Zoarces*. A *proteolytic enzyme* is present in the intestine of a number of cyprinids (*Al-Hussaini*, 1949) and in gold fish, *Carassius auratus* (*Sarlbhi*, 1951) but the source of the enzyme is not very clear. Probably, *trypsinogen* is formed in the pancreas and *erepsin* and *enterokinase* are secreted by the intestine. It has been suggested that the absence of stomach in fishes is compensated by the production of pancreatic trypsin and an intestinal enzyme of the same nature.

Amolase is produced in the pancreas of teleosts and an amyloclastic activity has been reported from the extracts of the intestinal mucosa of several species. *Maltase* and *lipase* are also present in the intestinal extracts of several species of fishes. *Sarbahi* (1951) found *lipase* in the extracts from liver and pancreas. The source of the production of these enzymes is, therefore, not clearly established. Studies on the digestive physiology of teleosts suggest a correlation to exist between the enzymes and the diet of the animal. *Al-Hussaini* (1949) found that the concentration of *carbohydrases* to be highest in predominantly herbivorous fish *Cyprinus* and lowest in the carnivorous, *Gobio*, while the concentration of *protease* was opposite to it.

5

Blood Vascular System

In many small invertebrates, circulatory system is not necessary. Distances are short, structures are simple, and transportation of materials within the body may be effected by diffusion and by such flow of internal fluids as may result from body movement. With greater size and complexity, a definite circulatory system is required. A comparison with human communities is a fair analogy. In a village, a transportation system is unnecessary; stores, school, and church and all close to the homes they serve. With growth of the community, this is no longer the case, and an organized transportation system is a public necessity. Among invertebrates, we see the development of two types of circulation *open* and *closed* systems.

In the former, a heart forms a pump whence blood is forced out through a series of vessels, the *arteries*, to various parts of the body. From the point at which these vessels terminate, however, the blood oozes back to the heart through tissue spaces; return vessels are absent. In the more advanced *closed system*—the type present in the vertebrates—the blood almost never leaves the vessels and thus lacks direct contact with the body cells that lie in the separate interstitial fluid. The blood serves the tissues by diffusion of materials through small, thin-walled terminal vessels, the capillaries, and returns to the heart again through a second series of closed channels, the *veins*.

Amphioxus shows an intermediate stage in development for although both outgoing and returning vessels and equivalents of capillaries are present, these all lack the continuous endothelial lining found in true vertebrates. The foremost functions served by the vertebrate circulatory system are those of transportation to the cells (via the interstitial fluid) of the materials needed for their sustenance and activity and of the removal of wastes. The tissues must be constantly supplied with oxygen from gills, skin, or lungs, and fed with a small but steady stream of food materials from intestine or from storage and manufacturing centers, notably the liver. Wastes must be removed: carbon dioxide, destined for gills or lungs; nitrogenous wastes, to be excreted in the kidneys; and excess metabolic water. A second important function of the circulation lies in the influence that it exerts over the maintenance of the stable and narrowly defined internal environment necessary for cellular welfare.

The constant circulation of the blood makes for uniformity of composition in the interstitial fluids of every region and aids in the maintenance of relatively uniform temperatures. The vascular system has still further functions. It aids in the struggle disease and in the repair of injuries; through the circulation of hormones, the blood acts as an accessory nervous system.

Components

The basic structural features of the vertebrate circulatory system are familiar and readily demonstrated. A muscular pump, the *heart*, forces the blood stream by way of *arteries* to gills or lungs and to varied tissues of the body. The uniest circulatory units are networks of *capillaries*, through whose walls exchange of materials with the interstitial fluid surrounding body of cells takes place. From the capillaries, blood returns to the heart via *veins*; lymphatic vessels, originating in the tissues and unconnected with the arteries, form a route by which fluid can reenter the circulatory system. The veins between two capillary systems are termed a *portal system* and are named for the organ supplied (*e.g.*, the hepatic portal system consists of the veins transporting blood from the gut to the liver).

BLOOD

The *blood*, filling the vessels of the circulatory system, may be regarded as a tissue—a tissue with very remarkable properties, but one that is, in certain aspects, comparable to the connective and skeletal, to which, ontogenetically, the blood system is closely related. Each of these tissues consists of cellular elements derived from mesenchyme and lying in a "matrix". In bone, the matrix is a hard substance; in connective tissue, it is gelatinous in consistency; in blood, there is a liquid matrix in which the cellular components float freely. In teleost fishes the volume of blood varies from 1.5 to 3% of the body weight.

Blood Plasma

The liquid part of the blood, plasma, is a watery fluid of complex composition, of which—mammals, for example—some 7 to 10 per cent of the total volume consists of materials in solution. The blood is to be regarded as essentially a part of the complex interstitial fluid enclosed within the walls of the blood vessels, and both have a practically identical salt content. In addition, however, the blood contains materials peculiar to itself in the form of special *blood proteins—albumin*, *globulin*, and *fibrinogen*—which make up the greater part of the nonaqueous fraction of the plasma. These molecules are so large that they are normally unable to pass through capillary walls and thus leave the blood stream. Because the salt concentration of the blood is comparable to that of the interstitial fluids lying outside the blood vessels the presence of these blood proteins (notably the *albumins*) raises the osmotic pressure of the blood above that of the interstitial fluids—a point of importance in capillary function. *Fibrinogen* makes up only a small percentage of the total, but is important in clot formation when a blood vessels is cut.

One type of *globulin* also takes part in the formation of the blood clot. Other globulins are sensitive to protein materials of "foreign" origin and will, for example, causes the clumping (*agglutination*) of red blood cells of another species or even (as in man) of individuals of another blood type if the blood are admixed. Certain globulins are active agents in diseases protection as the antibodies effective against invading organisms or chemicals. The salts and blood proteins are stable, permanent, plasma constituents. In addition, however, analysis of the blood reveals various materials in transit. A quantity of glucose is always present, as well as smaller amounts of fats and of amino acids, or their way to the cells of the body. Waste materials, on the way to kidney, gills, or lungs, are also present, particularly carbon dioxide (mainly as a bicarbonate), nitrogenous waste (usually in the form of urea or uric acid), and lactic acid. In more minute amounts, enzymes may be identified in the plasma, and endocrine secretions—the *hormones*.

Blood Cells

Haemoglobin

Haemoglobin is the respiratory pigment of the blood and its amount varies with the total number of erythrocytes present in the fish. Exceptions are to be found among teleosts, however, some small forms

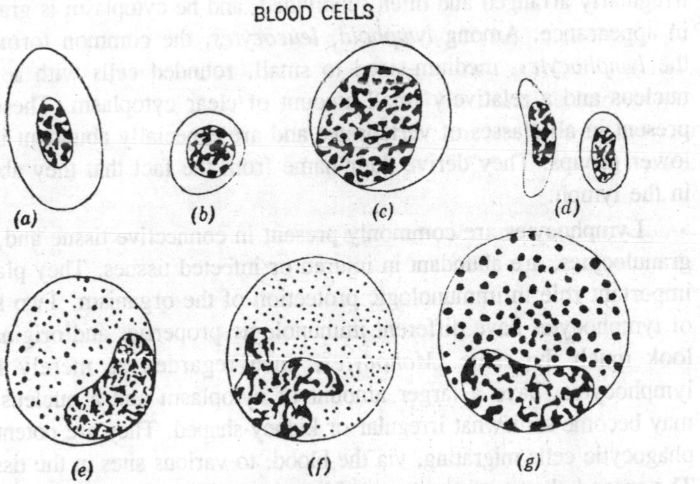

Fig. 5.1. Blood cells of fishes: (a) erythrocyte; (b) lymphocyte; (c) monocyte; (d) thrombocytes; (e) acidophilic granulocyte; (f) neutrophilic granulocyte; (g) basophilic granulocyte.

dwelling in very cold, well, oxygenated Antarctic waters have no haemoglobin or red blood cells.

Erythrocytes

The erythrocytes are nucleated, oval cells and vary from 7 to 36 microns in diameter in different species of fishes. Immature round red blood cells also occur in considerable numbers. Generally, erythrocytes vary from 20,000 to 3,000,000 per cubic millimeter of blood depending upon the habit of the fish. A sedentary fish, has much smaller number of erythrocytes than an active one.

White Blood Corpuscles

The white corpuscles characteristic of the blood and, to a lesser extent, the lymph streams are the *leucocytes*. They are much less numerous than erythrocytes. In amniotes, they almost never compose more than 1 per cent of the blood corpuscles, in fishes, however, the figure may rise to 10 per cent or over. The white corpuscles vary from 20,000 to 150,000 per cubic millimeter in different groups of fishes. Leucocytes vary greatly, in appearance and in types present, from one group of vertebrates to another, and it is difficult to establish any classification of them that will hold throughout.

In general, we may distinguish two major groups of white cells—*lymphoid types*, with a simple, single nucleus and a non-granular cytoplasm; and *granulocytes*, in which the nuclear materials are irregularly arranged and often subdivided, and he cytoplasm is granular in appearance. Among *lymphoid*, *leucocytes*, the common forms are the *lymphocytes*, medium-sized to small, rounded cells with a large nucleus and a relatively small amount of clear cytoplasm. These are present in all classes of vertebrates and are especially abundant in the lower groups. They derive their name from the fact that they abound in the lymph.

Lymphocytes are commonly present in connective tissue and, like granulocytes, are abundant in injured or infected tissues. They play an important role in immunologic protection of the organism. Two kinds of lymphocytes have different immunologic properties and origins but look much the same. *Monocytes*, once regarded as merely large lymphocytes, have a larger amount of cytoplasm and a nucleus that may become somewhat irregular or kidney-shaped. They are potentially phagocytic cells migrating, via the blood, to various sites in the tissues. The second characteristic group of leucocytes is that of the *granulocytes* or *polymorphonuclear leucocytes*. These, like the monocytes, are large cells.

The nucleus, as the alternative name, suggests, is irregular in structure and more or less subdivided into lobes. Most characteristic is the fact that the abundant cytoplasm is highly granular. In some cases, the granules are stainable with acid dyes: in other, basic dyes are effective; in still others, both types of dyes take hold to some degree. These three types of "polys" are hence reasonable termed, respectively, *acidophilic* (eosinophilic), *basophilic*, and *heterophilic* (neutrophilic) *granulocytes*. Heterophilic elements are the most abundant granulocytes in all vertebrate classes except, apparently, reptiles; acidophils are uncommon in humans but are widespread among vertebrate groups, basophils are generally still fewer in numbers and are seldom identified in fishes. Granulocytes appear to be essentially spherical while in the blood stream but are capable of ameboid motion and seem to be able to wriggle in and out of capillaries between endothelial cells. They tend to accumulate rapidly in injured, inflamed, or injected tissues.

The heterophilic granulocytes are capable phagocytes, "eating cells", and a major means of reducing bacterial infection. Eosinophils play a role in immune and allergic reactions. The blood cells are formed from the wall of the blood vessels in early embryonic stages but in adult teleost fish, spleen in the seat for their manufacture. In the chondrichthyes, white blood cells are manufactured in the organ of

Leydig, lying below the submucosa of the oesophagus. If the spleen is removed in these fishes, the erythrocytes are also produced in this organ. The spiral valve of the intestine in Chondrichthyes and Dipnoi, is also the seat for the production of several types of white blood cells.

HEART

The heart is more anteriorly placed than in other vertebrates, being situated directly behind and beneath the last pair of branchial clefts and internal to the ventral portion of the pectoral girdle. The organ is enclosed in a pericardial cavity, which, in the adult, is separated from the abdominal portion of the coelom by a transverse pericardio-peritoneal septum, and in the Lamprey (*Petromyzon*) is partially enclosed within a cartilaginous, cup-like modification of the hinder part of the branchial basket. In the *Ammocoetes-stage* of the Lamprey the pericardium is in communication behind with the general coelom, but the connection is lost in the adult.

In Elasmobranchs the two cavities are connected by a single pericardio-peritoneal canal, or by two such canals; and in *Chimaera*, and in the Sturgeon (*Acipenser*) and *Polyodon*, by a single canal. The heart consists of a *sinus venosus* which receives the venous blood from the body an, *auricle* and a *ventricle*, to which is added a *conus arteriosus* in the Elasmobranchs, certain Teleostomi (Crossopterygii, Chondrostei, and Holostei), and in the Dipnoi. Through these cardiac chambers the blood is forced in the order mentioned.

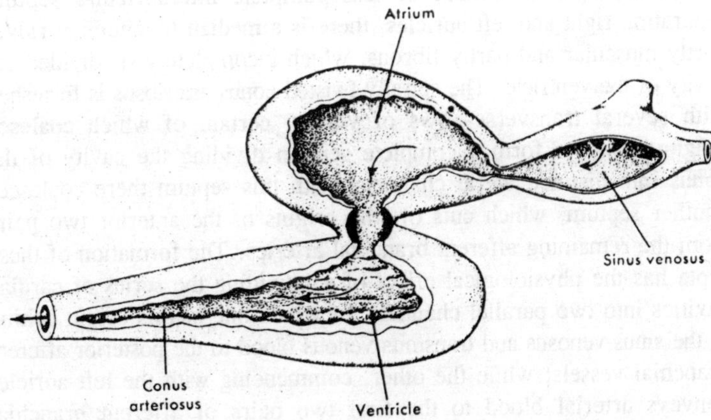

Fig. 5.2. Heart of Squalus acanthias opened to show chambers and blood flow. Arrow in sinus venosus is emerging from right common cardinal vein.

In the Dipnoi the auricle is subdivided by a more or less complete interauricular septum into a right and left auricle, the former receiving the venous blood from the sinus venosus, and the latter the aerated blood from the lung-like air bladder. The sinus venosus and the auricle have very thin walls; the ventricular walls, on the contrary, and very thick and in great measure are composed of a sponge-like network of muscular bundles which generally encroaches considerably on the ventricular cavity. Membranous valves, the sinu-auricular, and the auriculo-ventricular valves, are developed at the junctions of the sinus venosus with the auricle, and the auricle with the ventricle respectively. The conus arteriosus is muscular and contractile, and is interposed between the ventricle and the root of the ventral aorta.

Internally, the conus is provided with several transverse rows of pocket-shaped or semilunar valves. In Teleosts the conus is non-muscular and vestigial, and has but a single row of valves, corresponding to the most anterior of the multiple rows of valves in the Elasmobranchs. In these Fishes the vestigial conus is succeeded by a non-contractile, bulb-like dilatation, or *bulbus aortae*, of the root of the ventral aorta. In only a single Teleost, viz. *Albula*, one of the Albulidae, is the vestigial conus muscular, and at the same time provided with two rows of valves. In the Cyclostomata there is a bulbus with a single row of two valves but no true conus. In the Dipnoi (e.g., *Protopterus*) the heart, like the rest of the vascular system, exhibit certain interesting resemblances to the Ambhibian heart.

In addition to a more or less complete interauricular septum separating right and left auricles, there is a median longitudinal ridge, partly muscular and partly fibrous, which incompletely subdivided the cavity of the ventricle. The spirally-twisted conus arteriosus is furnished with several transverse rows of valves, certain of which coalesce longitudinally to form a complete septum dividing the cavity of the conus into distinct lateral channels; with this septum there coalesces another septum, which cuts off the origins of the anterior two pairs from the remaining afferent branchial arteries. The formation of these septa has the physiological effect of subdividing the series of cardiac cavities into two parallel channels, of which one has its origin behind in the sinus venosus and transmits venous blood to the posterior afferent branchial vessels; while the other, commencing with the left auricle, conveys arterial blood to the first two pairs of afferent branchial arteries. In *Neoceratodus*, however, the longitudinal septum in the conus is incomplete, and hence the blood which is sent to the anterior afferent vessels is mixed.

Working of the Heart

The venous blood flowing continuously towards the heart reaches the sinus and passes into the auricle by pushing apart the *semi-lunar valves*. During this, the pockets of the valves also become full of blood and the pressure due to the contraction of the auricle causes the valves to swell and adhere with each other, thus preventing the backward

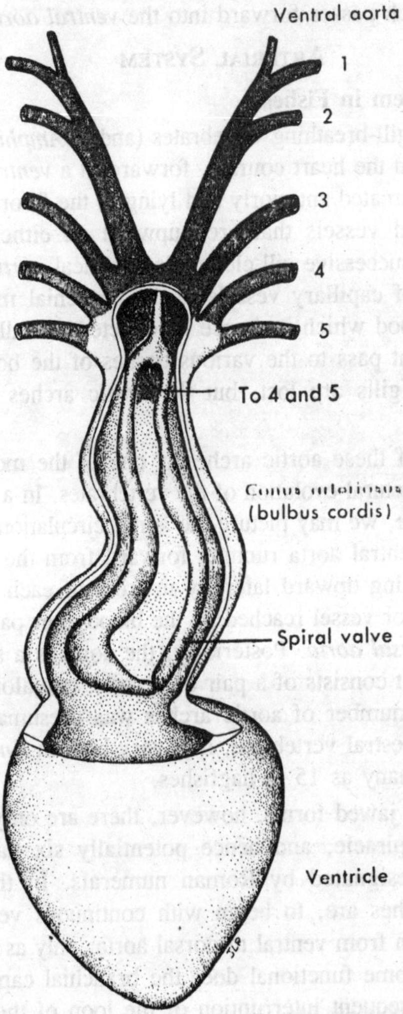

Fig. 5.3. Conus arteriosus and afferent branchial arteries (1 to 5) of the lungfish Protopterus.

flow of the blood. The blood now flows from the auricle into the ventricle by pushing apart the four *auriculoventricular valves*. As soon as the ventricular cavity is full, the valves also receive the blood, so that they bulge out and adhere with each other so as to effectively close the opening and thus, prevent the backward flow of the blood. The blood, now, pushes aside the *ventriculobulbar valves*, to enter the bulbus. Here again, the increased pressure inside the bulbus causes the valves to swell and close the passage, preventing backward flow of the blood, which passes forward into the *ventral aorta*.

Arterial System

Aortic Arch System in Fishes

In primitive, gill-breathing vertebrates (and in *Amphioxus* as well), and the blood from the heart courses, forward in a *ventral aorta*. This median trunk, bifurcated anteriorly and lying in the floor of the throat, gives off a paired vessels that arch upward on either side of the pharynx between successive gill clefts. Each typical *aortic arch* breaks up into a series of capillary vessels in the branchial membranes for aeration of the blood which is thence recollected dorsally into further arterial vessels that pass to the various tissues of the body and head. In tetrapods, the gills are lost, but the aortic arches are found in every embryo.

The history of these aortic arches is one of the most interesting chapters in the structural evolution of the vertebrates. In a diagrammatic primitive vertebrate, we may picture the aortic circulation as including, in addition to a ventral aorta running forward from the heart, paired aortic vessels passing upward laterally in front of each branchial slit or pouch. The major vessel reached by the blood after passage through the gills is the *dorsal aorta*. Posteriorly, the aorta is a single median trunk: anteriorly, it consists of a pair of vessels, one along either side of the head. The number of aortic arches was presumably high and variable in the ancestral vertebrate; it is very high in *Amphioxus*, and there may be as many as 15 in hagfishes.

In most living jawed forms, however, there are only five normal gill slits plus a spiracle, and hence potentially six pairs of aortic arches, usually designated by Roman numerals. In the vertebrate embryo, these arches are, to begin with continuous vessels passing without interruption from ventral to dorsal aorta, only as the branchial slits open and become functional does the branchial capillary system develop, with consequent interruption of the loop of the arch. In all vertebrates, the aortic arches develop in order; in the embryo, from

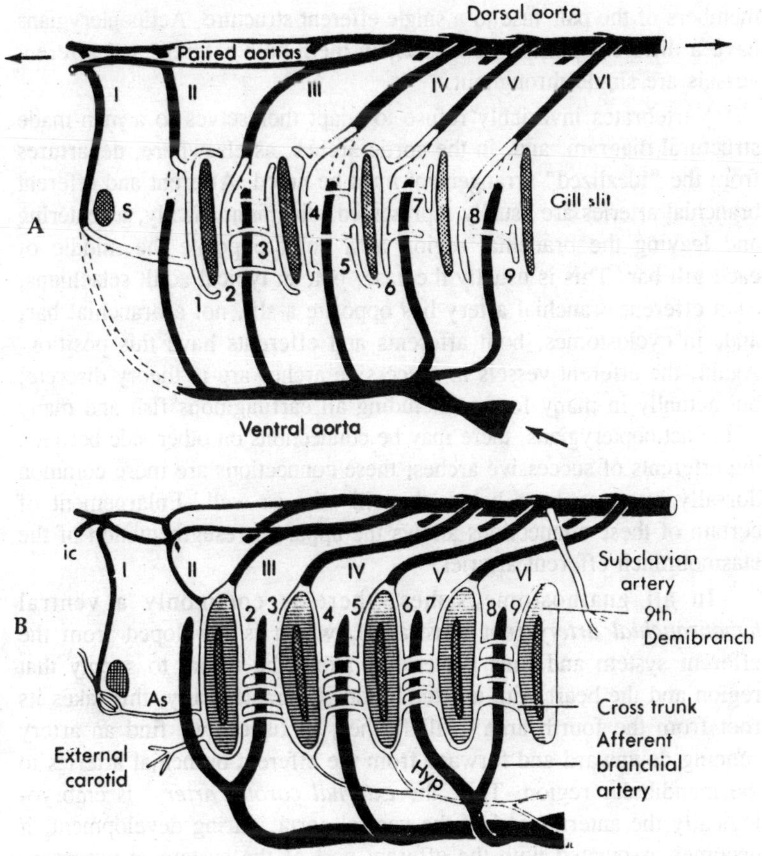

Fig. 5.4. Changes in embryonic aortic arches I to VI of Squalus during development, lateral view.

front to back. The first, or mandibular arch is, in the early embryo, the only passage from ventral to dorsal aorta.

In the cyclostomes, the gills are subspherical, pouchlike structures. Afferent branchial vessels ascend between successive pouches and divide to supply the adjacent halves of the neighbouring pouches; further afferent vessels supply the front half of the first pouch and the back half of the last. In all other living fishes, the branchial openings are slit shaped, and the afferent vessel of an aortic arch typically runs up between each successive pair of slits. In selachians and lungfishes, the afferent vessel continues as a single structure far up the gill, giving off capillaries as it goes. The efferent vessels of each gill are a pair, anteriorly and posteriorly placed; at the top of the gill slit, the two members of the pair fuse to a single efferent structure. Actinopterygians have a different pattern of vessels; in them both efferent and afferent vessels are single throughout.

Vertebrates invariably refuse to adapt themselves to a man-made structural diagram, and, in the aortic arches, as elsewhere, departures from the "idealized" arrangement must be noted. Afferent and efferent branchial arteries are usually represented, diagrammatically, as entering and leaving the branchial region at a point opposite the middle of each gill bar. This is usually the case; but, in typical adult selachians, each efferent branchial artery lies opposite a slit, not a branchial bar, and, in cyclostomes, both afferents and efferents have this position. Again, the efferent vessels in successive arches are in theory discrete; but actually in many forms, including all cartilaginous fish and many of the actinopterygians, there may be connections on other side between the efferents of successive arches; these connections are more common dorsally but may loop below the gill slits as well. Enlargement of certain of these connections allows the apparent resegmentation of the elasmobranch efferent arteries.

In all gnathostome fishes, there is commonly a ventral *hypobranchial artery* (not illustrated), which is developed from the efferent system and runs posteriorly along the throat to supply that region and the heart with aerated blood. Characterically, this takes its root from the fourth arch. Still further, in fishes, we find an artery running downward and forward from the efferent branchial arteries to the mandibular region. This, the *external carotid artery*, is embryologically the anterior end of the ventral aorta; during development, it becomes associated with the efferent part of the system, a necessary change if it is to supply oxygen to the lower jaw.

In cyclostomes, all the aortic arches are similarly developed; in jawed fishes, in which six arches are usually developed in the embryo, the posterior four usually attain a typical adult structure, but the first two may be highly modified. The first arch, the *mandibular*, runs dorsally in the embryo between mouth and spiracle; the second, the *hyoid arch*, lies between the spiracle and the first normal branchial slit. Presumably ancestral vertebrates had a well-developed arch I running up behind the mouth. But although a first aortic arch is found in the embryo, it is never found in typical fashion in the adult of any living jawed vertebrate (the numbering in cyclotomes is debatable so we omit them here). Its afferent portion almost universally disappears. In primitive fishes, there is a gill-like structure on the anterior margin of the spiracle. In theory, such a gill should be supplied by arch I.

But its blood supply comes, in the adult, from the efferent system of the second arch or from a branch of the dorsal aorta. The blood that reaches it has, thus, already been aerated.

The spiracular gill is therefore not a true gill, but a "*false gill*" or *pseudobranch*. The efferent portion of arch I persists in cartilaginous and a few bony fishes and runs up from the pseudobranch into the orbital region, enters the braincase to connect with the arteries beneath the brain, and supplies the eye with blood. The second aortic arch is somewhat variable but more persistent. It is generally present in typical fashion in Chondrichthyes and is also found in some bony fishes. But it is lost in most actinopterygians, and the same is true of the lungfish *Neoceratodus*. Arches III to VI are in general, as we have seen, both normally and fully developed in jawed fishes. In the lungfish *Protopterus*, an interesting exception is that arches III and IV run without break through the branchial region. This condition is associated with the reduced gills and developed lung of that genus and shows a partial parallel to developments in the amphibian relatives of the dipnoans. A few teleosts also have reduced the number of functional gills. The introduction of a lung in certain bony fishes makes, for most forms still in the fish stage, no great change in the circulatory system of the gills. The lung in *Polypterus* and lungfishes is supplied by an artery from the efferent portion of the sixth aortic arch or the aorta beyond this portion; hence the blood reaching it has already passed through the normal aerating device of the branchial capillaries.

Venous System

A generalized venous system consists of the following major streams: *cardinals* (anterior, posterior, and common), *renal portal*, *lateral abdominal*, *hepatic portal*, *hepatic sinuses*, and *coronary* veins. The hepatic portal system is derived from the embryonic subintestinal and the distal portion of embryonic vitellines, and the hepatic sinuses are derived from the vitellines between liver and heart. Two additional streams develop in lungfishes and tetrapods: a pulmonary stream from the lungs and a postcava from the kidneys. These eight channels and their tributaries drain the entire body—head, trunk, tail, and appendages. As development progresses, they are slowly modified by deletion of some vessels and addition of others. Modifications are few in lower vertebrates, more numerous in higher vertebrates.

Basic Pattern: Sharks

An adult shark is an ideal living swimming blueprint of the basic venous channels of vertebrates. In asmuch as very few changes occur in the embryonic channels later development, knowledge of the venous channels of sharks and how they develop is a good introduction the venous channels of other vertebrates.

Cardinal Veins

The sinus venosus receives all blood returning to the heart of sharks. Most of this blood, except that from the digestive organs, enters the sinus by a pair of *common cardinal veins*. These use the transverse septum as a bridge from the lateral body walls to the heart. They appear early in development, and remain essentially unchanged thereafter. Blood from the head other than the lower jaw is collected by a pair of *anterior cardinal* (*precardinal*) *veins* lying dorsal to the gills. The anterior cardinal pass caudad and empty into the common cardinals. The embryonic anterior cardinals, like common cardinals, remain essentially throughout life.

The earliest *embryonic posterior cardinal* (*postcardinal*) *veins* are continuous with the caudal vein. These embryonic postcardinals pass cephalad lateral to the developing kidneys and receive a series of renal veins from them en route. They then empty into the common cardinals. Their cephalic ends in adult sharks expand to become *posterior cardinal sinuses*. While these embryonic postcardinals are functioning, a new pair of postcardinal veins is forming between the kidneys from a subcardinals plexus. These new veins become confluent with the old postcardinal at the anterior end of the kidneys, and they, too, drain

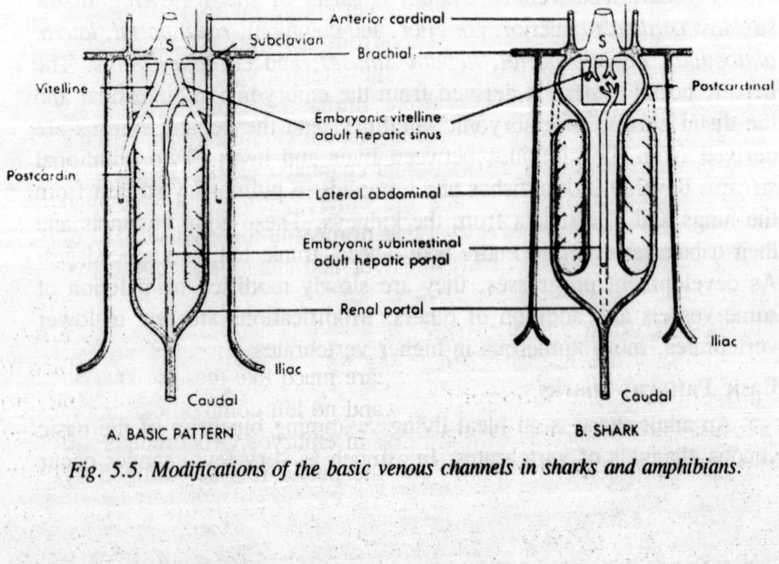

Fig. 5.5. Modifications of the basic venous channels in sharks and amphibians.

the kidneys. As more and more blood from the kidneys flows into the new veins the older postcardinals are lost anterior to the kidneys. Thereafter the name *posterior cardinal* is applicd to the newer vessels. In adults they drain chiefly the kidney, body wall, and gonads.

Renal Portal System

At an early stage in development in sharks some of the blood from the caudal vein continues forward beneath the gut as a subintestinal vein that drains the digestive tract. Later the connection of the caudal vein with the subintestinal is lost. When the old posterior cardinals are lost anterior to the kidneys, all blood from the tail thereafter enters the capillaries surrounding the kidney tubules (*peritubular capillaries*). The result is a *renal portal system*.

Lateral Abdominal System

Commencing at the pelvic fin from which it receives an *iliac vein* and passing forward in the lateral body wall on each side is a *lateral abdominal vein*. At the level of the pectoral fin it receives a *brachial vein*, after which the vessel to the depth toward the heart to enter the common cardinal vein. The part of the abdominal stream between brachial and common cardinal is the *subclavian vein*. In addition to collecting blood from the paired fins, the abdominal stream also receives a *cloacal vein*, a metameric series of *parietal veins* from the lateral body wall, and minor tributaries. The basic channel remains unmodified during subsequent development.

Hepatic Portal System and Hepatic Sinuses

Among the first to appear in any vertebrate embryo are paired *vitelline*, or *omphalomesenteric veins* from the yolk sack, whether functional or not, to the heart. One of the vitelline vein is soon joined by the embryonic *subintestinal vein* that drains the digestive tract. As the developing liver enlarges it encompasses the vitelline veins, causing them to the broken into many sinusoidal channels. Caudal to the liver one vitelline vein disappears, and the other, with the subintestinal vein becomes *hepatic portal system* that drains the digestive organs of the coelom, and the spleen. Between liver and sinus venosus the two vitelline veins become *hepatic sinuses*.

Other Fishes

The venous channels of other fishes are much like those of sharks. Cyclostomes have no renal portal veins and no left common cardinals, although two common cardinals develop in embryos. Abdominals are lacking in most ray-finned fishes, and the pelvic fins are drained by the postcardinals. In dipnoans the pelvic fins are drained by an unpaired ventral abdominal vein that ends in the sinus venosus, and the right postcardinal is missing. Blood from the swim bladders of all ray-finned fishes empties into the hepatic or common cardinal veins; in dipnoans it empties into the left atrium. Coronary veins in all fishes empty into the sinus venosus.

Lymphatic System

In addition to blood-vessels, fishes posses a lymphatic system, consisting of smaller vessels, *lymph-capillaries* or *lymph-spaces* distributed in the connective tissue of different parts of the body, and by their union ultimately forming larger lymph-vessels or *sinuses* which communicate with certain of the principal veins, the whole forming a series of channels for the collection of the blood-plasma which has exuded from the blood-capillaries for the nutrition of the tissues, and for its conveyance to the general venous system. The fluid in the lymphatics, or lymph, consists of dilute blood-plasma containing leucocytes but devoid of red corpuscles.

At the points where the larger lymphatics open into the veins, *lymph-hearts* may be developed. In the Eel (*Anguilla vulgaris*) there is a lymph-heart in the tail, which communicates by a valvular orifice with the smaller of the two caudal veins, and by its rhythmical pulsations propels the lymph into the vein. In *Silurus* there are two caudal lymph-hearts. Apart from the lymphoid tissue, which is so abundantly present in certain parts of the body, fishes appear to be devoid of the special "lymphatic glands" of the higher Vertebrates.

Blood-Glands

All the important blood-glands of other vertebrates have their representatives in fishes. Nothing is certainly known of the function of these organs in fishes, but from the general structural resemblance which they present to their equivalents in the higher vertebrates, it is perhaps not unreasonable to infer that they are similar in function. If this be so, the blood-glands of fishes are organs for leucocyte-formation and phagocytosis, involving the destruction and removal of effect red blood corpuscles; in addition, they may also be concerned with certain obscure chemical change of the composition of the blood, which have an important relation to general or local nutrition.

Spleen

This lymphoid organ is the largest of all the blood-glands, and, in the form of a compact or more or less lobulated body, is present in all fishes, and possibly in Cyclostomes. In position the spleen is usually in close proximity to the stomach to which it is attached by an extension round it of the peritoneal investment of that organ. Thus, in the Dog-Fish (*Scyllium*), the spleen is a large reddish body attached to the convexity of the U-shaped stomach, and, in addition, sends a long narrow lobe between the distal limb and the valvate portion of the intestine. In the Sturgeon (*Acipenser*), the organ is also large, but is attached to the left side of the commencement of the intestine. In the Cod (*Godus*) among Teleosts and spleen is much elongated and is situated on the dorsal side of the stomach. In the Dipnoi (e.g., *Protopterus*) the organ is probably represented by a large compact lymphoid mass, closely connected with the dorsal and lateral walls of the stomach.

Thymus

In the embryo of Elasmobranch and Teleost, the thymus has a multiple origin, being derived from a series of distinct epithelial thickenings, one of which is developed at the dorsal extremity of each of the gill-clefts except of the spiracle. These rudiments subsequently detach themselves from the epithelial surface and sink inwards, eventually fusing together on each side to form a single independent structure. Later, the epithelial mass thus formed becomes invaded by connective tissue, and by leucocytes which form lymph follicles, and the thymus gradually assumes the structure of a lymphoid organ. From its mode of development it has been suggested that the thymus owes its evolution to the metamorphosis and ingrowth of branchial filaments, but it is also noteworthy that each embryonic rudiment of the organ closely resembles, both in position and origin, one of the developing branchial tongue-bars of *Amphioxus*.

The abundance of leucocytes which it contains has also prompted the further suggestion that the origin of the thymus may be due to the necessity of providing for the phagocytic protection of the gills themselves from the ravages of harmful micro-organisms, fungoid spores, etc. as well as to aid in the removal of such portions of the gills as may have been injured. A thymus is probably present in all fishes, if not in the adult at all events in the embryo, but is always relatively in size. In Elasmobranchs the organ lies on each side above branchial arches and beneath the dorsal masculature; and in Teleostomi at the dorsal extremity of the last branchial arch, in close proximity to the musçous membrane of the branchial cavity. In a similar position in the Dipnoi (e.g., *Protopterus*) there are, on each side, two contiguous lobes of lymphoid tissues which apparently represent a thymus.

Lymphoid Tissue

In addition to certain of the ductless glands, and the local or diffused masses of their characteristic tissue already mentioned in connection with the alimentary canal, lymphoid tissue is often abundantly present in other parts of the body. There is, for example, a mass of this tissue on the heart of the Sturgeon (*Acipenser*). The anterior enlarged portion of the mesonephros, commonly termed the "*head-kidney*" of the Teleostomi is almost entirely composed of lymphoid tissue, which has replaced, whether at partially the proper renal structure; and from the presence of free red blood corpuscles and of crystals of oxy-haemoglobin and other derivatives of haemoglobin, it may be inferred that the "head-kidney", in common with the more orthodox blood-glands, performs a blood-destroying function. On the other hand, the example of the spleen, which is alike the seat of leucocyte formation and of blood-destruction, renders it unnecessary to reject the view that the "head-kidney" is an organ in which leucocytes or blood-corpuscles are formed. In but few Teleostomi is a purely lymphoid "head-kidney" entirely wanting, as, for example, in the Sun-Fish (*Orthagoriscus mola*).

As previously mentioned the Dipnoi are remarkable for the extraordinary development of lymphoid tissue, in as much as it forms a thick investing mass round the kidneys and gonads in addition to its exceptional abundance in the walls of the alimentary canal. The absence of ordinary lymphatic glands in fishes is well-known, and it is at least probable that, functionally, the want of these lymphoid organs may be compensated for by the superabundance of lymphoid tissue in other parts of the body.

6

EXCHANGE OF GASES

The mechanics and physiology of respiration are especially interesting in fishes, for there is considerable diversity of approach to the problem of acquiring and transporting oxygen, not only in fishes of different habitat, but also in fishes of different groups living in the same habitat. A much wider scope for activity is given to a scombroid, for example, than to a lamprey or a hagfish by the design of their respiratory and circulatory systems. A surprising number of fishes are facultative or obligate air-breathers, rising to the surface to gulp air into various kinds of respiratory chambers. Several of these, like the walking catfish (*Clarias batrachus*) of the southern states of America, synchronize these trips to the surface so that the water boils with a mass of fishes for a moment or two, and remains quite until the next excursion (a strategy which apparently confuses predators). Some, like mudskippers, spend much of their lives out of water, but the great majority of fishes spend all their lives in a dense viscous fluid which contains relatively little oxygen.

At the sea surface, for instance, air-saturated water contains (at 20°C), only around 3% of the oxygen in the same volume of air. Since oxygen only crosses cell surfaces by diffusion, this means that the gas exchangers for acquiring it must have a large area, and a relatively large expenditure of energy is needed to force water to flow around them. Alternatively, the fish has to accept a very low rate of oxygen uptake. What is more, and we need to bear this in mind when considering the design of these gas exchangers, the blood is almost always quite different in composition and osmolarity to the external water. The blood has to be intimate contact with the water to make the diffusion pathway short; so there must be special arrangements to circumvent or at least cope with water and ion flow across the exchanger.

Physics of Respiratory Medium

Water, as a medium, is less suitable for respiration than is air. For instance, it is about 1,000 times more dense and therefore more energy must be expended in passing it over the respiratory surfaces. Furthermore, any given volume of air will contain more oxygen than

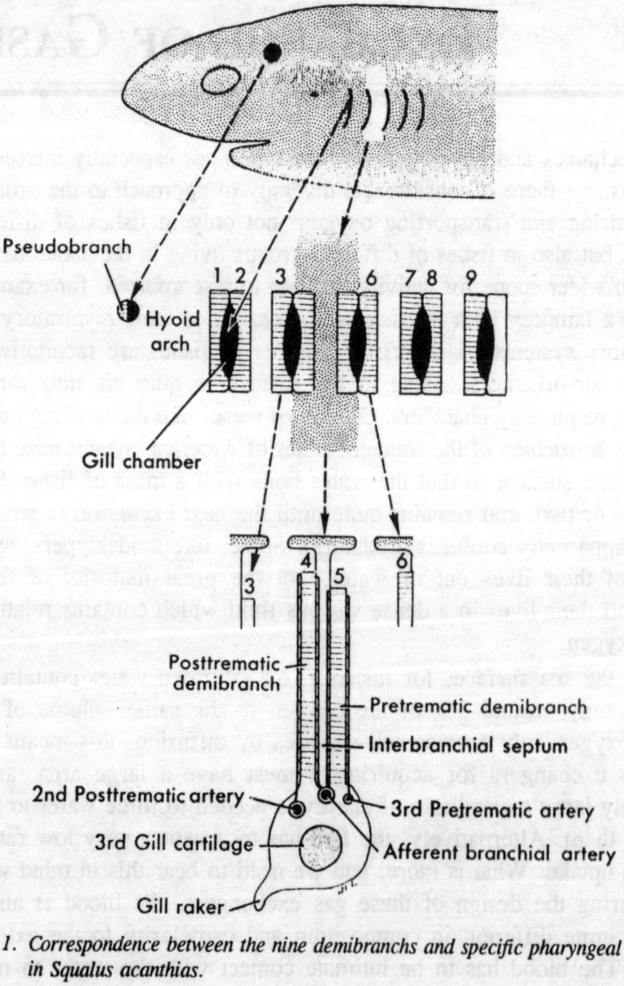

Fig. 6.1. Correspondence between the nine demibranchs and specific pharyngeal arches in Squalus acanthias.

the same volume of water under similar physical conditions. Yet fishes are particularly well adopted for respiration in water and, in fact, are able to utilize up to 80% of the oxygen contained in the water passing across the gills, whereas man is only able to use about 25% of the oxygen inhaled by the lungs. Two factors seem to be primarily responsible for the fishes, efficiency. One is the structure of the gills and the arrangement of blood circulation through them; the other is the fact that there is a continuous flow of water across the gills at all stages respiration.

Gills

Gills are composed of numerous *gill filaments* or *gill lamellae*, which are thin-walled extensions of epithelial surfaces. Each contains a vascular network. Blood is brought extremely close to the surface, thus facilitating the ready exchange of gases. In their aggregate, gills present a relatively large surface for rsespiratory exchange.

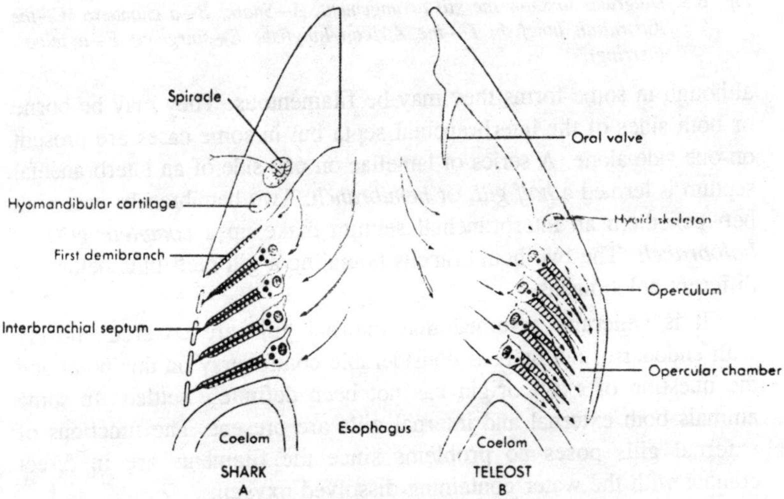

Fig. 6.2. Gills of a pentanchid shark and a teleost, frontal section, schematic. Arrow show routes of respiratory water.

Types of Gills

Gills are two general types, external and internal. *External gills* develop from the integument covering the outer surfaces of visceral arches. They are usually branched, filamentous structures covered entirely with ectoderm and are not related to the visceral pouches. *Internal gills* are usually composed of a series of parallel *gill lamellae*,

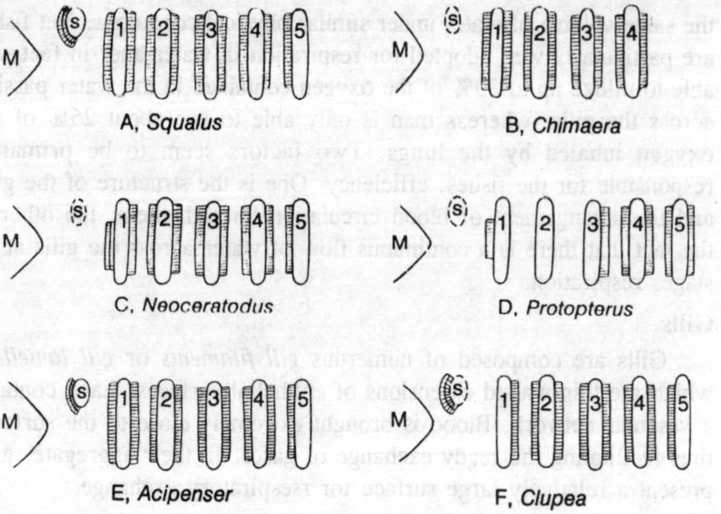

Fig. 6.3. Diagrams to show the gill arrangement. A—Shark; B—a chimaera; C—the Australian lungfish; D—the African lungfish; E—sturgeon; F—a teleost (herring).

although in some forms they may be filamentous. They may be borne or both sides of the interbranchial septa but in some cases are present on one side alone. A series of lamellae on one side of an interbranchial septum is termed a *half gill*, or *hemibranch*. Two hemibranchs enclosing between them an interbranchial septum make up a *complete gill*, or *holobranch*. The two hemibranchs bounding a gill cleft thus belong to different holobranchs.

It is generally assumed that internal gills are covered entirely with endoderm, but there is considerable controversy on this point and the question of exact origin has not been definitely settled. In some animals both external and internal gills are present. The functions of external gills poses no problems since the filaments are in direct contact with the water containing dissolved oxygen.

When internal gills are used in respiration, in most cases water, containing dissolved oxygen, is taken into the mouth and passes through the internal gill slits into the gill clefts. As the water passes over the gill lamellae, oxygen is taken from the water and carbon dioxide is released. The water then passes through the external gill slits to the outside. In lampreys and certain marine teleosts it has been determined that internal gills, in addition to their respiratory function, play an important role in the excretion of salt from the blood. In teleosts, such nitrogenous wastes as ammonia have been shown to be excreted via glandular cellular elements present in the gill lamellae. The gills may thus supplement the kidneys as excretory organs.

Comparative Anatomy of Gills

Fishes. In fishes and tetrapods a series of skeletogenous *visceral arches* encircles the pharynx. In tetrapods these become greatly modified, but in fishes they serve primarily to support the gills. They are located between the gill clefts, one behind the other, at the bases of the interbranchial septa. The *first* is called the *mandibular arch*: the *second* is the *hyroid arch*. The remaining visceral arches are referred to by number (3, 4, 5, 6, etc.).

The first *gill pouch* or *cleft* lies between the mandibular and hyoid arches and is often referred to as the *hyomandibular cleft*. In fishes it is either modified to form a *spiracle* or is closed altogether. The ancestral placoderms had a full-sized gill slit between the mandibular and hyoid arches. Two main types of *internal gills* are to

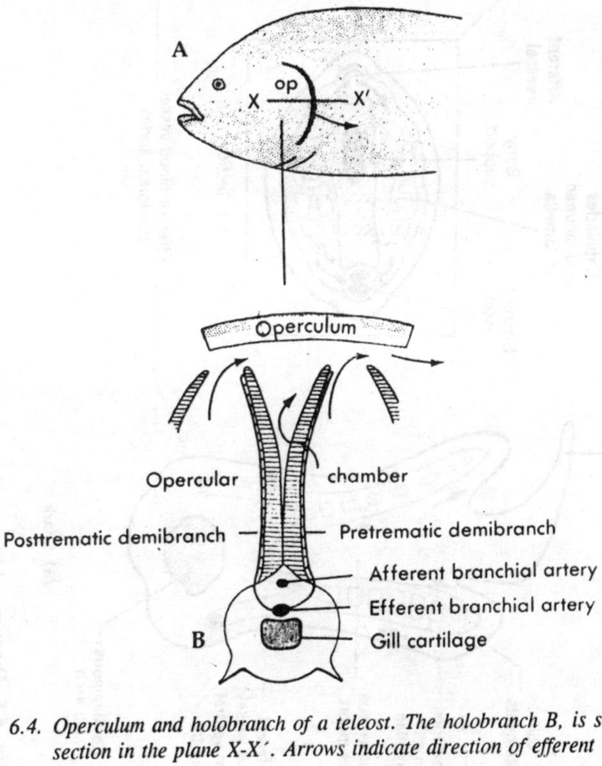

Fig. 6.4. Operculum and holobranch of a teleost. The holobranch B, is seen in cross section in the plane X-X'. Arrows indicate direction of efferent water flow.

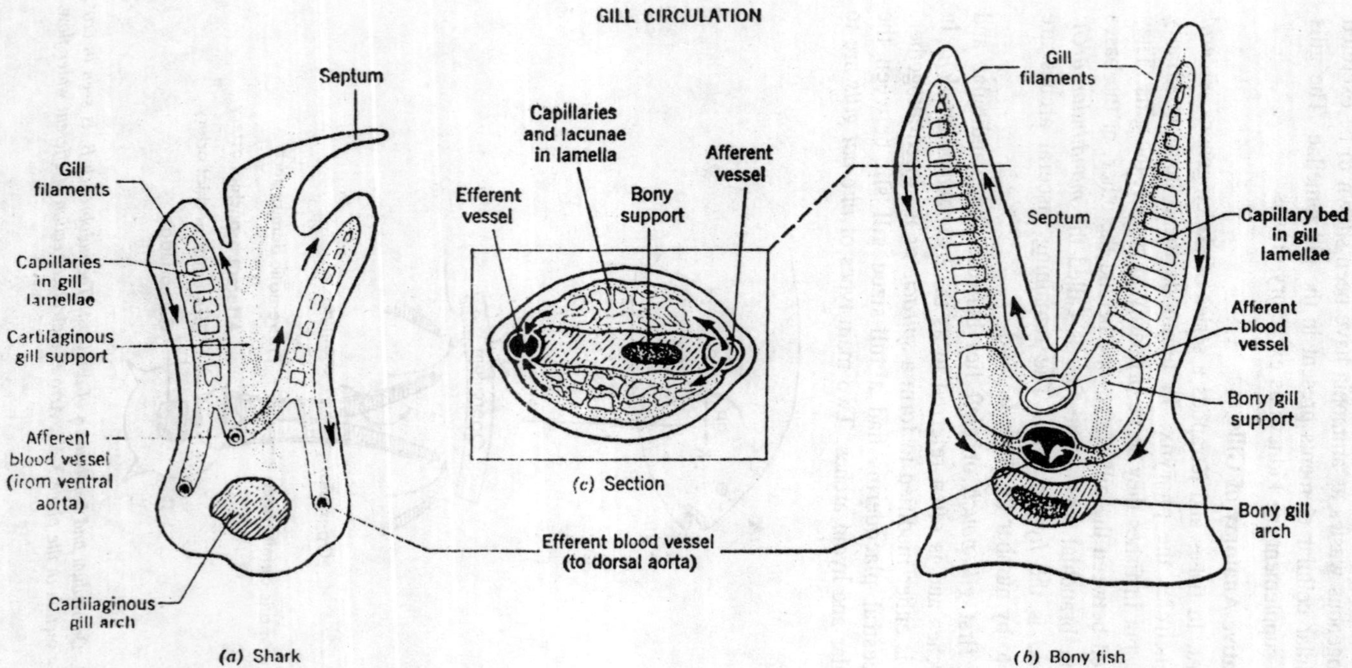

Fig. 6.5. Diagrams of horizontal sections through gills of a shark (a) and a bony fish (b) with detailed cross section (c) near apex of (b). Arrows indicate direction of blood flow.

be found among fishes. The first and more primitive type is typical of elasmobranchs. In this group the interbranchial septa are exceptionally well developed and extend beyond the hemibranchs which are closely applied to them. Each bends posteriorly at its distal end in such a manner that a row of separate *external gill slits* is formed. Thus the interbranchial septa, in addition to separating the gill clefts, serve to protect the gills themselves.

In the frilled shark, *Chlamydoselachus*, the interbranchial septa are continued to an unusual degree and project over the external gill slits. In the dogfishes and sharks the external gill slits and lateral in position, but in adult skates and rays they are ventrally located been through they are in a lateral position during the early part of embryonic life. Small cartilaginous *gill rays*, or *branchial rays*, in the form of a single row project from each visceral arch into the interbranchial septum to which they give support. Rigid, comblike *gill rakers* usually project from the gill arches. Their function is to prevent food and foreign bodies from entering the gill clefts. The second type of gill is found in the remaining fishes. In these the interbranchial septa are reduced to varying degrees to form slender structures from which the hemibranchs protrude into a single *branchial*, or *extrabranchial*, *chamber* located on each side between the operculum and gills.

The distinction between internal and external gill slits is less obvious in these. Two rows of gill rays are usually present, rather than one as in the Elasmobranchs. Each row supports a hemibranch. Beginning with the He locephali, the septum of the hyoid arch is enormously developed and grows caudad as a cutaneous fold, covering the gill clefts posterior to it. The gill cover, or *operculum*, which is bony fishes is strengthened by a number of thin *opercular bones*, protects the gill in the branchial chamber which thus opens to the outside through one gill aperture. In *Latimeria* the operculum is poorly developed and partially ossified. In a few forms as in the reel the opening of the branchial chamber to the outside is small and round, but it is usually long and slitlike.

In one group of fishes, the symbranchiates, there is a single, midventral opening. Although in Elasmobranchs, chimaeras, sturgeons, and some others a true hemibranch receiving unoxygenated blood is located on the posterior side of the hyoid arch, this is lacking in *Polypterus*, *Amia*, *Polyodon*, *Teleosts*, and *Dipnoi*. Teleosts have four complete holobranchs on each side. Since in fishes no aortic arch courses through the tissue posterior to the last gill cleft, no hemibranch

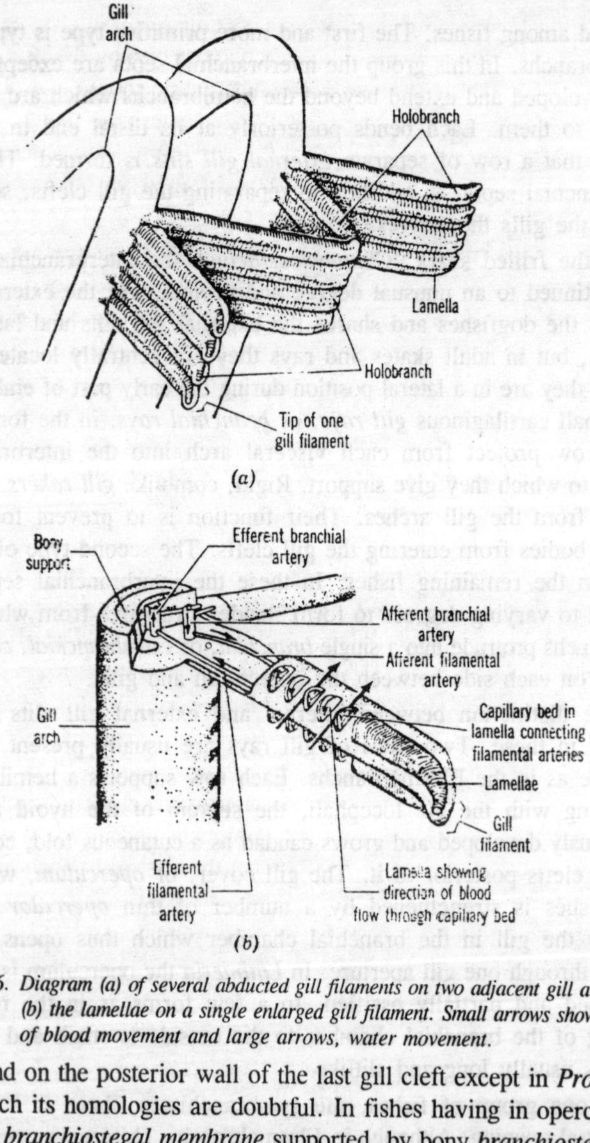

Fig. 6.6. Diagram (a) of several abducted gill filaments on two adjacent gill arches, and (b) the lamellae on a single enlarged gill filament. Small arrows show direction of blood movement and large arrows, water movement.

is found on the posterior wall of the last gill cleft except in *Protopterus* in which its homologies are doubtful. In fishes having in operculum, a pliant, *branchiostegal membrane* supported by bony *branchiostegal rays* usually extends from the inner surface of the operculum to the body wall. It serves as a one-way valve which permits water to leave the branchial chamber through the opercular opening but does not allow it to enter. Other folds, from the maxillary and mandibular regions, form *oral valves* located just within the mouth opening. These permit water to enter the mouth but not to leave it. Gill respiration in these fishes is brought about by raising the opercula and closing the branchiostegal folds. This causes a change in pressure within mouth and pharynx, so that the oral valves are opened and water enters the mouth. Lowering the operculum, with a resulting increases in pressure, causes the oral valves to close and the branchiostegal valves to open. Oral valves seem to be lacking in *Latimeria*.

As water passes over the gills, emerging through the opercular slit, oxygen is taken up, and carbon dioxide liberated. Respiration in fishes thus involves a series of muscular contractions in the walls of mouth and pharynx, bringing about a flow of water. It is of interest that some fishes use the stream of water emerging from the branchial chamber as a propelling force in locomotion. Certain rapidly swimming fishes keep their mouths open while swimming, thus forcing a stream of water to enter the mouth, bathe the gills, and leave the branchial chamber without restoring to muscular contractions and opercular movements. In most, however, the act of inspiration causes water to enter the mouth and pharynx, and expiration results in its expulsion.

According to *J.F. Daniel* (1934) respiration in Elasmobranchs is brought about by the interaction of a complicated series of buccal and pharyngeal muscles which, in the absence of specific valves, in most cases, regulate the passes of water over the gills. When water enters the mouth the external gill slits close, and when the slits open, the mouth closes. In one species, *Heterodontus huncisci*, the number of respirations has been observed to be approximately 35 per minute when the animal is at rest. In most Elasmobranchs and in a few other fishes (*Acipenser*, *Polyodon*, and *Polypterus*) the first gill pouch has become

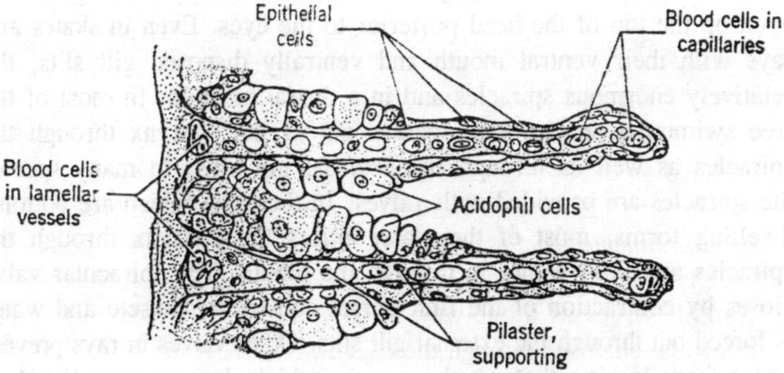

Fig. 6.7. Section through two adjacent lamellae of a gill filament.

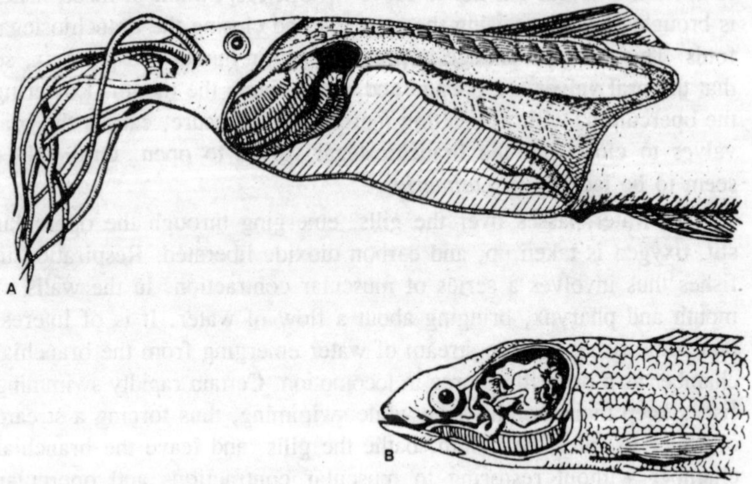

Fig. 6.8. Accessory respiratory organs in teleosts. A—Saccobranchus; B—Anabas.

modified and opens to the outside by means of a *spiracle*. In other fishes the spiracle has been lost or is merely represented by a blind pocket in this region of the pharynx.

Rudimentary gill lamellae may be located on the anterior wall of the spiracle. Since the blood supply to these lamellae consists of oxygenated blood, they do not perform a respiratory function, and the term *false gill*, or *pseudobranch*, is customarily applied to them. The blood in the pseudobranch, however, is actually exposed to oxygenation a second time, so that the brain and eyes, to which this blood is diverted, receive an unusually high percentage of oxygen. This is undoubtedly of some advantage to the animal. The spiracles generally open on the top of the head posterior to the eyes. Even in skates and rays with their ventral mouth and ventrally disposed gill slits, the relatively enormous spiracles and in a dorsal position. In most of the free swimming sharks water passes out of the pharynx through the spiracles as well as through the external gill slits. In many species the spiracles are provided with valves. In the rays, which are bottom-dwelling forms, most of the water *enters* the pharynx through the spiracles and little comes in through the mouth. The spiracular valve closes by contraction of the first dorsal constrictor muscle and water is forced out through the external gill slits. Other valves in rays prevent water from leaving through the mouth, which does not usually close entirely. Under certain conditions rays have been observed to reverse the usual procedure and to "spout" water *out* through the spiracles.

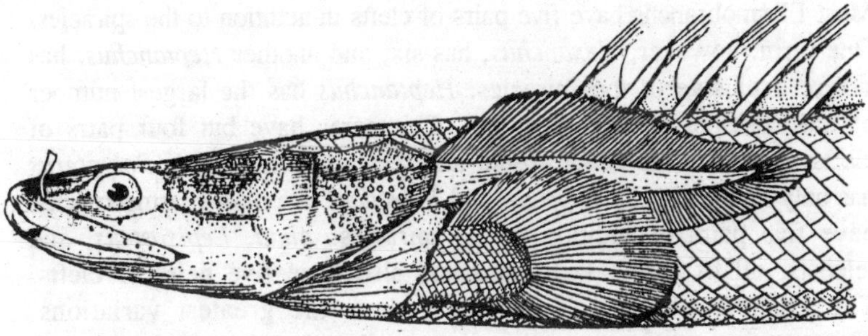

Fig. 6.9. Larval form of the primitive African ray-finned fish, Polypterus; the large external gill arises from the top of the hyoid arch and extends back above the pectoral fin.

This seems to be a protective device enabling the animal to eject foreign matter that has entered the pharynx with the incurrent stream of water.

It is of interest that in one of the ovoviviparous elasmobranchs: *Petroplatea micrura*, certain villous projections from the uterine wall of the mother enter the spiracles of the developing young, supplying nutriment directly to the walls of their digestive tracts. Although in most fishes a *true* hemibranch, receiving unoxygenated blood is lacking on the posterior side of the hyoid arch, a modified *opercular gill*, or *pseudobranch*, receiving oxygenated blood may be present. Opercular gills of this type are found in *Amia*, the Dipnoi, *Latimeria*, and many teleosts. Such opercular gills probably represent the spiracular pseudobranch which, upon closure of the spiracle, has shifted its position. *Lepidosteus* is peculiar in having both dorsal and ventral gills on the inner surface of the operculum.

The dorsal lamellae belong to a pseudobranch, whereas the ventral gill is a true hemibranch. The number of gills and gill clefts varies among fishes. There is some confusion as to just what each represents since different authors use different numbers in designating them. Throughout this volume we shall consider the mandibular arch to be the *first* visceral arch, the hyoid the *second*, and shall refer to the remaining arches as 3, 4, 5, 6, etc. The hyomandibular cleft between the mandibular and hyoid arches is actually the first cleft. However, since it is either modified to form a spiracle or is closed altogether, and because it never bears true gills, we shall refer to it only as the spiracle, spiracular cleft, or hyomandibular cleft.

The cleft between the hyoid and third visceral arches is the first *typical gill cleft* and, therefore, will here be designated as gill cleft.

Most Elasmobranchs have five pairs of clefts in addition to the spiracles. One form, however, *Hexanchus*, has six, and another *Heptanchus*, has seven, exclusive of the spiracles. *Heptanchus* has the largest number of gill clefts of any gnathostome. Chimaeras have but four pairs of clefts, the spiracles being absent and the last cleft closed. *Polypterus* has only four pairs but *Acipenser* and *Polyodon*, like most Elasmobranchs, have five pairs in addition to the spiracles. *Amia*, *Lepisosteus*, and teleosts, all of which lack spiracles, also have five pairs of clefts. The Dipnoi, which lack spiracles, show the greatest variations. *Epiceratodus* and *Protopterus* have five pairs of clefts, but *Lepidosiren* has only four.

All the branchial arches of *Epiceratodus* bear functional gills. In *Lepidosiren* the first slit has been closed, and only three holobranchs are present. In *Protopterus* no lamellae border the first two clefts or the anterior wall of the third. A hyoidean pseudobranch lies anterior to the first gill cleft. The gill filaments of *Lepidosiren* and *Protopterus* are coarse and of little importance in oxygen uptake. They are important, however, for carbon dioxide elimination. The presence of external gills is rare among fishes. In the larva of *Polypterus* a single pair of external integumentary gills is present in the region of the hyoid arch. Each gill consists of a narrow main, central portion which bears a double row of filaments. Larval Dipnoi possess four pairs of external cutaneous gills located on the visceral arches. These external gills disappear in adult life, although vestiges may remain. In certain elasmobranchs during embryonic life the gill filaments are so long that they extend out the pharynx through the spiracles and function in the manner of external gill filaments. It is believed that they may even aid in the absorption of nutriment.

Blood Supply to Gills

The fine structure of the secondary folds allows the blood to come into very intimate contact with the water; indeed the two fluids are separated only by a very thin membrane through which the oxygen and carbon dioxide may readily pass. The folds are well supplied with blood, the flow of which is so directed that the blood and water streams pass one another in opposite directions. The arrangement of '*counter currents*' is found in many parts of the animal body where an efficient exchange of dissolved substances is required between two fluids. In the gills, the countercurrents ensure that blood which is leaving the gills almost fully oxygenated meets water entering with its full oxygen content, while blood deficient in oxygen and entering the gills meets water from which much of the oxygen has already been extracted. In this way there is always a greater amount of oxygen in the water than in the blood it meets so that oxygen continues to pass into the blood throughout its passage through the gills. The efficiency of the countercurrent system is shown by the fact that the uptake of oxygen through the gills falls from 50% to 9% if the direction of water flow is experimentally reversed.

Larval Gills

Larval gills are of three kinds—external gills that are outgrowths from the external surface of one or more gill arches; filamentous extensions of internal gills that project through gill slits to the exterior; and the internal gills of late anuran tadpoles, which are hidden behind the larval operculum. External gills usually develop before gill slits open and before any opercular fold has started to develop. They can

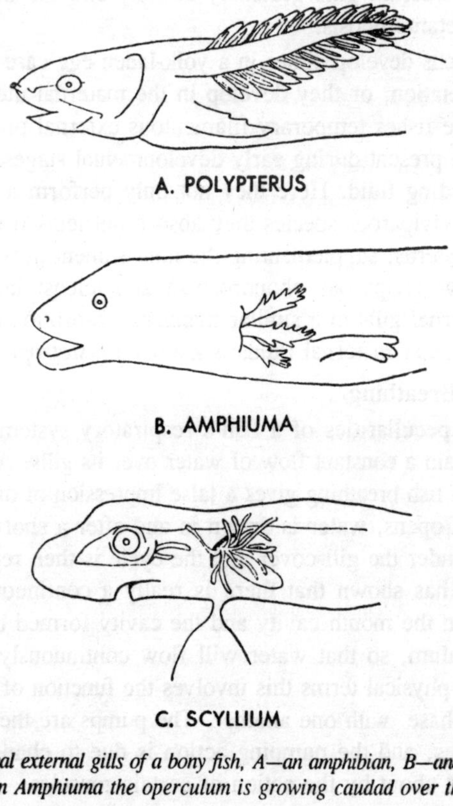

Fig. 6.10. Larval external gills of a bony fish, A—an amphibian, B—and an elasmobranch, C. In Amphiuma the operculum is growing caudad over the gills.

be waved about by branchial muscles and can often be retracted. They develop in the embryonic or larval stages of most dipnoans (*Ceratodus* is an exception), in all amphibians (including apodans), and in a few ganoid fishes such as sturgeons and *Polypterus*. The latter has only one pair.

Although external gills are larval structures, there are perennibranchiate urodeles and a perennibranchiate lungfish, *Protopteurs*. The latter has four pairs of larval gills and retains three pairs in a reduced state throughout life. External gills develop in anuran tadpoles on pharyngeal arches III to V. Later, when pharyngeal pouches II to V rupture to the exterior, the pouch linings become folded to form a set of internal gills. Thereupon, a fleshy operculum grows backward from the hyoid arch over the gill region, enclosing external gills and internal gills in an opercular chamber that retains only a single excurrent pore, this being located at the posterior edge of the left operculum. Thereafter, the external gills gradually atrophy and the internal gills function until metamorphosis.

Elasmobranchs develop either in a yolk-laden egg case attached to underwater vegetation, or they develop in the maternal uterus. In the embryos of these fishes temporary filamentous external projections of internal gills are present during early developmental stages, projecting into the surrounding fluid. Here they not only perform a respiratory function but, in viviparous species they absorb nutrients from the fluid in the maternal uterus, supplementing the nourishment provided by the yolk sac. A few viviparous chondrostean and teleost larvae utilize filamentous external gills in a similar manner. Absorption of nutrients from uterine or other maternal fluids is known a *histotrophic nutrition*.

Mechanism of Breathing

One of the peculiarities of a fish's respiratory system is the fact that it can maintain a constant flow of water over its gills. A superficial examination of a fish breathing gives a false impression of discontinuous flow: the mouth opens, water is drawn in and after a short interval is expelled from under the gill-cover and the cycle is then repeated. But recent research has shown that there is really a continuous pressure gradient between the mouth cavity and the cavity formed between the gills and operculum, so that water will flow continuously across the gill surfaces. In physical terms this involves the function of two pumps slightly out of phase with one another. The pumps are the mouth and opercular cavities, and the pumping action is due to changes in their volumes brought about by the action of certain muscles.

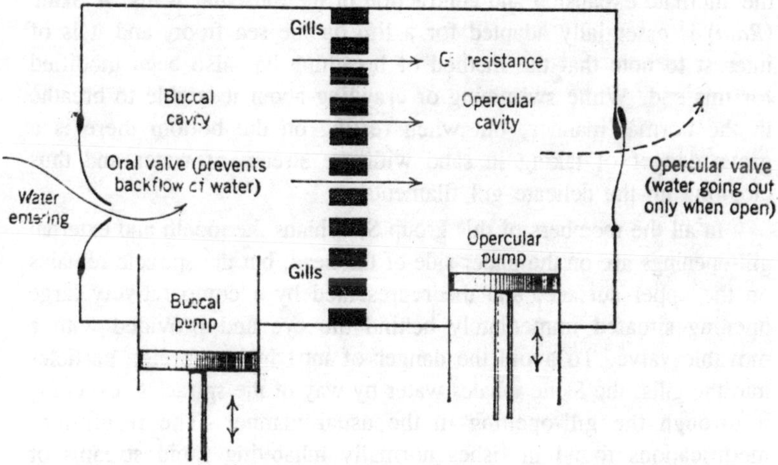

Fig. 6.11. Diagram of mechanism for gill ventilation by means of two pumps; the stage shown is shortly after the commencement of inspiration.

There are, of course, numerous slight modifications to this basic system, modifications usually associated with the different habits of the various species. For example, in some fast swimmers like the Mackerel and Tunny neither pump functions when the fish is swimming and they rely on the water forced through the slightly open mouth by their passage through the sea. In the case of the Mackerel the oxygen requirements are so high that the fish is obliged to swim continuously if sufficient water is to pass over the gills. In yet other species pumping movements are absent during swimming but start once the fish comes to rest. A fish can modify its breathing rate to suit its oxygen requirements and the basic rate varies from species to species according to their habits. If the water is deficient in oxygen the respiratory rate increases, as it will do if the fish is excited or very active. Temperature will also affect respiratory rates, but probably because of its effects on the solubility of oxygen. Most fishes breathe in the manner described, but there are some which, owing to their peculiar mode of life, been obliged to modify this process in accordance with their change in habits.

The Lamprey (*Petromyzon*), for example, has adopted a parasitic habit, and spends a good deal of its time attached to other fishes by means of its sucker-like mouth. It is quite obvious that while in this position it would be impossible for the animal to inhale water through the mouth without losing its hold. Thus, water is often taken into and expelled from the branchial sacs by their external openings, through the alternate expansion and contraction of the muscular walls. A Skate (*Raia*) is essentially adapted for a life on the sea floor, and it is of interest to note that the method of breathing has also been modified for this end. While swimming or crawling about it is able to breathe in the normal manner, but when resting on the bottom there is a grave danger of taking in sand with the stream of water and thus clogging up the delicate gill filaments.

In all the members of this group Selachians the mouth and external gill-openings are on the under side of the head, but the spiracle remains on the upper surface, and the represented by a comparatively large opening situated immediately behind the eye and provided with a movable valve. To avoid the danger of introducing foreign particles into the gills, the Skate inhales water by way of the spiracles, expelling it through the gill-opening in the usual manner. The respiratory modifications found in fishes normally inhabiting rapid streams or mountain torrents, where they are in the habit of fixing themselves to stones and other objects to avoid being swept away by the current of water, and dealt with in a subsequent chapter.

Gills in Excretion

In addition to their respiratory function the gills play an important part in the excretion of certain waste products and in the maintenance of the fish's salt balance. Both marine and freshwater Bony Fishes secrete some nitrogenous waste products through the gills, in the form of ammonia and urea. It has been estimate that in the Carp (*Cyprinus*) and the Goldfish (*Carassius*) six to ten times as much nitrogenous matter is excreted through the gills than by the kidneys. Certain cells in the gill filaments are able either to take up or excrete chlorides. The importance of this mechanism for maintaining the salt balance of the body need not be stressed, and it is not surprising to find that 'chloride' cells and particularly well developed in fishes with a wide salinity tolerance.

Gills and Food Capturing

In those fishes whose food consists of minute creatures swimming about in the water there is obviously a danger of some of these escaping by way of the pharyngeal openings and perhaps clogging or injuring the delicate filaments. This danger is lessened by special structures known as gill-rakers which take the form of a double row of stiff appendages on the inner margin of each hoop-like gill-arch. These rakers, by projecting across the pharyngeal openings, serve to strain the water which is to bathe the gills, and to prevent and solid particles from passing over with it. Generally, the front row of rakers on each arch interlocks with the hinder row on the adjoining arch, and the two together form and effective sieve.

In the Pike (*Esox*), feeding almost entirely on other fishes, the gill-rakers are represented merely by bony knobs, which may serve to block the passage of larger food particles. In the Herring-like fishes (Clupeidae), on the other hand, the food consists of plankton, and the rakers are very numerous and take the form of long, slender, and close-set bristles. In many filter-feeders the mechanism is made even more perfect, each primary gill-raker giving off secondary and tertiary branches, the whole apparatus having the appearance of the finest gauze. The form and number of the gill-rakers may differ considerably even in two closely related species, but this difference may generally be correlated with a difference in the normal diet. The two species of Shad (*Alosa*) found in seas and rivers provide an excellent example: the Allis Shad (*A. alosa*), having about eighty rakers on the lower limb of each arch in the adult fish, whereas the Twaite Shad (*A. fallax*) has only thirty. The Allis Shad feeds largely on small crustaceans, although it takes a certain number of larval fishes. The Twaite Shad does not eat crustaceans to nearly the same extent, but is much more destructive to the young fry of other fishes.

Other Organs of Respiration

In addition to the gills, the skin of many fishes help in respiration. In *Periopthalmus*, tail is used for breathing. Some species possess larval gills which are filamentous outgrowths from the gills. In some species as *Salmo* and *Misgurnus* the gills lamellae and long, filiform structures and extend out of the gill slits in larval stages. *Polypterus* larva also has a pair of external gills which disappear in the adult. In some fishes, air bladder is modified as a lung for respiration. In *Amia* and *Lepidosteus*, the air bladder is highly vascular and sacculated internally and serves for respiration. In the Dipnoi, air bladder is completely modified as a lung. In other teleosts, air bladder may serve as a reservoir of oxygen. Besides these structures, some species possess accessory respiratory organs to help in breathing.

ACCESSORY RESPIRATORY ORGANS

In some tropical fresh water fishes, special structures called the accessory air breathing organs are present in addition to the gills. Such structures are generally developed in response to the exceptional environmental conditions which include life in foul water or life out of water for short periods. The air breathing organs enable the fish to tolerate oxygen depletion in the water or to live out of water for short periods. These additional respiratory structures are rarely developed in marine fishes and have been reported in *Chiloscylium indicum*, *C. griseum* (*Geroge*, 1953), and a sole fish of the genus *Achirus* (*Breder*, 1941). Air breathing organs are generally found in fishes living in shallow stagnant, fresh water of tropical regions where deoxygenation of water occurs. These organs may also develop in fishes inhabiting torrential streams of the hills, that are liable to dry up during summer.

Several types of accessory respiratory organs have been evolved in different types of fishes. In *Mastacembelus* however, the unmodified gills secrete a large amount of mucous to keep them moist and facilitate gas diffusion for sometime when the fish is out of water. Although gills serve as the chief organs of respiration in most fishes, other respiratory mechanisms may be present in addition. The skin of the eel is also important in respiration, an unusual condition among fishes. Vascular areas in the skin, in some forms, may also aid in respiration.

Some fishes possess pockets, projecting from the branchial chambers, in which water may be carried. Other accessory respiratory organs in fishes may enable the animals to utilize oxygen taken from the air. In a few tropical Teleosts curious labyrinthiform organs are developed in connection with certain of the branchial arches, and serve as accessory breathing organs. In the Indian "Climbing Perch" (*Anabas scandens*), of the family Anabantidae, the organ consists of three or more concentrically-arranged bony laminae, with wavy, crenulated margins, attached by a common bony base to the upper extremity of the fourth branchial arch, and enclosed in a special dorsal enlargement of the branchial cavity. The vascular membrane which invests the laminae is abundantly supplied with venous blood by a branch of the fourth afferent branchial artery, the equivalent efferent vessel joining the dorsal aorta.

Essentially similar organs are found in several genera of Osphromenidae (e.g. *Polyacanthus*, *Osphromenus*, and *Trichogaster*). A simpler form of respiratory organ of somewhat the same type occurs in the Indian family Ophiocephalidae. In these Fishes there is, on each side, an accessory branchial cavity, situated above that which contains the gills, but freely communicating with it. The cavity is lined by a thickened and puckered vascular membrane, but otherwise contains no special respiratory structures. In the Siluroid genera *Clarias* and *Heterobranchus* the accessory organ takes the form of branched, arborescent and highly vascular structures, developed as outgrowths from the dorsal extremities of one or two branchial arches, and enclosed

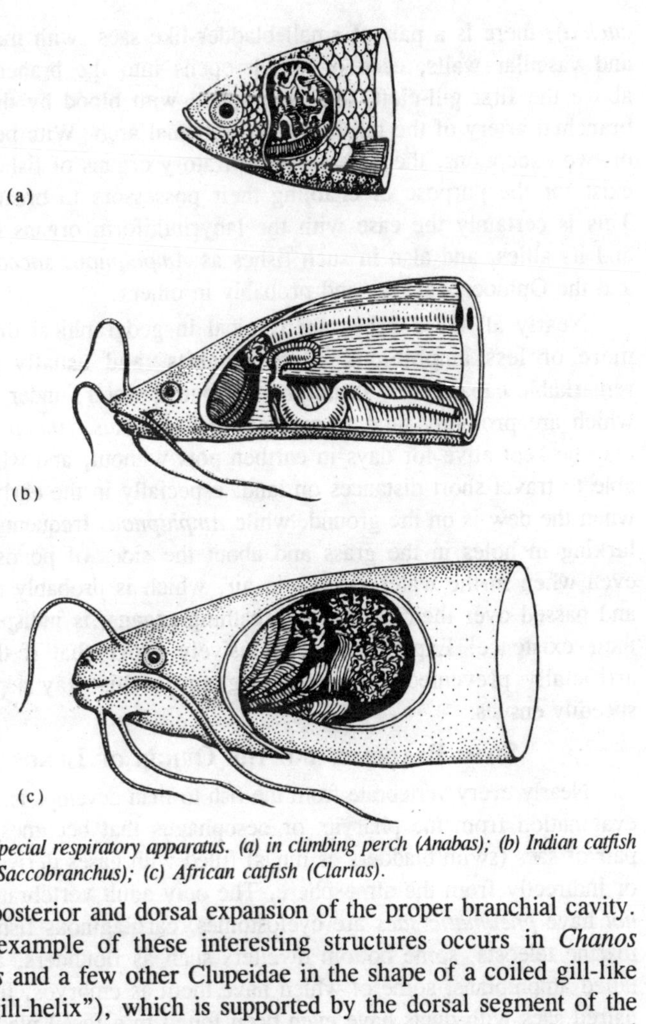

Fig. 6.12. Special respiratory apparatus. (a) in climbing perch (Anabas); (b) Indian catfish (Saccobranchus); (c) African catfish (Clarias).

within a posterior and dorsal expansion of the proper branchial cavity. Another example of these interesting structures occurs in *Chanos salmoneus* and a few other Clupeidae in the shape of a coiled gill-like organ ("gill-helix"), which is supported by the dorsal segment of the fourth branchial arch, and enclosed in a similarly curved caecal extension of the branchial cavity. Each gill derives its blood from the fourth afferent branchial artery, the corresponding efferent vessel joining the fourth efferent branchial artery.

A similar spirally-coiled "*gill-helix*" is found also in *Heterotis chrenbergii*, amongst the Osteoglossidae, and in several species of Characinidae. In other Teleosts the accessory breathing organ assumes the condition of paired lung-like outgrowths of the branchial cavity. Thus, in one of the Symbranchidae, the Indian "Cuchia Eel" (*Amphipnous cuchia*), there is a pair of small bladder-like sacs, with membranous and vascular walls, each of which opens into the branchial cavity above the first gill-cleft, and is supplied with blood by the efferent branchial artery of the gill-less first branchial arch. With perhaps one or two exceptions, the accessory respiratory organs of fishes seem to exist for the purpose of enabling their possessors to breathe in air. This is certainly the case with the labyrinthiform organs of *Anabus* and its allies, and also in such fishes as *Amphipnous saccobranchus*, and the Ophiocephalidae, and probably in others.

Nearly all these fishes are tropical in geographical distribution, more or less amphibious in their habits, and usually possess a remarkable capacity for sustaining life out of water, under conditions which are promptly fatal to ordinary fishes. Thus, *Anabas scandens* may be kept alive for days in earthen pots without, and when free is able to travel short distances on land, especially in the early morning when the dew is on the ground, while *Amphipnous* frequents marshes, lurking in holes in the grass and about the sides of ponds. In fact, even when in the water, access to air, which is probably swallowed and passed over their accessory breathing organs, is indispensable to their existence. Experiments conclusively prove that if the fish is artificially prevented from obtaining air in this way asphyxiation speedily ensues.

Swim Bladders and the Origin of Lungs

Nearly every vertebrate from the fish to man develops and unpaired evagination from the pharynx or oesophagus that becomes one or a pair of sacs (swim bladders or lungs) filled with gases derived directly or indirectly from the atmosphere. The only adult vertebrates that do *not* have *pneumatic sacs* are cyclostomes, cartilaginous fishes, a few marine teleosts, some bottom dwellers such as flounders, and a few tailed amphibians, some of which have them as embryos. Imprints of paired sacs with ducts have even been found in a fossil placoderm. It can be said with considerable confidence that vertebrates that fail to develop pneumatic sacs have probably lost the genetic capability to do so. After evaginating, pneumatic sacs may retain their unpaired connection with the foregut, or the duct may close. Pneumatic sacs hereafter will be called lungs in tetrapods and swim bladders in fishes, regardless of their function. Swim bladders may be paired or unpaired.

The pneumatic duct, when presents in adults, connects to the oesophagus ventrally in dipnoan and chondrosteans but has shifted toward the dorsal side in holosteans and teleosts. Infrequently the pneumatic

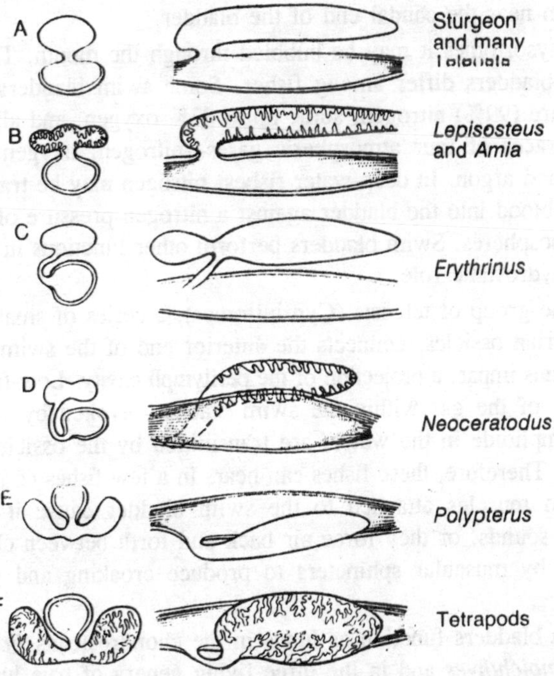

Fig. 6.13. Diagrammatic cross sections and longitudinal sections of the swim bladder or lung in various fishes and in tetrapods.

duct connects to the pharynx or stomach. Swim bladders are retroperitoneal, lie close to the kidneys, and bulge more or less into the coelom. The walls contain elastic tissue and smooth muscle, and the lining is relatively smooth except in the true lungfishes. Fishes are said to be *physostomous* when the duct is open, *physoclistous* when it is closed. Dipnoans, chondrosteans, holosteans, and a few teleosts are *physostomous*. Among the latter are catfish, carp, eels, herring, pickerel, and salmon.

Only a few *physostomous* fishes use the swim bladder for respiration. Swim bladders in teleosts serve chiefly as *hydrostatic organs*. The volume of gas in the bladder can be reflexly regulated at its source, thereby altering the specific gravity of the fish and increasing or decreasing its buoyancy. The gas in the hydrostatic swim bladder usually comes from the blood. It is actively transported into the chamber of the bladder from a network of small arteries (*rete mirabile*) in the lining of the bladder and named the *red gland*. The gland is supplied from the coeliac artery, and its associated tortuous veins empty into the hepatic portal vein. The gas is reabsorbed in an area of modified epithelium near the caudal end of the bladder.

In physostomes it may be bubbled through the mouth. The gases in swim bladders differ among fishes. Some swim bladders contain almost pure (99%) nitrogen, some up to 87% oxygen, and all contain at least traces of four atmospheric gases—nitrogen, oxygen, carbon dioxide, and argon. In deep-water fishes, nitrogen may be transported from the blood into the bladder against a nitrogen pressure of as high as 10 atmospheres. Swim bladders perform other functions in addition to their hydrostatic role.

In one group of teleosts (Cyprinifomes), a series of small bones, the Weberian ossicles, connects the anterior end of the swim bladder and the sinus impar, a projection of the perilymph cavity. Low-frequency vibrations of the gas within the swim bladder, evoked by waves of similar amplitude in the water, are transmitted by the ossicles in the inner ear. Therefore, these fishes can hear. In a few fishes contractions of striated muscles attached to the swim bladder cause it to emit thumping sounds, or they force air back and forth between chambers separated by muscular sphincters to produce croaking and grunting sounds.

Swim bladders function as lungs in the chondrosteans *Polypterus* and *Calamoichthyes* and in the three living genera of true lungfishes (dipnoans), except that *Neoceratodus* uses its single swim bladder in this manner only when the oxygen content of the water is low. In these air-breathing fishes, the swim bladder is lined with low septa and may exhibit thousands of tiny air sacs. Like tetrapod lungs, they are supplied by arteries arising from the sixth embryonic aortic arch; and in dipnoans the venous return is to the left atrium, as in tetrapods. An oropharyngeal pump forces air from the oropharyngeal cavity into the swim bladder.

Expulsion of air from the bladder to the oropharynx is a result of the vacuum created by lowering the oropharyngeal floor while the mouth and nares are closed, elasticity of the bladder, contraction of smooth muscles in its walls, and pressure on the body wall exerted by the surrounding water. It is subsequently bubbled to the exterior through the mouth. The striking similarities between swim bladders and lungs suggest that these are the same organs. In the Devonian when the fresh water was warm and periodically stagnant, and therefore low in dissolved oxygen, aerial respiration may have made the difference between extinction and survival. At that time, placoderms and most crossopterygians had pneumatic ducts. We can only speculate on which came first, respiratory or hydrostatic swim bladders, but these two conclusions seem well founded: they were functioning in aerial respiration long before vertebrates ventured onto land, and closure of the pneumatic duct in *physoclistous fishes* is probably a mutation from a more primitive, open duct condition.

Significance of the Accessory Respiratory Organs

It has been shown that a higher percentage of oxygen is generally in the accessory respiratory organs than in the gas bladder of fishes. Most of the fishes possessing such structures are capable of living in highly deoxygenated water. They have been observed to gulp in air from the surface and to pass it to the accessory respiratory organs. If prevented from reaching the surface, these fishes die due to *asphyxiation*. All these show that the accessory organs are capable of maintaining life of the fish. It has also been shown that the absorption of oxygen in the air breathing organs, is much greater than the excretion of carbon dioxide. Hence, most of the CO_2 is excreted in the gills and the chief function of the accessory organs is the absorption of oxygen required for the sustenance of life.

Role of Blood in Respiration

Although in members of certain lower phyla there is a direct exchange of gases between the cells and the environment, in vertebrates and some other forms the blood serves to bring oxygen from the environment to the cells and to carry carbon dioxide from the cells to the environment. It is the presence of the pigment *haemoglobin* that given vertebrate blood its unusual capacity or carrying oxygen. Haemoglobin is present in the blood of all vertebrates and even in certain invertebrates. In vertebrate blood it is confined to the red blood corpuscles, or *erythrocytes*. It is interesting that the muscles of birds and mammals also contain a form of haemoglobin called *myoglobin*. There are minor differences in the hemoglobins of *different forms* and they vary somewhat in their properties.

The importance of haemoglobin in connection with respiration lies in its ability to combine with oxygen through the medium of iron which it contains. Two atoms of oxygen, under favourable conditions, unite with each atom of iron in the haemoglobin complex. Thus $Hb_4 + 4O_2 = Hb_4O_8$. Haemoglobin which has combined with oxygen in this manner is spoken of as *oxyhemoglobin*. It is scarlet red in colour and a very unstable compound, since it given up its oxygen so readily.

After having given off oxygen it is known as *reduced haemoglobin*, or simply *haemoglobin*, and has a purplish-blue colour. The red corpuscles with their contained hemoglobin also play a prominent role in transporting carbon dioxide. When carbon dioxide leaves the cells and tissues of the body it enters the blood-stream and diffuses into the red blood corpuscles. Here most of the carbon dioxide combines with water to form *carbonic acid* as followed: $CO_2+H_2O=H_2CO_3$. An enzyme, *carbonic anhydrase*, present in the red cells, facilitates the reaction.

The carbonic acid thus formed combines with potassium (formerly in combination with haemoglobin) to form *potassium bicarbonate*. Bicarbonate ions then diffuse out into the plasma in exchange for chloride ions which enter the cell. In the plasma the bicarbonate ions combine with sodium to form *sodium bicarbonate*. Thus, carbon dioxide is transported by the plasma as well as by the red corpuscles. At the respiratory surface oxygen enters the red blood corpuscles and the above processes are reversed. Carbon dioxide is then liberated as a free gas.

Phases of Respiration

In animals in which the blood is used in carrying oxygen and carbon dioxide, two phases are involved in respiration. They are referred to as *external* and *internal respiration*. The term *external respiration* is used to denote the exchange of gases between the blood and the environment. This usually takes places in the capillaries of the gills or lungs, but in some cases other structure such as the skin is also utilized. In gills the capillaries are almost in direct contact with the water in which oxygen is dissolved. In lungs they are practically in contact with air that has been taken into the cavities of the lungs. Even though lungs are situated within the body, nevertheless the term external respiration is used.

During external respiration the blood takes up oxygen and loses most of the carbon dioxide which it is carrying. *Internal respiration* refers to the gaseous exchange between the blood and the tissues or cells of the body. Oxygen is taken from the blood, which in turn gains carbon dioxide given off by the tissues. The principles involved in internal and external respiration follow the "laws of gases" and are physical rather than biological in nature. In vertebrates, respiratory organs are present which serve to facilitate external respiration. Certain requirements are demanded by respiratory organs in order that they may function properly:

1. A large, vascular surface area must be provided so that an ample capillary network may be exposed to the environment.
2. The membrane surfaces through which gaseous exchange occurs must be moist at all times and be thin enough to permit the passage of gases.
3. Provision must be made for renewing the supply of the oxygen-(containing medium air or water) which comes in contact with the respiratory surface and for removing the carbon dioxide which has been given off from that surface.
4. Blood in the capillary network must circulate freely.

7

Removal of Nitrogenous Wastes

Animals eliminate some metabolic wastes through the gut and the skin, but most are eliminated through special excretory organs, the kidneys. In elimination fishes and other aquatic animals have a particular problem in their gills and oral membranes are permeable both to water and salts. In the ocean the salinity of the water is more concentrated than that of the body fluids of the fish, and water is drawn out, but salts tend to diffuse inward; hence, marine fishes drink sea water. In contrast, in fresh water, fishes lose salts and take up water through the gills because their internal salt concentration is greater than that of their surroundings. Many nitrogenous wastes of fishes pass through the kidneys which also assist in water-salt balance (homeostasis) by the excretion or retention of certain minerals. The gills also take a prominent part in waste excretion, eliminating mainly ammonia. Before considering problems of osmoregulation in fishes whose body fluids are either more or less concentrated than the water in which they live, let us first look at hagfish, where the body fluids are similar to sea water, and there is little or no osmotic gradient across the permeable surfaces.

Hagfishes and the Origin of the Glomerular Kidney

In hagfish, there are kidneys of the usual vertebrate mesonephric type although the persistent pronephros remains rather enigmatic. Hagfish blood is isosmotic or slightly hypertonic to sea water; blood sodium and chloride levels are similar to those in sea water, although divalent ions are significantly lower. Tissue sodium and Chloride levels are high, but even so there is a greater difference between blood and tissue ions than in other fish and osmotic balance between blood and tissues is maintained by high levels of intracellular amino acids.

Experiments with tritiated water have shown that hagfish are freely permeable to water, much more so than are other fish where the blood is osmotically different to the medium and it is not surprising that hagfish are unable to tolerate fresh water and are killed by 25% sea water. There is no doubt that the acraniates arose in the sea, and the situation in hagfish (which are all marine) seems to show that they too have always been marine, and have inherited their 'sea-water' blood composition from marine protochordates. Yet lampreys (which either live entirely in fresh water, or have a fresh water larval stage in the life history) are quite different to hagfish in their ionic composition and osmoregulation, as we shall see, and investigations on hagfish urine composition by *McInerney* have shown unexpected sodium-concentrating mechanisms in the kidney.

Munz and *McFarland* (1964) concluded that hagfish lacked any mechanism for re-absorbing sodium from the glomerular filtrate (a necessary trick for any animal living in fresh water), and suggested that this was good evidence for the marine origin of the group. But *McInerey's* careful experiments on hagfish set up in water with low salt content (made isosmotic with sea water by adding sucrose) showed that they responded to low external sodium by increasing renal sodium uptake and it is difficult to think that this 'surplus' capacity of the sodium recovery mechanism would be needed in sea water. *McInerney* suggests that it could be considerable as a left over from an original freshwater ancestry. This view would fit with Homer Smith's famous hypothesis of 1932 that the glomerular kidney first arose in fresh water as a device for excreting water.

Although the two living agnathan groups are so different from each other their relationships are hard to ascertain, both could be considered as being originally freshwater. But the alternative view that the glomerular kidney first arouse in sea water, as a device to regulate ions by producing a filtrate which could be selectively altered, and that marine fish possessing glomerular kidneys were 'pre-adapted' to enter fresh water now seems more probable, and both hagfish and lampreys are today considered to have arisen in the marine environment. The earliest fossil lampery. *Mayomyzon* of the Pennsylvanian, is associated with a marine fauna, and *Jamoytius* (*Ritchie*, 1968) which may be ancestral to both living agnathan groups is also marine. Further investigations of the sodium recovery mechanism of hagfish are obviously desirable.

Lampreys

All lampreys develop in fresh water, passing several years as filter-feeding ammocoete larvae buried in mud. Some species remain in fresh water after metamorphosis, but others pass down streams and rivers to the sea, where they feed and grow until they return to fresh water to breed. They therefore have similar problems of osmoregulation to such *anadromous fish* as salmon, and it is very striking (considering the remote affinity of agnathans and teleosts) that both groups adopt similar solutions. In fresh water, the larval ammocoete and the adult have blood that is more concentrated than the water, the animals thus tend to gain water. They excrete large amounts of urine, apparently larger amounts than do freshwater teleosts (it is hard to sample urine without stressing the animals, and values for urine production are open to objection on several grounds). A high rate of urine production enables the animals to get rid of the water which enters via the permeable surfaces but if there is not to be concomitant loss of ions, there must be selective absorption mechanisms in the kidney so that the ions (or as high a proportion of them as possible) in the glomerular filtrate and reabsorbed as the fluid passes down the nephron to the collecting duct. Lampreys have efficient mechanisms for reabsorbing sodium and chloride but are less efficient at reabsorbing potassium.

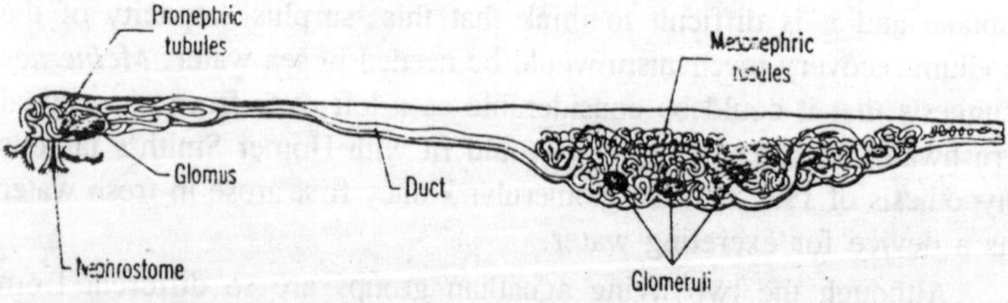

Fig. 7.1. Kidney of a lamprey (Petromyzon) larva of 22 millimeters showing relationship of pronephric and mesonephric portions.

Isotope studies of *ammocoete* larvae have shown that the potassium and sodium lost via the urine is replaced via active uptake mechanisms in the gills, and that the animals can maintain a stable blood composition even in very dilute solutions (5—30μm Na/1). Here then is the first appearance of extrarenal routes for ion uptake, complementing mechanisms in the kidneys; such mechanisms are more familiar in freshwater teleosts. In the sea, lampreys face the reverse problem of losing water. However, adult lampreys are only rarely

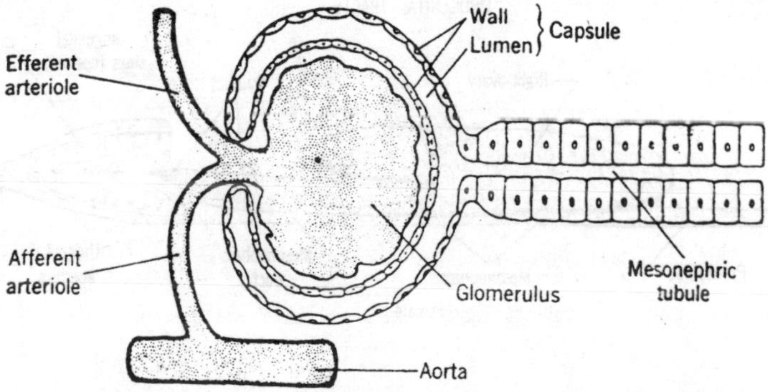

Fig. 7.2. Diagram of a renal corpuscle (glomerulus plus capsule) at the end of a uriniferous tubule as in a fish mesonephros.

caught at sea (they can usually be seen attached to any basking shark closely examined, but they drop off when the shark is caught), and so they are very few analyses of the blood of marine lampreys. When they are caught as they enter rivers, the marine osmoregulatory mechanism has already changed so that the animals cannot live in sea water; some are able to tolerate 50% sea water and it is on such already partially adapted animals the mechanism of marine osmoregulation has been examined.

Pickering and *Morris* found that, like marine teleosts, marine lampreys drink sea water, thus absorbing water and monovalent ions. These are excreted by specialized 'chloride' cells on the gills, and divalent ions entering the body fluids are excreted by the kidneys. The blood is more dilute than sea water, and although most of the water drunk is absorbed, little urine is produced and excess water is lost extrarenally. The distant relationship between Agnatha and teleosts suggests that they must have independently evolved similar mechanisms for osmoregulation, and have had a similar history of secondary return to the sea. Before considering this, let us look first at teleost osmoregulation.

Structure of the Kidney of Teleosts

Teleosts have radiated widely in the Sea and in fresh water; many are able to cope either with migrations from one to the other, as to live in estuaries and brackish water of varying salinity. That is why teleosts have different types of kidneys. The kidney's are paired, elongated structures placed above the alimentary canal and are close to the vertebral column. The teleostean kidney is generally divided into two portions, the head kidney and the trunk kidney, but in many

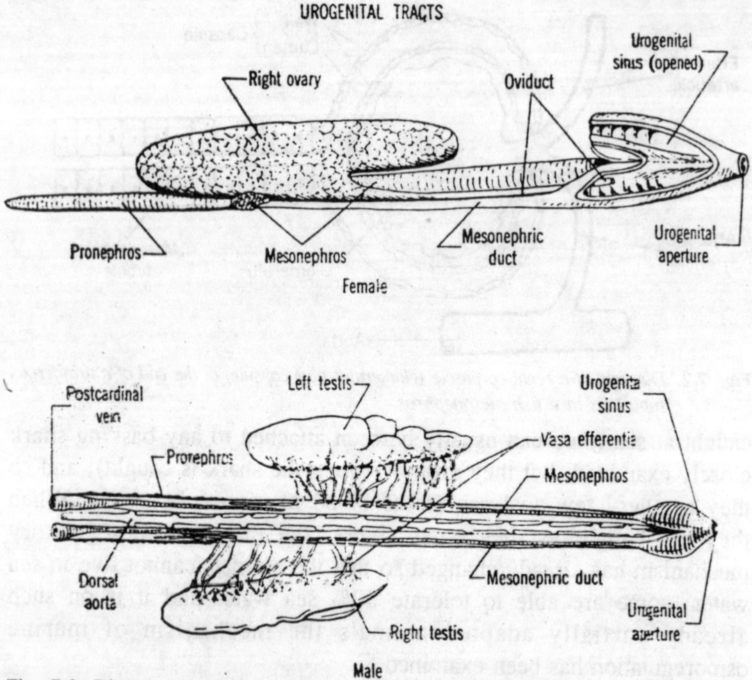

Fig. 7.3. Diagram of relationships of kidney, gonads, urinary ducts, urogenital sinus, and blood vessels in a male and female holostean fish gar (Lepisosteus).

species, these regions can not be distinguished by external examination. Generally, there an no conspicuous differences in shape between the two sexes. According to *Ogawa* (1961), the marine teleostean kidney can be divided into five types:

1. In some fishes both the kidneys are completely separate as in Lophiidae.
2. In seahorse and pipe fishes only the extreme posterior portion of the kidney is fused. There is no distinction of head and trunk kidney.
3. In marine catfishes and eels the middle and posterior portions of the kidneys are fused. There is a clear distinction between the trunk and head kidney.
4. In clupeidae, the two kidneys are completely fused and there is no clear distinction between trunk and head kidney.
5. In some fishes the posterior portions of the kidney are fused. The anterior is represented by two slender branches. Head and trunk kidneys are clearly distinguishable. Example: Most of marine fishes belonging to Pleuronectidae, Cottidae, Scopelidae etc.

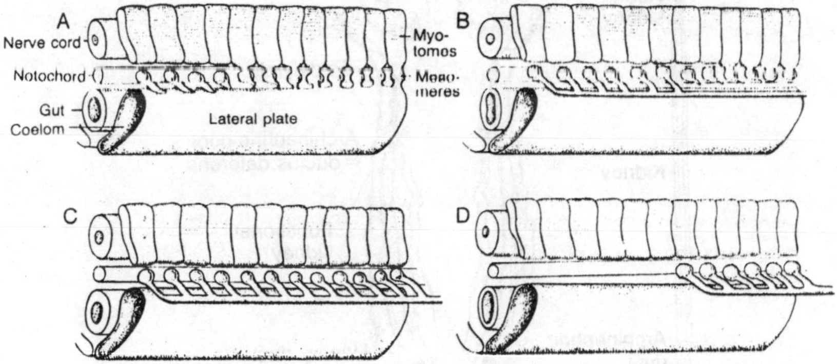

Fig. 7.4. Diagrams of the anterior part of the trunk of an embryo (skin removed) to show the development of the archinephric duct.

Histologically, the trunk kidney is made up a large number of nephrons, each consisting of a renal corpuscle or the Malpighian body and the tubule. The intertubular space is full of lymphoidal tissue which is unevenly distributed. The head kidney is generally made up of lymphoid, hematopoietic, interrenal and chromaffin (supra renal) tissue and is devoid of renal corpuscles and the troubles.

In some species, however, a few nephrons and collecting ducts may be present in the lymphoid tissue of the head kidney, but the glomeruli and the renal corpuscles are usually absent. The head kidney is therefore, not excretory in function. It is derived from the pronephric hematopoietic tissue but in some teleosts, the mesonephros grows forwards extending into the head kidney. The renal elements found in the head kidney of some teleosts are very likely mesonephric. A typical nephorn of a fresh water teleost consists of the following parts:

1. A renal corpuscle containing a well vascularised glome rulus.
2. A ciliated neck region of variable length, connecting the renal corpuscle with the tubule.
3. An initial proximal segment with prominent brush border and numerous prominent lysosomes.
4. A second proximal segment with numerous mitochondria but less well developed brush border.
5. A narrow ciliated intermediate segment which may be absent in some species.
6. A distal segment with relatively clear cells and elongated mitochondria.
7. A collecting duct system.

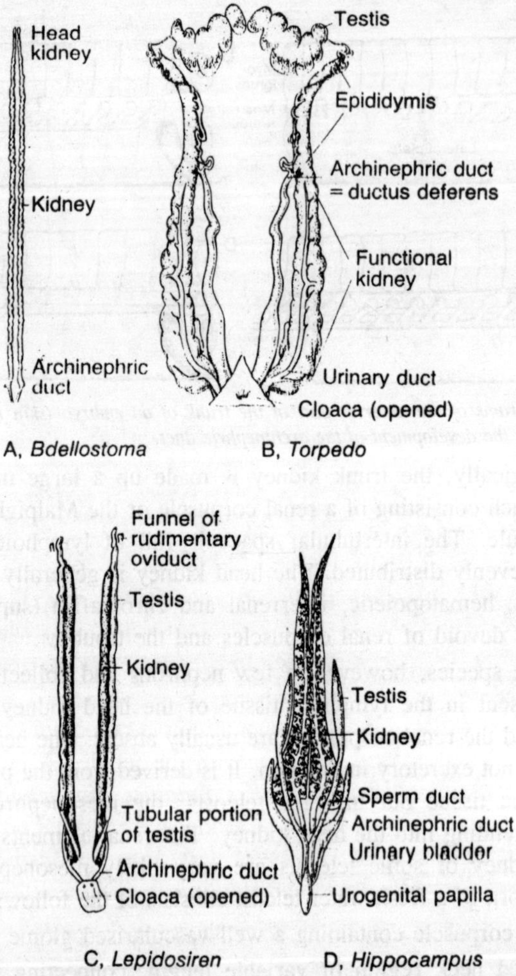

Fig. 7.5. Urogenital systems in ventral view of males of A, the slime hag Bdellostoma; B, the elasmobranch Torpedo; C, the lungfish Lepidosiren; D, a teleost, the seahorse Hippocampus.

The renal corpuscles are numerous, generally spherical in shape and each contains a highly vascular glomerulus in the fresh water fishes. In the marine fishes, however, there is a great variation in number, shape and size of the renal corpuscles. Large and well vascularised renal corpuscles are rarely found. The kidneys of many marine fishes generally have less well developed glomeruli than the fresh water fishes. They contain only a few small glomeruli which are

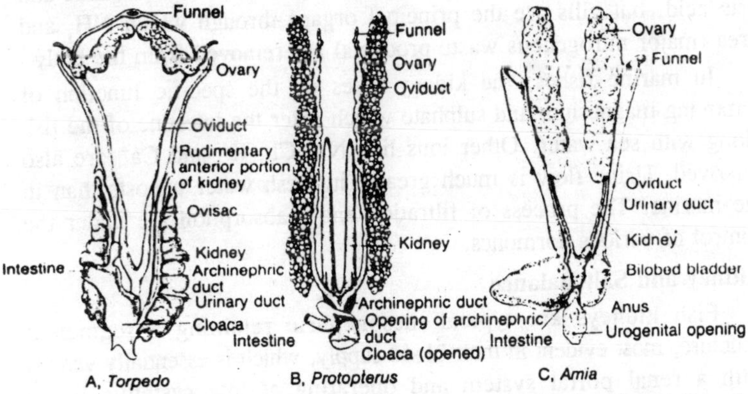

Fig. 7.6. Urogenital systems in ventral view of females of A, the elasmobranch Torpedo; B, the lungfish Protopterus; and C, the primitive actinopterygian Amia. In Torpedo, the shell gland is not developed.

poorly vascularised and are probably non-functional (*Therapon jarbua*, *Arius*, *Myoxocephalus*). Only a few medium sized glomeruli may be present along with the smaller ones. in some species, the glomerulus has a central avascular core surrounded by capillary loops which effectively reduces its functional area (e.g., *Chirocentrus*, *Tricanthus*, *Myoxocephalus*).

In the goose fish, *Lophius*, the glomerulus become non-functional in the adult, by losing its connection with the tubule. A number of marine fishes (*Opsanutau*, *Porichthys notatus*, *hippocampus* and *Syngnathus*) have completely aglomerular kidney. A typical nephorn of a marine glomerular teleost consists of:

(i) renal corpuscle containing the glomerulus,

(ii) a neck segment which may be very short,

(iii) two or three proximal segments,

(iv) a variably present intermediate segment,

(v) collecting tubule.

A distal segment is probably absent. The glomerulus and the Bowman's capsule together act as ultra filters where the blood is filtered under high pressure. The excretory fluid flows through the renal tubule and is driven by the net filtration pressure, and the movement of the long cilia of the neck segment and cilia of the cells of other segments. Na^+ and Cl^- are almost completely reabsorbed from the ultrafiltrate. Glucose is also largely removed from the filtrate. The kidney of fresh water fishes may also serve as an important excretory organ for minor nitrogenous compounds such as creatine and uric acid, but gills are the principal organs through which NH_3 and urea (major nitrogenous waste products) are removed from the body.

In marine fishes, the kidney takes up the specific function of removing magnesium and sulphate which enter the intestine of the fish along with sea water. Other ions like N^+, Cl^-, K^+ and Ca^+ are also removed. Urine flow is much greater in fresh water teleosts than in the marine. The process of filtration and reabsorption are under the control of various hormones.

Kidney and Salt Balance

Fish kidneys are typical mesonephric retaining a segmental structure, most evident in their blood supply, which is essentially venous with a renal portal system and operating at low systemic blood pressures not above 20 mm Hg. A particularly interesting feature of the fish kidney is that the nephrons in different fish are extraordinarily diverse, so that the function of the different nephorn segments can be inferred by comparing different fish. Perhaps the most striking modification of the nephorn occurs in some marine teleosts (23 species of 7 families at the present count) where the glomerulus is reduced or even completely lost. Study of such aglomerular kidneys has been fruitful for renal physiology. In this way tubular secretion was first demonstrated, and insulin was found to be the tracer of choice for the glomerular filtration rate (GFR). As we should expect, urine and indeed kidney function are different in freshwater and marine teleosts. Freshwater teleosts produce between 0.1 and 1.4 ml of urine hr^{-1} $100^{-1}g$ body wt., whilst in most marine species, urine production is around 1/10 of these values. The relative clearance of water via the kidney is up to 30% in fresh water, and practically nil in sea water, as the urine in fresh water is much more dilute than the plasma; in sea water it is more or less isosmotic with the plasma.

Wide simultaneous variations in GFR and urine production take place in normal fish without change in urine concentration, and it seems that recruitment of glomeruli under varying conditions plays an important role in the function of the kidney. Even small variations in systemic blood pressure (since it is always low), can 'shut down' glomeruli, and coupled with linked changes in tubular absorption greatly change urine production without change in urine concentration. Measurements of GFR in single nephrons of the trout, for example, have shown that in fresh water, 45% of the nephrons are filtering, whereas in sea water, only 5% filter. Obviously, such as flexible system has the great merit that it allows more or less automatic regulation of urine production as euryhaline fish move from the sea into fresh water of *vice-versa*, and the osmotic loads change.

Tubular Structure and Function

Although there is a wide spectrum of tubular structure in different fish, well described by *Hickman* and *Trump* (1969), three basic types are all we need recognize here. The proximal segment next to the glomerulus (where it occurs) is in two parts, one equivalent to the mammalian proximal segment, the other concerned with the secretion of divalent ions; the distal segment being concerned with water and monovalent ion absorption, and in elasmobranchs, with urea absorption. In fresh water, manovalent ions are conserved by tubular absorption, whilst in marine fish, divalent ions are secreted by the tubule. Although *Hickman* and *Trump* remark that it would scarcely do the marine teleost kidney justice to consider it purely a magnesium sulphate pump, this is certainly one of its main roles. $Mg2^+$ and SO_4^- enter the plasma via the sea water drunk (sea water contains around 50 mM Mg^{2+} and 25 mM SO_4^{2-}) and are excreted exclusively through the kidney. Na^+, Cl^- and other ions are also excreted, but extra-renal routes are more important for these ions.

As soon as marine euryhaline fish, such as the southern flounder *Platichthys flesus*, enters diluted sea water, Mg^{2+} secretion by the distal tubules drops dramatically, without notable change in plasma Mg^{2+} concentration and then, moves slowly, the tubules become less permeable to water, and Na^+ is more completely absorbed so that the urine osmolarity declines.

In elasmobranchs, through not in *Latimeria*, 95% of the urea in the glomerular filtrate is reabsorbed by the distal tubule perhaps by a counter-current system where the initial thinner part of the distal tubule is applied to the proximal tubule. After the distal region of the tubule, collection ducts of differing sizes in different species pass the urine to a urinary bladder, which in some species at least, plays an important role in modifying the ionic content of the urine. In *Lophius* and *Platichthys* for example, water is reabsorbed when the fishes are kept in sea water, and in trout fresh water ions are resorbed and conserved. These processes are under the control of the pituitary hormone *prolactin*.

Osmoregulation

Most fishes are confined to either salt or fresh water and have a limited salinity tolerance. They are known as *stenolaline species*. In the sea, the concentration of salt in the water is more than in the body fluids of the fish, so that water tends to diffuse out. Hence, marine fishes tend to drink water. Fresh water fishes, however, lose salts, and water enters into the body through semipermeable membranes of the buccal cavity and the gills. This is because of a higher concentration of salts within the body of fresh water fishes, than that of the surroundings. Kidneys and the gills play an important part in the excretion of nitrogenous wastes and in maintaining water-salt balance.

Fresh Water Fishes

Osmoregulation in fresh water fishes can be studied under the following sub-headings:

Nitrogenous excretion

The larva of lamprey (cyclostomes) has a functional pronephric kidney. Pronephric kidney develops in the anterior part of body and consists of 1-13 uriniferous tubules which open into pericardial cavity by nephrostome which has got one glomerulus. Coelomic fluid are swept into nephorstomes at the site of glomerulus (circulatory component of pronephric tubule). Here the substances to be conserved or reabsorbed into blood and the remaining watery filtrate passes by duct to the urinogenital sinus. In *Amia* (Holostiai) nephrostome opens in the anterior coelom. Here the excess of water is filtered from the blood through glomeruli and salts and sugar are reabsorbed into the blood stream through epithelium of kidney tubules and released into the surrounding capillaries.

Teleosteriai

A typical kidney of teleosts consists of a glomerular with open neck segment, proximal convoluted tubule which is developed into two cytologically differences:

A. Destal convoluted tubule

B. Initial collecting tubule.

Urine is filtered into the glomerulus and passes into the kidney from where ultimately to the exterior. According to the *"Smith*–1939" some nitrogenous wastes are excreted from the gill surface in the form of ammonia and urea. He found that the carp 6-10 times as much nitrogen excreted by gills as by the kidney. On the basis, he concluded that kidney excretes creating and uric acid and gill excretes ammonia and urea.

Lung Fishes (Dipnoi)

Structure of kidney tubule of Dipnoi is similar to the tubule of other fresh water fishes. Although second part of the proximal tubule

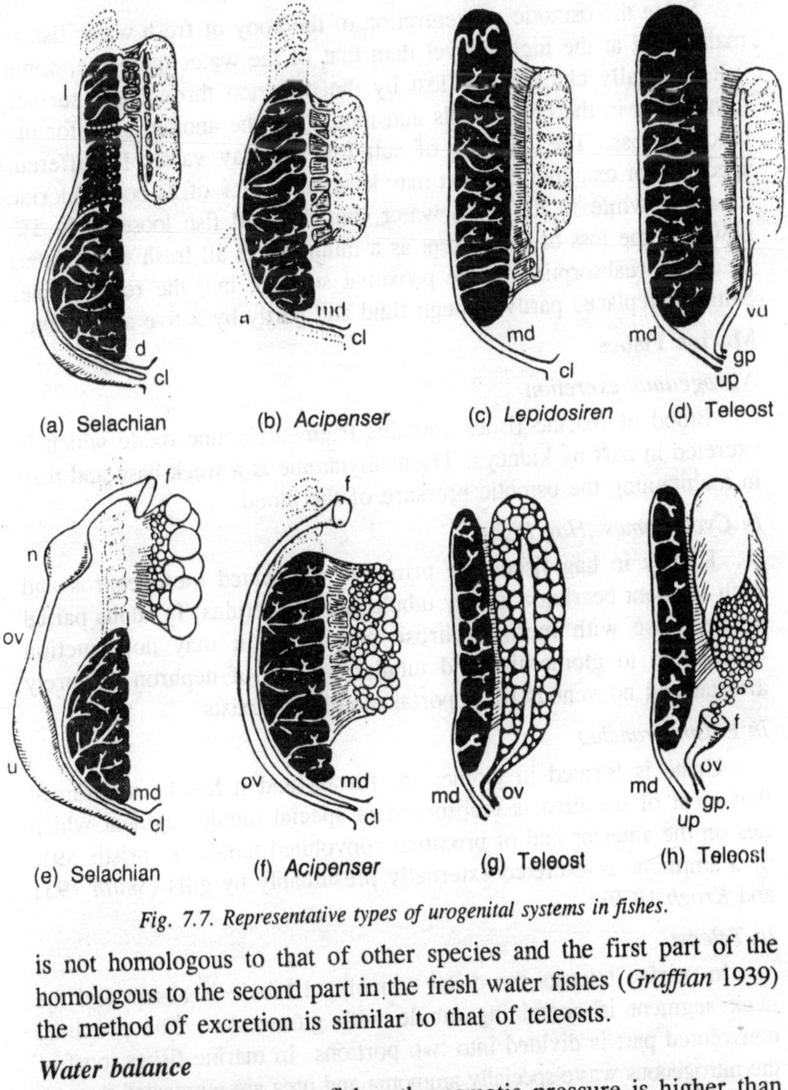

Fig. 7.7. Representative types of urogenital systems in fishes.

is not homologous to that of other species and the first part of the homologous to the second part in the fresh water fishes (*Graffian* 1939) the method of excretion is similar to that of teleosts.

Water balance

In all the fresh water fishes the osmotic pressure is higher than that of the surrounding water and this is a tendency for water to diffuse into the animal body where there is a water permeable membrane (gills, oral membrane and internal surfaces). The thick skin greatly reduced water permeability although small amount of water do enter a fish through the skin. All fresh water fishes have glomerular kidneys which help in filtering out excess of water.

Salt balance

Since the osmotic concentration of the body of fresh water fish is maintained at the higher level than that of the water therefore, some salts specially chloride are lost by the diffusion through the surface tissues and in the urine. Gills and mouth are the another sites for the chloride loss. The quantity of salt lost per day varies in different species. For example, Salmon may loose upto 17% of its body chloride in a day while in the fresh water, but the gold fish looses only 5% however, the loss of salt is kept as a minimum in all fresh water fishes by active reabsorption in the proximal segment into the renal tubule. Salts are replaced partly through fluid and partly by active absorption.

Marine Fishes

Nitrogenous excretion

Blood of marine fishes contains triemethylamine oxide which is excreted in part by kidneys. Triemethylamine is a week base and may in maintaining the osmotic pressure of the blood.

In Cyclostomes (Hag Fishes)

Kidney in hag fishes are primitive segmented mesonephros and each segment bearing a kidney tubule and glomerulus. In adults paired pronephrose with meso nephrose persist but it may not function Circulation to glomerular and tubular portion of nephron is purely arterial and no venous renal portal circulation exists.

In Elasmobranches

Urine is formed in glomerular filtrate and it has been assumed that most of the urea is reabsorbed in special tubule segment which lies on the anterior end of proximal convoluted tubule. In pristis 89% of a ammonia is excreted externally presumably by gills (*Smith* 1931 and *Krogh* 1939).

In Teleost

In marine teleosts the distal convoluted tubule is absent and the neck segment is containing tubule with glomerulus. The proximal convoluted part is divided into two portions. In marine fishes most of the nitrogenous waste specially ammonia and urea are eliminated through the gills. Urine eliminated through kidney consist of cretin, createnin, nitrogen compounds and trimethyle amine oxides.

Water balance

In contrast to fresh water environment marine fishes live in a medium which is hypertonic to body fluids and tissues and thus tend to loose water and gain salts through their osmotic membranes. To counteract the water losses, marine fishes bring sea water and thus increase the salt contents in the body fluid. Water balance in these fishes has been described by *"Smith* 1953" as follows: Water which is absorbed by gills dilutes the blood and the urine flow increases, consequently the excretion of urea increases and thereby reducing the blood area concentration which in turn reduces the urine flow and starts new cycle of urea retention. In this manner the fish is always supplied with enough free water to fulfil the urinary requirement.

Salt balance

Blood of cyclostomes has a ostomotic concentration which approaches that of the sea water though with the low chloride contents being isotonic to their surroundings, the hagfishes do not have to drink water and use whatever water they derive from their food for the formation of their own urine. Marine bony fishes eliminate their sodium salt which comes mainly from food and sea water through the gills and gut. Out of which only traces leave through the urine. Univalent ions specially the chloride passes from intestine to blood stream the leave through the gills whereas ions of magnesium, Ca, sulphate and other ions balance remain in the intestine and the alkaline nature of the fluid promotes their change to insoluble oxide or hydroxide which are passed out with the fishes. Chloride or salt cells in marine fishes eliminate excess of chlorine ion whereas, in fresh water fishes chlorine or salt absorb chlorine.

Osmoregulation in Anadromous Fish

Anadromous fishes are those which moves from salt water to fresh water far spawing purposes. The Salmon spends its adult life into the sea and migrates upto the rivers to spawn. The eggs develop to the fish like or larger fish and then they return to seuta water. *'Sawyer-*1955' worked on Peteromyzon, and showed that fish migrating from sea water to fresh water assumes very quickly the function of typical fresh water fishes and dies after bleeding. Marine lampreys can regulate the body fluid until the time of the anadromous migration and after migration to fresh water they can not resist even of the sea water.

Osmoregulation in Catadromous Fish

Catadromous fishes migrate from fresh water to salt water for the purpose of spawing romo. For ex-eels. Eel migrates as an adult conteanable fresh water to spoon in sarggasso in sea in the atlantic ocean. The larvae produced here drift about 1-3 years as stenohaline marine animals. The larvae than metamorphose into young eels which become euryhaline and eventually enter rivers and fresh water. These young ones display good osmotic regulation during migration from fresh water to sea water. The renal unit of the kidney of the eels contain glomerulus, neck segment proximal and distal convoluted tubules and initial collecting tubule, when transfer from fresh to salt water and the rivers eels show significant changes in weight for several hours, presumably because of the osmotic flow of water into (fresh water) or out of (sea water) the fish. The quantity of water passing through the surface of the eel in small under these conditions as most of the exchange is either by mouth or by urine excretion.

Osmoregulation in Eurihaline Fishes

The eurihaline fishes are able to tolerate a wide range of salinity in water although some such species do not appear to have any preference in their movements while other species undertake short spawing migration to fresh or breakish water. If the fish is transferred to more concentrated water there has been a temporary loss of weight (water) and increase in salt concentration but necessary physiological adjustment is met within a hour and weight and concentration return to normal. On the other hand fishes go from sea to fresh water they gain water and loose salt and must accumulate by increased urine flow and salt retention. Structures permitting adjustment to salinity.

Body surface

Fishes possess permeable body surfaces and help in making the adjustment to the new medium which is an adaptation.

Kidney

Kidney of fresh water fishes has many glomeruli (10,000) which are larger than those of the typical marine ray fin with glomerular kidney. These characteristics are related to salt contents of the environment and therefore to the volume of the urine excreted.

Gills

The gills are the sites of the active ions absorption which is necessary to replace the salt losses or to supplement the minerals of the food.

Endocrine Control of Osmoregulation

The amount of urine and salt balance is regulated in fishes by endocrine secretion. Hormone may influence the kidney either by an increase or decrease blood pressures which alters the filteration rate into the capsule of the renal corpuscles and thus affect amount of fluid excretion. Hormones may also affect renal excretion by specific action on tubulcs cells to change permeability and reabsorption rate by specific substances. In fishes where gills and kidney shared the osmoregulation processes, hormones also influence filteration or absorption process at the gills.

Adrenal cortical hormones which include sodium chloride play an important role in the regulation of gills and kidney function of fishes. Other endocine tissues such as thyroid, suprarenal body and hypothalamus in fish help in regulating the mineral balance. Thyroid injections of the gills may induce it to leave the water for lengthen periods. It appears that thyroid hormones support from influencing other changes. During the hormones metamorphosis affect the water and mineral balance of fishes not only to adapt to temporary life outside of water but also in changes from fresh water to salt water (migratory).

8

INTEGRATED SYSTEM

The nervous system of vertebrates evolved from the irritability nature of protoplasm. Its characteristic feature is the responsiveness. It includes brain, spinal cord, peripheral nerves and autonomic nerves innervating body organs. The rudiment of the feature brain and spinal cord first appears in the embryos of some Cyclostomes (e.g., *Bdellostoma*), of Elasmobranchs, and of Chondrostei, (e.g., *Acipenser*), and of *Neoceratodus* among the Dipnoi, in the form of a tubular medullary canal pinched off from the epiblast of the dorsal surface of the body. By a somewhat different method, but with the same final result, a medullary canal is formed in other Cyclostomes (e.g., *Petromyzon*), in the Holostei and Teleostei, and in *Lepidosiren*, from a solid ingrowing keel of epiblast which subsequently becomes tubular. Later, the medullary canal in the head enlarges, and becomes divided by two transverse constrictions into three vesicles, the primary *fore-*, *mid-*, and *hind-brain*, leaving the rest of the canal to form the spinal cord.

BRAIN ARCHITECTURE

Although the superficial aspects of brain anatomy are significant, an adequate understanding of the working of the brain can no more be gained from them than a knowledge of the working of a telephone system can be gained from an acquaintance with the external appearance and room plan of the telephone exchange. What is of importance in a telephone system is arrangement of switch-boards and writing; in a brain, the centres act as switchboards, in a sense, and tracts of fibres from the wiring between them. Possibly the brain wiring was primitive much like that of the spinal cord with fibres criss-crossing to interconnect all areas. Generally, however, there is a strong tendency

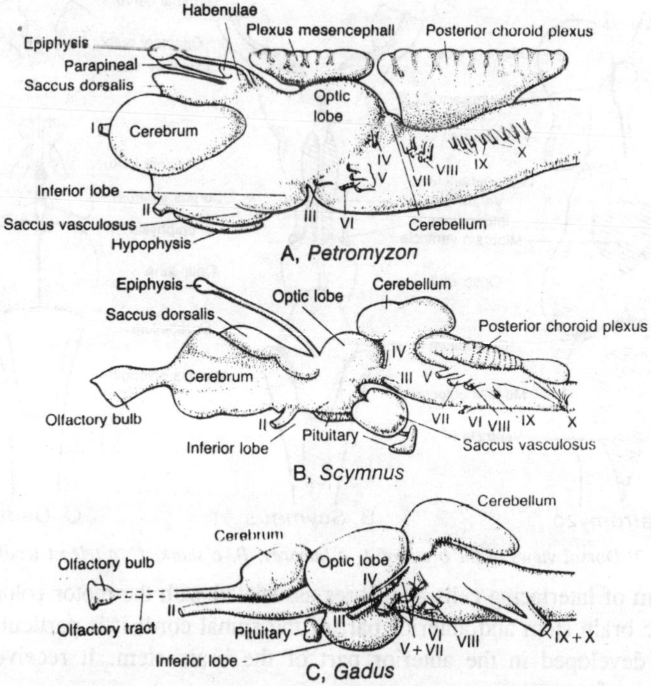

Fig. 8.1. Lateral views of brain of A, a lamprey; B, a shark; C, a codfish.

for the clastering together of nerve cells of specific functions in centres and the assembling of fibres with like connections into definite (through often very hard to see) bundles. Although certain special centres have special names, most are termed *ganglia* or *nuclei* (making an unfortunate, duplicate, biologic use of this latter word).

Nuclei or centres, in a broad sense range in size from tiny clusters of cells embedded in the gray matter of the brain stem and discernible only on microscopic study, to such massive structures as the cerebellum or cerebral cortex Bundles of fibres connecting nuclei with one another are in general termed *tracts*; the fibres constituting such a tract are, of course, axons of neurons whose cell bodies lie in the nucleus of origin. A tract is generally given a compound name, the two parts of which designate its origin and termination; thus the corticospinal tract carries impulses from the cortex of the cerebral hemispheres to or toward the motor cells of the spinal cord. Despite the general tendency of brain cells and fibres to organize into clean-up centres and tracts, a primitive condition persists in the *reticular formation*. This is a

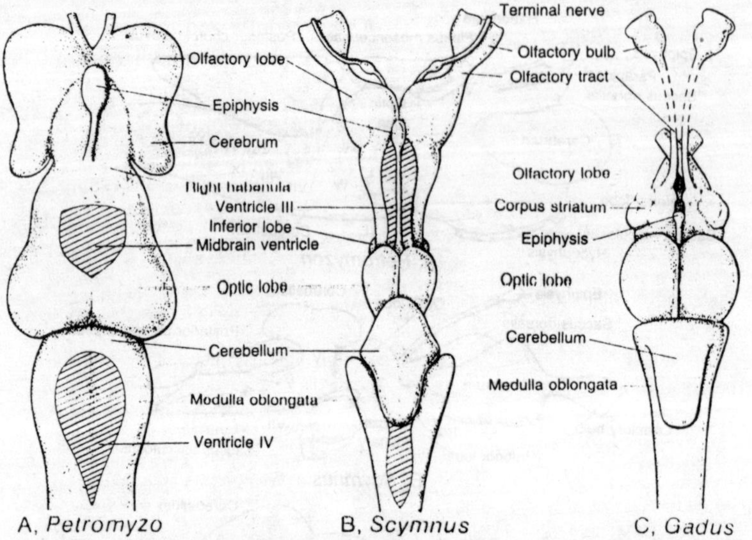

Fig. 8.2. Dorsal views of the brain of A, a lamprey; B, a shark; C, a teleost (codfish).

system of interlacing cells and fibres associated with the motor columns in the brain stem and anterior part of the spinal cord; it is particularly well developed in the anterior part of the brain stem. It receives a variety of sensory input.

This system is important in motor coordination. Still further, the reticular network appears to function in carrying stimuli downward from the anterior regions of the brain to the motor centres of the medulla and cord. The brain is constructed on a bilaterally symmetric pattern; in consequence, bryo, is simpler than its specialized outgrowths, the cerebral hemispheres and cerebellum. Even in the stem, however, the anterior part appears to have been the seat of complex nervous centres from the beginning of vertebrate history. It is in the hind-brain in the medulla oblongata that we find the closest structural approach to the cord.

Further, with the medulla and the adjacent parts of the midbrain are connected all the cranial nerves except the atypical ones from nose and eye and the terminalis. The medulla oblongata is, particularly in lower vertebrates, closely comparable to an anterior section of the spinal cord—enlarged, however, through the expansion of the central canal to form the fourth ventricle, and with the columns of gray matter of either side widely separated dorsally as a consequence. For most of its length, the roof of the medulla is thin, membranous, and infolded to form the *posterior choroid plexus*; anteriorly, the medulla is covered by the cerebellum. In the gray matter of the medulla and the posterior part of the midbrain, there is a series of columns of nuclei of gray matter basically similar to those present in the cord. In that region, we have noted the presence of a dorsal column, containing association centres for the reception and distribution of sensory impulses, and a ventral column containing motor neurons; we have further noted evidence that both columns may be subdivided into somatic and visceral components.

A similar situation is found in the medulla and the posterior part of the midbrain, although the widely expanded ventricle makes for a somewhat different appearance. A horizontal groove, the *sulcus limitans*, running along the inner surface of the brain stem on either side, separates a dorsal sensory region from a ventral motor region. Further, as in the cord, each of these two can be subdivided into somatic and visceral components.

Brain in Different Groups of Fishes

Cyclostomes

Lamprey. In the lamprey the olfactory nerves are paired and relatively short. They extend back from the walls of the midline nasal capsule to enter the medial anterior aspect of the olfactory bulbs. These bulbs are the larger part of the telencephalon. Behind each is a small cerebral lobe. The olfactory bulbs and cerebral lobes have small ventricles which connect with the medial third ventricle, a good part of which lies in the telencephalon. The diencephalon has a membranous roof which is evaginated outward as a large *saccus dorsalis*. Anteriorly to the saccus is a small paraphysis; on top of the saccus rests a *parietal organ* and above this a pineal organ. Both of these organs are developed as photoreceptors.

The parietal organ has a nerve extending down to the left *habenula*; the nerve of the pineal organ extends down and back on the right side to the posterior commissure. The anterior end of the diencephalon is not marked by a velum transversum. Ventrally the optic nerves cross without exchange or interlacing of fibres. The hypothalamus has small inferior lobes and an indistinctly differentiated "*saccus vasculosus*." The floor of the infundibulum the pars nervosa, is attached to the underlying anterior lobe of the pituitary. The mesencephalon has bilateral optic lobes between which is a membranous, evaginated chorioid plexus. This plexus is fused to the plexus overlying the fourth ventricle in the myelencephalon. A small transverse cerebellum and a tapered myelencephalon, much like those of the amphibian, lie behind the optic lobes.

Myxinid. The brain of the myxinid is quite unlike that of the lamprey. It lacks any chorioid plexus of extensive membranous roofs. This lack is apparently due to late changes in development and is not primitive. The posterior part of the brain (myelencephalon) is large, while the anterior three segments are relatively reduced. From front to back there are paired olfactory bulbs, cephalic enlargements, diencephalic and mesencephalic lobes. The metencephalon cannot be distinguished or in lacking. There is a swelling above the midline groove of the diencephalon formed by the fused anterior and habenular commissures. Behind these is a small epiphyseal evagination. Ventrally there is a small infundibulum ending in an undifferentiated *saccus rasculosus*. There are small bilateral *lobi inferiores*. The anterior lobe of the pituitary is small and inconspicuous. The pars nervosa of the infundibulum is undifferentiated.

Chondrichthyes

Shark. The nervous system of the shark, *Squalus*, is typical of this group. The olfactory bulbs lie close to the olfactory capsules; the olfactory nerve is made up of many small bundles of fibres entering the bulb directly. The bulb is connected with the cerebral hemispheres by a long olfactory tract. The length of the tract is determined by elongation of the head. The cerebral hemispheres are relatively large than those of teleosts and have thick walls. The third ventricle extends forward for half the length of the telencephalon. Anteriorly the cerebral lobes and joined across the midline, and the lateral ventricles extend anterolaterally into the olfactory bulbs.

Within each hemisphere the lateral ventricle expands as a distinct chamber. Posterior the root between the cerebral lobes is membranous and is extended outward as a distinct paraphysis. The posterior end of the telencephalon is marked by a transverse velum above and by the preoptic recess below. The diencephalon has a membranous roof which is developed as a dorsal sac; the velum transversum gives rise to a chorioid plexus. There is a long, thin epiphysis extending upward from the posterior end of this roof. The epiphysis is attached in the region of the habenulae, which lie just anterior to the posterior commissure.

Ventrally there is an optic chiasma and the hypothalamus is drawn out and back as a distinct infundibulum ending in a saccus vasculosus. The infundibulum is attached below to the pituitary. In the region just above the saccus vasculosus there are distinct, bilateral inferior lobes.

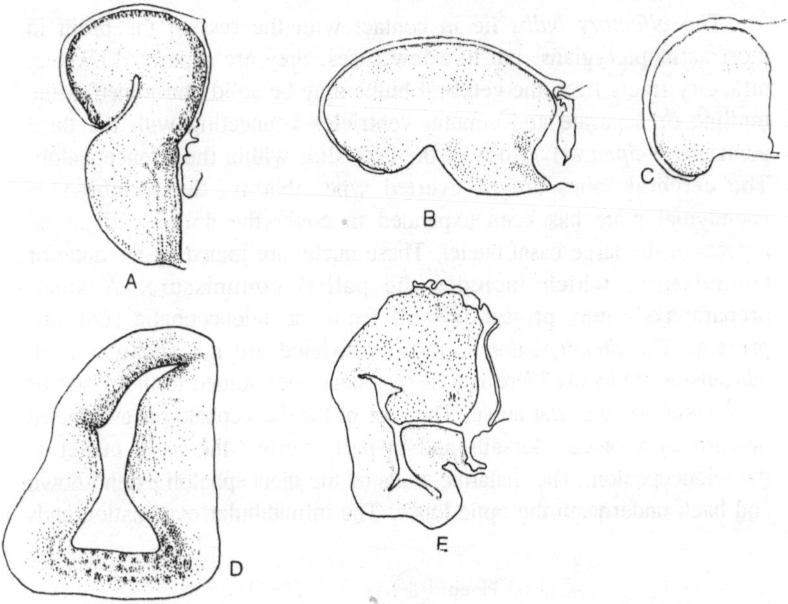

Fig. 8.3. Transverse sections through one cerebral hemisphere of various Osteichthyes; the midline is on the right of each drawing. A, Polypterus, a chondrostean, B, Amia, a holostean; C, Carassius, a teleost; D, Protopterus, a lungfish; E, Latimeria, a crossopterygian.

Behind the diencephalon the optic lobes are large bilateral. The floor and the sidewalls of the mesencephalon are thick as in the other groups. The cerebellum is large and has a small cavity (metencoel) extending upward into it and into its various lobes.

Posterolaterally there are folded restiform bodies which project forward. There is no *valvula cerebella* in this group. The *myelencephalon* has a large chorioid plexus in its roof with digitations extending down into the fourth ventricles; the lateral walls and floor are not as well developed as in the teleosts.

Holocephalan. The brain of *Hydrolagus* agrees in detail with that of *Squalus* differing in that the diencephalon (and part of the telencephalon) is much elongated. This modification of the brain is related to the large eyes, whose sockets are separated only by an interorbital septum. The optic lobes are small and the cerebellum is not distinctly divided into bilateral halves. The lateral-line lobes are larger and extend further back. The saccus vasculosus is comparable.

Actinopterygians

The *olfactory bulbs* lie in contact with the rest of the brain in most actinopterygians, but in a few cases, they are separated by long olfactory tracts from the cerebral bulbs may be solid and joined at the midline or separate and contain ventricles connecting with the third ventricle (*Acipenser*). Much of the latter lies within the telencephalon. The cerebral lobes are of everted type, that is, the membranous ependymal plate has been expanded to cover the dorsal and lateral aspects of the large basal nuclei. These nuclei are joined by the anterior commissure, which includes the pallial commissure. A small preparaphysis may project upward from the telencephalic roof just present. The *diencephalon* is much restricted are not apparent from the outside. Dorsally there is a large pineal body joined to the posterior commissure by a pineal nerve. The roof of the diencephalon is evaginated upward as a saccus dorsails and in part overlies the basal nuclei of the telencephalon. The thalamic walls of the diencephalon extend down and back underneath the optic lobes. The infundibular evagination ends

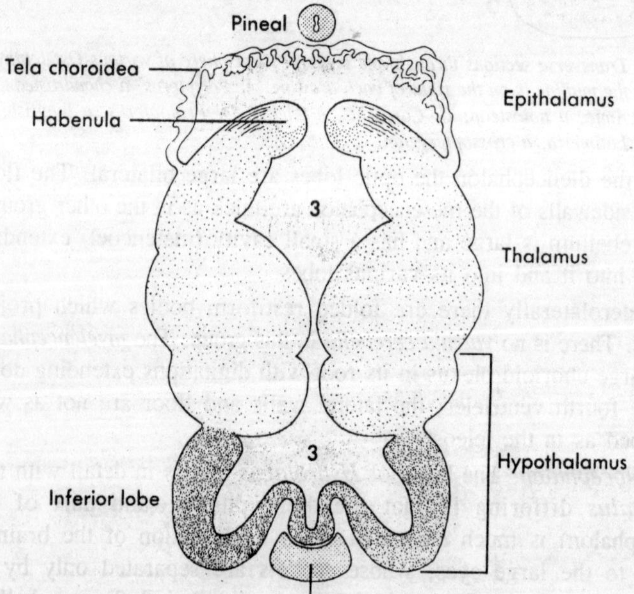

Fig. 8.4. Diencephalon of a shark in cross section, generalized.

in a large *saccus vasculosus*. This is usually thought to be sensory in function. It is now assumed that they are endocrine.

The pars nervosa and the anterior lobe of the pituitary form a large mass lying somewhat anterior to the saccus vasculosus. Dorsolateral to the succus are the bilateral inferior lobes of the hypothalamus. The optic nerves form a chiasma anterior to the inferior commissure of the thalamus; in this decussation the two nerves cross one above the other; one nerve passing through the other or both nerves interlacing. The *mesencephalon* is represented by large optic lobes, which are essentially hollow; anteriorly, they touch the telencephalon and posteriorly the cerebellum. The mesencephalic brain stem is not distinguished by any particular structures from the rest of the lateral wall and floor. The occulomotor nerve emerges from this part of the stem. The *metencephalon* is a large cerebellor mass dorsally. This mass is extended forward as a tongue between the optic lobes; this extension is the *valvula cerebelli*. The valvula is folded upon itself in the salmon and projected into the cavity of the mesencephalon. The decussation of the trochlear nerve passes below the posterior end of the valvula.

In the mormyrid fishes the valvula is extended outward above the optic lobes as folded bilateral lobes. These cover much of the brain. The stem portion of the metencephalon is not marked from the rest of the brain stem, but from it arises the fourth and perhaps a part of the fifth nerve. The membranous roof of the myelencephalon is modified into a chorioid plexus. The vagal lobes are highly developed in some teleosts and may meet above the fourth ventricle forming a tuberculum impar (*Lampanyctus*). From the sides and ventral wall of the medulla arise nerves V to X. The seventh and eighth nerves are represented by a common group of roots. The tenth has two main roots: one lateral-line sensory root the other a sensory and motor root serving the branchial arches and the viscera. The sixth nerve arises ventrally from the motor part of the myelencephalon. Posterior to these, the spino-occipital nerves and finally the spinal nerves arise.

The spino-occipital nerves exit through the cranial wall. The first spinal lies between the skull and the first vertebra. In the medulla there is a pair of giant neurons called *Mauthner's cells*. The cell bodies lie at the vertical of the roots of the seventh and eighth nerves near the midline, the axons cross and descend through the spinal cord just below the central canal. These giant fibres are observed in the larvae of salamanders as well as actinopterygians. They function in the co-ordination of swimming. In the adult of some teleosts these cells are lacking. It would be difficult to describe all of the many variations of brain pattern observed among the actinopterygians. The most diagnostic feature appears to be everted telencephalic lobes. Other features are the usual lack of evaginated olfactory bulbs and the occurrence of a valvula projected into the roof or cavity of the mesencephalon. The development of the lateral line system in this group.

Choanate Fishes

Dipnoan. The brain of the dipnoan *Protopterus* is not unlike the amphibian. The telencephalon of this group is peculiar in that the olfactory bulbs like dorsal to the anterior ends of the elongate cerebral lobes. A "preparaphysis" arises from the roof anterior to the velum transversum. The *diencephalon* is relatively small with a large saccus dorsalis of chorioid tissue forming its roof. On this is a *pineal body*, whose stem extends down and back toward the posterior commissure. The habenular nuclie are prominent and lie to either side of the pineal stem. The habenulae are joined by a (superior) commissure. The hypothalamus has slight inferior lobes. The optic lobes are fused to form a single midline mass. The cerebellum is a narrow transverse ridge as in the amphibian.

Neoceratodus is somewhat different in that the cerebral hemispheres are smaller and with membranous dorsal and medial walls. This membranous part forms a thick, folded, glandular choroid plexus which continues as the roof of the diencephalon. In the young, the dorsal and medial walls of the lobes are nervous, the adult condition is achieved by an eversion suggesting that of the actinopterygian. The optic lobes are slightly separated.

Actinistian. The brain *Latimeria* is distinct from that of the dipnoans. It is extremely small as compared with the cranial cavity (a parallelism to the situation in *Acipenser*); except for the olfactory bulbs it lies entirely behind the dorsum sella, in the occipital part of the skull. The olfactory bulbs lie in contact with the olfactory capsules and are connected by very long stalks with the rest of the telencephalon. The cerebral lobes have a restricted ependymal septum as in the young of *Neoceratodus*. The diencephalon has a pineal organ above, and an infundibulum with bilateral inferior lobes and a saccus vasculosus ventrally.

The *saccus vasculosus* is an irregular evagination of the rear wall of the infundibulum above the level of the pituitary and is only slightly developed in this fish. It is perhaps represented by the posterior recess of higher forms. Such a saccus is well developed in actinopterygians and sharks. The pituitary of *Latimeria* is drawn out anteriorly in the direction of the sella turcica. The optic lobes form a single medial mass behind which is a large cerebellum resembling that of the shark or actinopterygian (*i.e.,* with large facial lobes). The cerebellum lacks a *valvula* but has distinct auricles (*restiform bodies*).

Spinal Cord

The central nervous system of fishes loses much of its structural complexity in the transition from brain to spinal cord. Already in lampreys (Petromyzontiformes), but more pronouncedly in sharks and rays (Elasmobranchii), a cross-section of the cord shows a central matter of the cord of bony fishes (Osteichthyes) roughly resembles a letter X with paired dorsal and ventral horns. The dorsal horns receive somatic and visceral sensory fibres and the ventral horns contain motor nuclei, with centres for dorsal or ventral musculature lying in median and lateral positions respectively.

Peripheral Nervous System

It includes cranial and spinal nerves as given under the following heads:

Cranial Nerves

It is usual to describe the cranial nerves of Cyclostomes and Fishes as consisting of ten serially disposed pairs, *viz.* the *olfactory* (I), *optic* (II), *oculomotor* (III), *trochlear* (IV), *trigeminal* (V), *abducens* (VI), *facial* (VII), *auditory* (VIII), *glossopharyngeal* (IX); and the *vagus* (X). Like the spinal nerves, the cranial nerves collectively include somatic sensory (general cutaneous) and motor fibres, and also visceral sensory and motor fibres, all of which have their own special centres in the brain, but the proportions of these nerve components differ greatly in different nerves. Certain preoral nerves (III, IV, and VI) are exclusively *somatic motor*; others (I and II) are special *sensory nerves* for the olfactory and visual organs; but most of the other cranial nerves include several components, and are therefore *mixed nerves*. Besides these components some cranial nerves include also a quasi-independent system of nerve-fibres, which converge from certain cutaneous sense-organs to an independent centre in the medulla oblongata, the *tuber acusticum*, and is probably derived from the general cutaneous system of nerve components.

Such nerve fibres, including also the auditory nerve, which has its origin from the same centre, constitute the *lateralis system*. Perhaps the most striking feature in the postoral cranial nerves is the predominance of the visceralis or sympathetic system over the somatic. Omitting the lateralis fibres and a relatively few somatic sensory fibre,

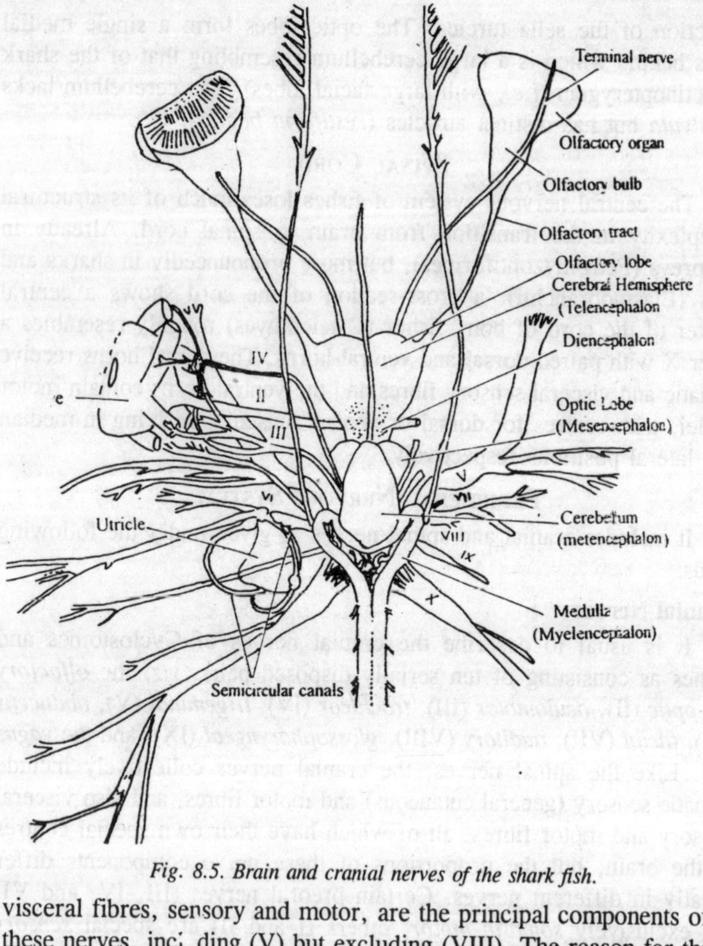

Fig. 8.5. Brain and cranial nerves of the shark fish.

visceral fibres, sensory and motor, are the principal components of all these nerves, inci. ding (V) but excluding (VIII). The reason for this is to be found in the fact that splanchnic or visceral muscles in relation with the jaws and branchial arches have usurped the place of somatic muscles in the muscular system of the head. For developmental and other reasons the olfactory and optic nerves stand in a category of their own, and the same may be said of the third, fourth and sixth nerves, which innervate the muscles of the eyeball.

The remaining nerves, all of which have their origin in the medulla oblongata, possess certain features in common, and as they are related to the gill-clefts in such a way that each forks over a cleft, they may be conveniently distinguished as "*branchial*" or "*branahiomeric nerves*". A typical branchial nerve consists of (1) a *principal ganglion* near the origin of the nerve from the brain; (2) a *main trunk* which gives off; (3) a somatic sensory branch or *dorsal nerve* to the skin; (4) a *palatine nerve* (visceral sensory) to the oral or pharyngeal mucous membrane; (5) an *epibranchial ganglion* which is associated with a transitory embryonic *epibranchial sensory organ* at the dorsal border of a branchial cleft; (6) a *pre-branchial* nerve (visceral sensory), skirting the anterior margin of a cleft in its ventral course; and (7) a *post-branchial* branch (visceral motor) similarly related to the hinder margin.

The first six cranial nerves resemble those of the higher Craniates in their mode of origin from the brain, in the physiological nature of their component fibres, and in their peripheral distribution, and therefore, they need not be specially referred to here. Comparing this nerve with a typical branchial nerve it would seem that the *profundus* and *superficialis ophthalmic* nerves and dorsal nerves; the *maxillaris* and *mandibularis*, pre- and post-branchial branches, respectively, relation with the modified gill-cleft which forms the mouth, while the branch to the oral surface represents a *palatine* nerve. The most important of the distinctive features in the cranial nerves of Fishes are to be found in the relations of nerves (VII), (IX), and (X) to branchial clefts, and in the lateralis system of nerve components and its association with the lateral line sensory organs. The seventh of *facial* nerve is an exceptionally interesting nerve. Besides the usual components of a typical branchial nerve certain of its so-called branches are wholly or largely derived from the lateralis system. For this reason the nerve may be said to consist of two portions, the *facial proper*, or those fibres which constitute the facial nerve in air-breathing Craniates, and the *lateralis branches* which solely innervate lateral line sense-organs, and are therefore peculiar to aquatic forms.

The facial proper has a ganglion (the *facial* or *geniculate ganglion*) on its root, and on entering the orbit after traversing the cranial wall it gives off a *palatine* nerve. Just over the spiracle a *pre-branchial* nerve the representative of the *chorda tympani* of Mammals, leaves the main trunk, and passes ventrally in relation with the anterior wall of the spiracle to its ultimate distribution in the walls of the mouth cavity. The main trunk, now called the *ramus hyomandibularis*, then pursues a ventral course behind the spiracle as a post-branchial nerve, and certain of its mainly motor branches which pass downwards in connection with the hyoid arch supply the muscles of that arch, and, if an operculum is present, the opercular muscles as well.

The lateralis portion of the facial includes the following principal branches each of which may have a ganglion on its root; (1) an *ophthalmious superficialis*; (2) a *buccolis nerves* with its *ramus oticus*; and (3) *external mandibular* nerves which course in the ramus hyomandibularis. The addition of the great *lateralis* nerve, which is usually described as the later branch of the tenth nerve and of the eighth or *auditory* nerve which supplies the auditory organ, completes the enumeration of the main factors of the lateralis system. The ninth or *glosso-pharyngeal* nerve, perhaps the most typical of all the branchial nerves, has per- and post-branchial branches which enclose the hyobranchial cleft. Its palatine nerve usually extends forwards and anastomoses with the corresponding branch of the seventh, thus forming a connection (*Jacobson's* region of *gray substance* consisting primarily of nerve cells and a surrounding area of *white matter*, the nerve fibres. These fibres are sheathed, or myelinated, and are collected in bundles according to their function and connections. Centred in the gray matter of the cord is the central canal.

As in other vertebrates, the gray (*anastomosis*) between the two cranial nerves. In some Elasmobranchs and Teleosts fibres derived from the dorsal branch of the ninth nerve innervate a few sense-organs of the lateral sensory canal of thc head, and hence that nerve sometimes contains lateralis fibres. The tenth or *vagus* is a compound or mixed nerve. Besides the great lateralis nerve generally associated with it, the *vagus* includes as many typical branchial nerves as there are branchial clefts behind the hyo-branchial cleft, and in Elasmobranchs and in *Chimaera* these nerves have independent origins from the medulla oblongata. Each nerve has the typical structure, a ganglionated trunk which forks over a gill-cleft into the usual pre- and post-branchial branches and palatine branches to the pharyngeal walls.

In the Dipnoi the lateralis nerve is connected with the superficial ophthalmic branch of the seventh nerve by a commisural nerve which curves across the outer face of the auditory capsule. A somewhat similar anastomosis is also present in *Petromyzon*. The *vagus* also includes a large *ramus intestinalis*, which in Elasmobranchs, at all events, has a distinct ganglionated root. The nerve forms characteristic plexuses on the oesophagus and stomach, and in Cyclostomes its branches may extend nearly the whole length of the intestine.

In Ganoids and Teleosts there is an interesting nerve known as the "*lateralis accessorius*". It is a compound nerve, and owes its formation to the union of somatic sensory fibres derived in succession from dorsal branches of the (V), (VII), (IX), and (X) nerves and also from the corresponding branches of a variables number of spinal nerves. The finer branches of the nerve are distributed to the skin of one or more of the fins, or even, as in *Gadus*, to all the fins, especially to the numerous "end buds" which are present on those organs. In many Fishes a variable number of the anterior spinal nerves (*spinoccipital*) perforate the occipital region of the skull. They probably represent the ventral roots of the ordinary spinal nerves of the region.

Table 8.1. Cranial Nerves of Fishes and Their Functions

Nerve Number	*Name*	*Peripheral Connection*
0	*Terminal*	Olfactory bulb
I	*Olfactory*	Mitral cells; olfactory bulb; nucleus olfactorious anterior
II	*Optic*	Retina of eye
III	*Oculomotor*	Superior, inferior, anterior rectus and inferior oblique muscles of eyeball
IV	*Trochlear*	Superior oblique muscle of eyeball
V	*Trigeminal* divided into deep *ophthalmic*, *maxillary*, and *mandibular* branches	Anterior portion of head, upper and lower jaw regions
VI	*Abducens*	Posterior rectus muscle
VII	*Fascial*	1. Dorsal group: postorbital lateral line; taste buds on body, where present
	and	
VIII	*Acoustic*	2. Orbital group: supra and infra-orbital lateral line canals; taste buds on snout
		3. Acoustic group: inner ear; temporal lateral line canals.
		4. Branchial group: operculo-mandibular oral and jugal line canals; taste inside mouth; taste and touch from opercular, hyoid and mandibular regions; head muscles antagonistic to those under control of mandibular branch of V.
IX	*Glossopharyngeal*	1. Dorsal group: part of temporal lateral line canal; dorsolateral skin sensibility and proprioception, first gill slit region
	often fused with	2. Branchial group: taste organs in dorsal and ventral pharyngeal mucosa, muscles of first gill slit (third gill arch)
X	*Vagus*	1. Supratemporal branch: skin and lateral line organs of supratemporal region
		2. Dorsal recurrent branch: joins with body taste but innervation of VII

3. Body lateral line branch: from body lateral line organs
4. Viceral branch: nerves to and from internal organs
5. Branchial branches: to and from four posterior gill slits or equivalent region

Central Connection	Function
Forebrain; lamina terminalis	Somatic sensory, tip of snout? vasomotor?
Small centres of forebrain	Special sensory, carries smell impulses
Optic tectum of midbrain	Special sensory, carries visual impulses
Anterior brain under optic tectum (Mesencephalon)	Somatic motor, innervates four of the six striated eye muscles and muscles inside eye
Anterior brainstem under cerebellum (Mesencephalon)	Innervates one of the six striated muscles of the eye
Lateral column of medulla oblongata region of brainstem frequently overlaps with centres of VII (Myelencephalon)	Mixed somatic sensory and motor functions. Thermal and tactile sensibility of skin of anterior portion of head; proprioreceptors of head muscuture.
Ventral brainstem behind origin of V	Innervates sixth striated muscle which moves the eyeball
Elongate series of connected nuclei in middle brainstem (Medulla oblongata; Myelencephalon)	A. truly mixed nerve; special and somatic sensory visceral and motor functions. Brings lateral line acoustic and gravity impulses o brain centres, as well as taste and touch impulses from head and mouth; proprioception and innervation of some head muscles
Series of ganglia, or roots, mainly in dorsal and median end region of brainstem (Medulla oblongata; Myelencephalon)	A. mixed nerve; taste and lateral line sense, in afferent components; innervates anterior gill muscles (first gill slit)
Series of ganglia posterior to but fused with roots of nerve IX; often overlapping with roots of first occipito-spinal nerves at posterior end of brainstem (Medulla oblongata, Myelencephalon).	A. mixed nerve; motor control of bulk of gill muscles, special fibres from taste and lateral line organs, visceral motor and sensory to and from internal organs, parasympathetic autonomic nervous system.

Spinal Nerves

The spinal nerves are segmentally arranged. In the lampreys and hogfishes (Cyclostomata), the exits from the cord of dorsal and ventral roots of the spinal nerves do not correspond to one another; only among the hogfishes (Myxinidae) and there true mixed spinal nerves like those which occur, with few exceptions, throughout the jawed fishes—the sharks and rays (Chondrichthyes) and bony fishes (Osteichthyes). The jawed fishes in general, have spinal ganglia for the neurons of the sensory or dorsal root nerves although in many families such as the minnows (Cyprinidae), cods (Gadidae), perches (Percidae), and drums (Sciaenidae), some afferent fibres come to the cord also from supramedullary and inframedullary ganglia. From sharks (Squaliformes) to bony fish (Osteichthyes), there is an increasingly fibre differentiation and subdivision of ascending and descending fibre tracts between the brain and the spinal cord.

In the cearobins (Triglidae), the long and separate anterior rays of the pectoral fins carry special receptors, probably of touch; sensory nerves from these rays are marked on the spinal cord by separate swelling corresponding to their dorsal root nerves, and their fibre projections also reach the medulla and the midbrain. *Ipsilateral* and *contralateral reflexes* of the spinal nerves in one segment of the cord can occur in elasmobranchs; *plurisegmental reflexes* and common in sharks and teleosts. In view of the important role of the trunk in fish locomotion the question of the nervous control of swimming movements has received considerable attention.

There is controversy as to whether a fish requires outside stimulation in order to make coordinated swimming movements or whether swimming is due to the automatic activity of cord and medullar centres, only to be coordinated by peripheral stimuli. *Von Holst*, after performing many experiments with dogfish sharks (Squalidae), eels (*Anguilla*), loaches (Cobitidae), and minnows (Cyprinidae), suggested that swimming is due to endogenous rhythm from the posterior medulla, in turn passed on to the cord but influenced at all levels by outside "setting" stimuli. *Lissmann* and *Gray*, mainly on the basis of experiments with certain cat sharks (Scyliorhinidae), maintain that there is no automatic swimming centre but that swimming movements are of reflex nature and depend on the coordination of afferent impulses in spinal cord and medulla.

Autonomic Nervous System

In the autonomic nervous system of lampreys and hogfishes (Cyclostomata), there is no ganglionated sympathetic chain.

Nevertheless in these fishes a diffuse system of nerves and some plexuses of sympathetic function have been described. In the lampreys and hogfishes, both the heart and the organs of digestion are innervated by the vagus. In the sharks and rays (Elasmobrachii), sympathetic ganglia, are found along the sides of the vertebral column, they are irregular and do not extend into the head region. The smooth muscles of the gut and arterial walls are innervated by them. However, there is no connection between them and the nerves which supply the skin such as in mixed communicating rami of bony fishes (Osteichthyes) and higher vertebrates.

In all shark-like fishes the hearts and the internal organs of digestion are abundantly supplied by vagal fibres. In the bony fishes, segmentally arranged ganglia form a sympathetic chain as far forward as the trigeminal nerve (cranial nerve V); most preganglionic fibres for sympathetic ganglia of the head arise from the spinal cord of the trunk region and pass forward. Sympathetic connection to the skin nerves are found in the highest bony fishes (Teleostei). The *parasympathetic system* is almost exclusively represented in portions of the widely ramifying branches of the vagus (cranial nerve X).

A parasympathetic innervation of eye parts (in cranial nerve III) has already been mentioned. The sympathetic cardio-accelerating fibres of higher vertebrates are absent in all fishes and consequently vagal stimulation has a powerful inhibitory action on the sinus venosus and the auricle but no effect on the ventricle.

The stimulation of branches of the vagus leading to other internal organs as well as the administration of *cholinergic* (parasympatomimetic) and *adrenergic* (sympatomimetic) drugs has produced conflicting results, especially in action on the gas bladder and certain portions of the alimentary canal. The nerves from the paling centre run in the spinal cord and leave it at various in different fishes (in the minnow *Phoxinus*, segments 12 to 18; in the salmon-trout genus *Salmo*; segment 26; and in the searobins, *Trigla* segment 3). The melanophore nerves then go to the subvertebral sympathetic chain where they divide into an anterior bundle leading to the colour cells of head and nape, and a posterior bundle, to the trunk and the tail. Little is known about melanophore control in cyclostomes; however, certain elasmobranchs have at least some sympathetically innervated melanophores. The presence of a darkening centre in the midbrain and of parasympathetic melanin disperser nerves has been postulated.

9

SENSE ORGANS

The receptor are basically meant to receive both physical or chemical stimuli from their external as well as internal environments. *Extero receptors* which are greatly developed, receive impulses related to the external environment similarly entero receptors which are poorly developed, receive impulses related to the internal environment. *Algesi receptors* for pain belong to this category which is probably not experienced as a sensation by fishes. In fishes various types of receptors or sense organs have been studied. These are:

CUTANEOUS SENSE ORGANS

These organs, the most remarkable and certainly the most characteristic of the sense-organs of Cyclostomes and Fishes, are bud-like groups of epidermic cells in relation with the ends of sensory nerve fibres. Each consists of a central core of sensory cells provided with terminal cuticular sensory hairs, and surrounded by a zone of supporting and mucous-secreting cells which leave the hairs exposed at the apex of the bud. Two kinds of these organs can be distinguished, which differ in their innervation and in their position in the skin. Of the two, the so-called *end-buds* are the more primitive. They occupy a superficial position in the epidermis, and their sense-cells are as long as the supporting cells. They are present in Cyclostomes and Elasmobranchs, and especially in Teleosts, where they are irregularly distributed over the surface of the body, on the fins, lips, and barbels, and also in the epithelium of the mouth and pharynx.

In the Dipnoi they are limited to the oral cavity, and in the higher Craniates they becomes taste buds. Their somatic sensory nerves are derived from the (VII), (IX), and (X) cranial nerves and the lateralis

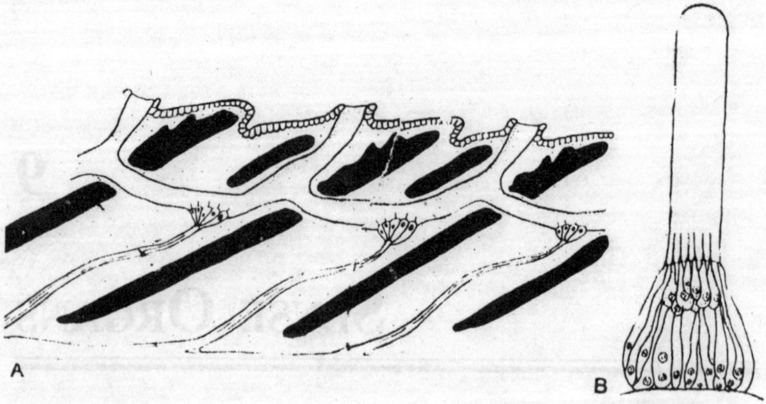

Fig. 9.1. Lateral lines of fish. A, part of the main lateral line canal of the perch (Perca) showing three pores to the surface and three neuromast organs. B, Diagram of a single neuromast organ showing the sensory processes and the gelatinous cupula.

accessorius. In the second type, usually called "*nerve-eminences*," the sensory cells are shorter than the supporting cells, and they are always innervated by the lateralis system. When first developed in the embryo they are quite superficial, like end-buds, but later the epidermis in which they lie sinks inwards so as to line a series of pits, closed sacs, tubes, open grooves, or closed canals. *Pit-organs*, so abundant on the head and trunk of Teleosts are simple epidermic pits with insunken nerve eminences, disposed in groups or in lines (accessory lateral lines) or irregularly distributed. The "*Spalt-papillen*" of Elasmobranchs are pit-organs in which the orifice of the pit is reduced to a slit. The

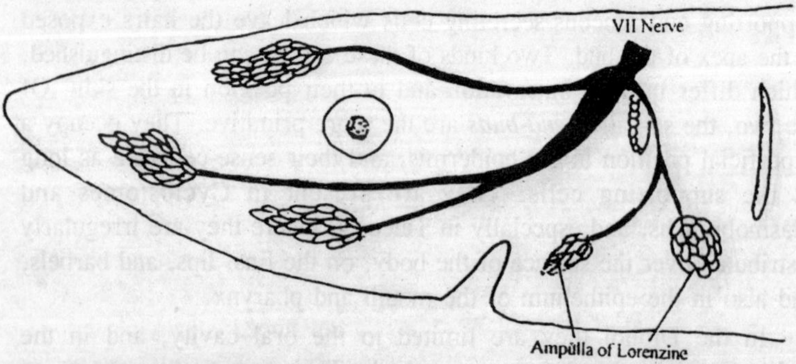

Fig. 9.2. Location and innervation of the ampullae of Lorenzini in the dogfish shark. The ampullae are innervated by fibres from the eighth cranial nerve distributed via the seventh cranial nerve.

more deeply seated *Savi's vesicles* on the ventral surface of the Torpedo, and the *nerve sacs* of Ganoids, and similar organs converted into closed sacs and pinched off from the rest of the epidermis. *Lorenzini's ampullae* or mucous canals, which are found in definitely located groups on the lateral and upper surfaces of the head in Elasmobranchs, may perhaps be compared to pit-organs prolonged inwards to form subcutaneous tubes each of which terminates in a radially-separate, chambered dilatation or ampulla, containing groups of sensory cells.

Lateral Line System

The *lateral line system* is a set of sense organs restricted to the fishes and aquatic stages of the amphibians. Although it has connections in the brain with the afferent nerve pathways from the auditory system of the inner ear, it is a different system. It was already present in fossil *Ostracoderms*, and is thought to have originally evolved from primitive cutaneous sense cells. Fossil evidence from *Pteraspis* (Agnatha) onwards permits some tracing of lateral-line evolution, but embryological evidence is less clear because in some fishes the lateral-line arises from the otic or auditory placode, whereas in other, as for example an European minnow (*Leuciscus*), several separate lateral-line placodes form on head and body. In most of fishes the lateral line organs are properly organized and be arranged structurally as follows:

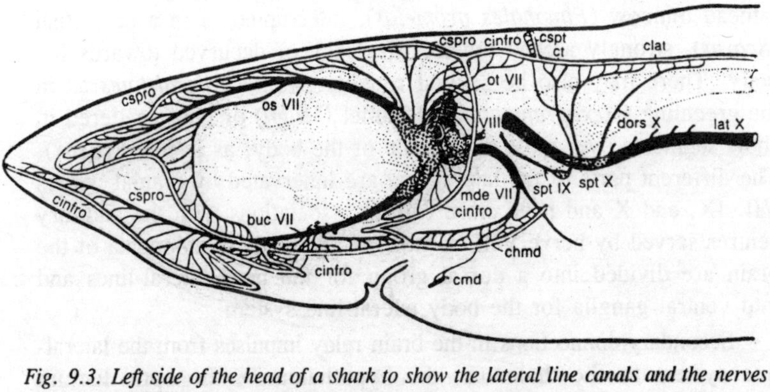

Fig. 9.3. Left side of the head of a shark to show the lateral line canals and the nerves supplying them.

1. In partially open or closed, variously developed canals located: (*a*) on the head where there are typically three main canals, the supraorbital, the infra-orbital, and the hyomandibular with branches, extensions, or connections such as the supratemporal commissural

canal that connects the canal system of the two sides of the body. (*b*) on the body where there is typically one canal extending from the cleithrum of the shoulder girdle to the base of caudal fin.

2. Free, arranged in groups on the head and in lines along the body (the elritze, *Phoxinus*; the loach, *Cobitis*).
3. Sunk, in pits, in groups or in several lines on the head or along the body (many elasmobranchs, cyclostomes) on the bands crossing the ventral surface (elasmobranchs, cyclostomes).

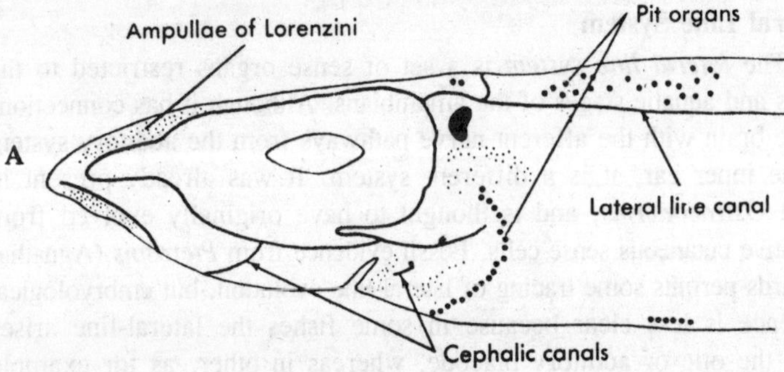

Fig. 9.4. Neuromast organs of Squalus acanthias.

Both the body lateral-line canal and that of the head exhibit several modifications. For example, the *body canal* may be extremely short as in the bitterling (*Rhodeus amarus*), intermediate in length as in the fathead minnow (*Pimphales promelas*), interrupted as in a parrotfish (*Scarus*), strongly arched towards the back of decurved towards the belly. There may also be several roughly parallel *lateral-lines* as in the greenling (*Hexagrammus*) and mullet (*Mugil*) or ones scattered in short segments over most of the side of the body, as in pikes (*Esox*). The different parts of the lateral-line are innervated by cranial nerves VII, IX, and X and have close central associations with the auditory centres served by nerve VIII. Centres in the medulla oblongata of the brain are divided into a dorsal group for the head lateral-lines and into ventral ganglia for the body lateral-line system.

Secondary connections in the brain relay impulses from the lateral-line system to the thalamus, the hypothalamus, the optic tectum (especially in sharks, Squaliformes) and the cerebellum (particularly in the Mormyridae). The giant cells of Mauthner with their final common path as wholesale movers of trunk and tail musculature, have axon tracts which bring them in direct contact with lateral-line branches of nerves VII, IX, and X as well as with the optic tectum, the sacculus

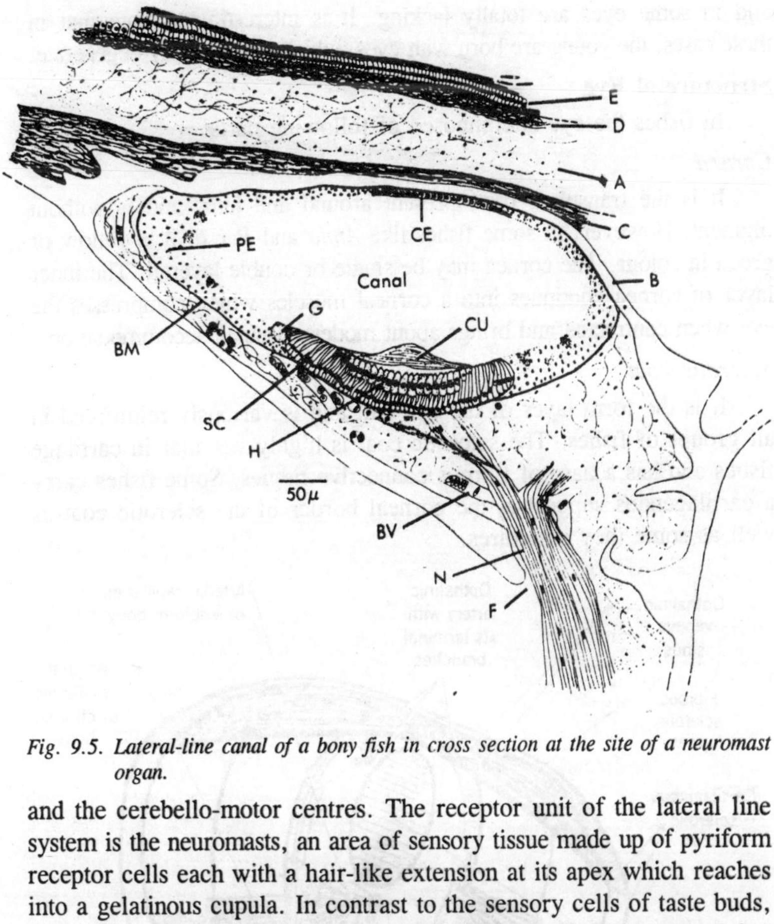

Fig. 9.5. Lateral-line canal of a bony fish in cross section at the site of a neuromast organ.

and the cerebello-motor centres. The receptor unit of the lateral line system is the neuromasts, an area of sensory tissue made up of pyriform receptor cells each with a hair-like extension at its apex which reaches into a gelatinous cupula. In contrast to the sensory cells of taste buds, which they resemble, they do not reach the base of the neuromast but are cushioned from below by nonsensory, supporting cells.

Eye or The Organs of Vision

The eyes are the most noticeable of sense organs in a fish. They are prominent, round, oval or eliptical structures situated on either side of the head. The eye of a fish is typically vertebrate and thus resembles our own. Fish eye is peculiar in the absence of eyelids, eyelashes and tear glands. Some sharks have a thin fold or skin, the *nictitating membrane* which can be drawn over the eye ball. In some bony fishes, the surrounding rim of the eye ball overgrows into a thick transparent sheeth, called the '*adipose lid*' Fishes which inhabit subterranean waters or caves have been found to have lost their vision and in some eyes are totally lacking. It is interesting to note that in these cases, the young are born with eyes, which subsequently degenerate.

Structure of Eye

In fishes the eye is composed of following parts:

Cornea

It is the transplant part, present around and lens, and is without pigment. However, in some fishes like *Amia* and *Perca*, it is yellow or green in colour. The cornea may be single or double layered. The inner layer of cornea continues into a corneal muscles which compresses the eye when contracted and brings about modest distance accommodation.

Sclerotic coat

It is the form layer of the eye ball and is variously reinforced in all groups of fishes. The sclerotic coat is highly vascular in cartilage fishes and has a base of fibrous connective tissues. Some fishes carry a cartilagenous support at the corneal border of the sclerotic coat as well as some they structures.

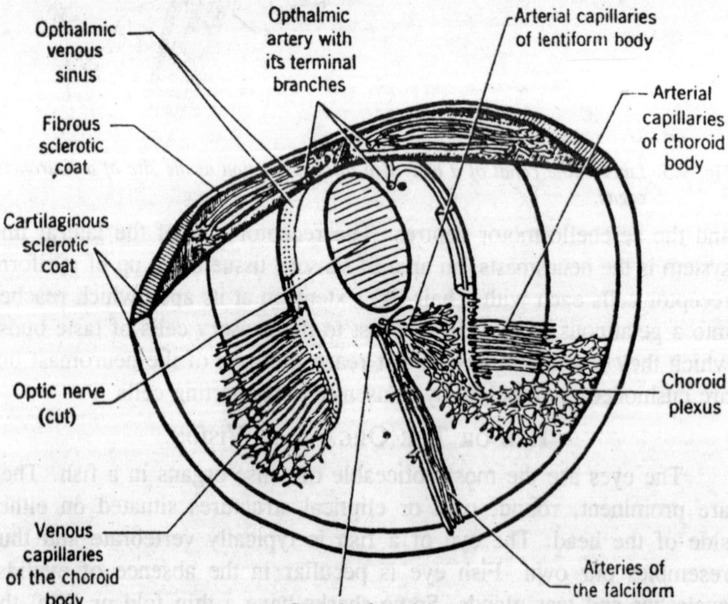

Fig. 9.6. Arrangement of blood vessels in the choroid body in the eye of a trout (Salmo).

Choroid coat

It is the middle layer developed from the embryonic mesoderm. It carries much melanin towards the posterior pole of the eye. The shiny nature of most shark eyes is due to the reflecting material in a well-developed choroid coat or *tapetum lucidium*. The tapetum functions to collect a maximum amount of light for the retina by scattering reflection from guanin crystals. The anterior part of the choroid just behind the cornea is modified into ciliary body having ciliary muscles and iris. Iris forms the pupil and controls the amount of light which reaches the retina. It has smooth muscles for the purpose. However, some fishes contain no muscles in their iris.

Retina

It is the innermost layer which is derived from embryonic ectoderm. It is restricted to the posterior 2/3 part of the eye ball

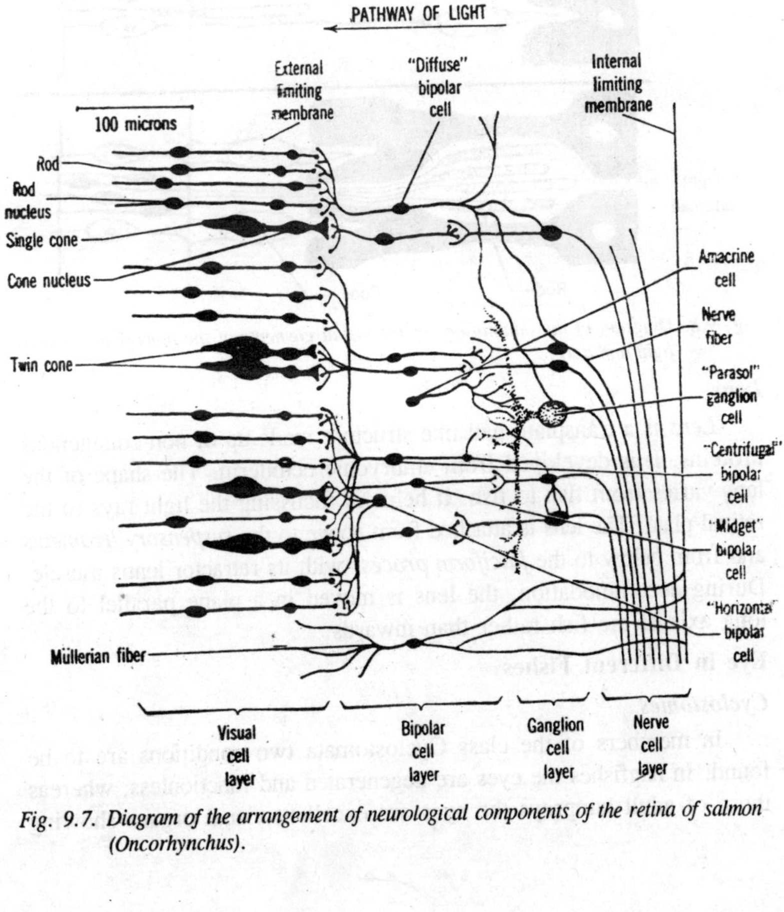

Fig. 9.7. Diagram of the arrangement of neurological components of the retina of salmon (Oncorhynchus).

behind lens. Fishes have typical vertebrate retina made up of several partially interlocking cell layers. As light hits the retina after travelling through the lens and the humours, the retinal layers encountered in order are: (a) fibres leading to the optic nerve, (b) their ganglionic cells, (c) bipolar cells which in turn, are in contact with, (d) the photoreceptor cells, proper, namely the *rod* and *cone cells* of the retina. Rods become functional in dim light whereas cones are stimulated by the strong light rays.

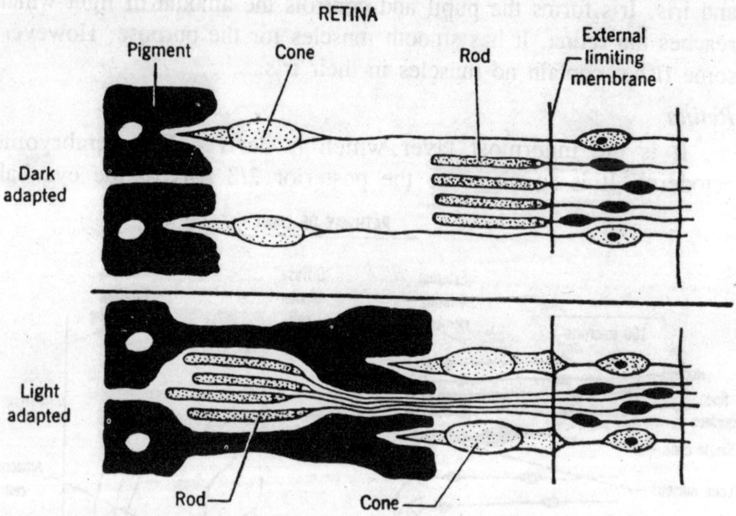

Fig. 9.8. Diagram of the adaptations of the visual elements of the bony-fish retina to light and dark.

Lens

Lens is a transplant ball like structure made up of non-collagenous proteins. It is developed from embryonic ectoderm. The shape of the lens varies from fish to fish. It helps in focussing the light rays to the retinal plate. The lens is attached from above to the *suspensory ligaments* and from below to the *falciform process* with its retractor lentis muscle. During accommodation, the lens is moved in a plane parallel to the long axis of the fish rather than inwards.

Eye in Different Fishes

Cyclostomes

In members of the class Cyclostomata two conditions are to be found. In hagfishes the eyes are degenerated and functionless, whereas those of adult lampreys the eyes are excellent visual organs showing

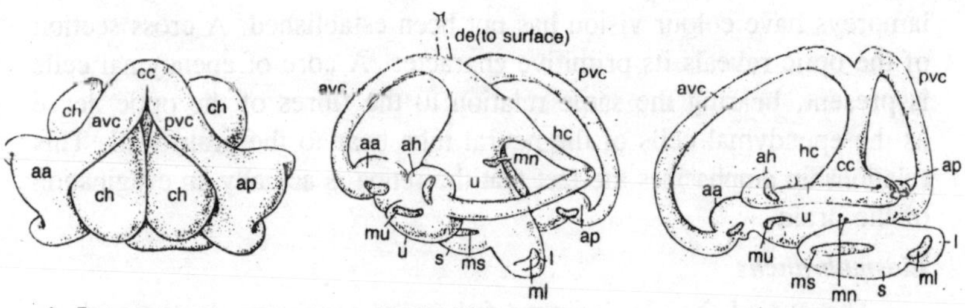

Fig. 9.9. Membranous labyrinth of A, lamprey; B, shark; C, teleost.

no signs of degeneracy. The hagfish eye is extremely small, being scarcely more than a millimeter in diameter. It lies entirely beneath the skin and is insensitive to light. The optic nerve is degenerate. Other nerves, as well as muscles and lens are lacking. Cornea, sclera, and choroid are not differentiated, and pigment is absent.

The retina is folded so as to fill entirely what would ordinarily be considered to be the chamber of the vitreous humor. A lens placode begins to form in the embryo, but it soon flattens out and disappears. The eye of the lamprey, although primitive, is nevertheless a well-developed structure. It offers no clue as to the possible evolutionary origin of the vertebrate eye. From the time that the *ammocoetes larva* burrows into the mud, until metamorphosis, the eyes are undeveloped and functionless. At the time of metamorphosis the eye develops markedly and the skin covering it becomes transparent. Among the important features are a flattening of the outer surface of the eyeball, a thin membranous sclera composed of dense connective tissue, a thin cornea *not* fused to the skin, lack of suspensory ligament and ciliary apparatus, fixed size of the pupil, and a permanently spherical lens, held in place by the vitreous humor. Eyelids are lacking.

In the lamprey the abducens nerve supplies both the external and inferior rectus muscles, whereas in other forms the latter is controlled by the oculomotor nerve which also supplies the superior rectus, internal rectus, and inferior oblique. Accommodation is brought about by flattening the cornea, when then pushes the lens *toward* the retina. This is accomplished by contraction of a special *corneal muscle*. Although both rod and cone cells are present in most forms, the rods for outnumber the cones. Both rods and cones are cone-shaped but differ in certain other respects. *Walls* (1942) is of the opinion that cones are phylogenetically older than rods and that the rods of lampreys are not related to the rods found in other vertebrates. Whether or not lampreys have colour vision has not been established. A cross section of the optic reveals its primitive character. A core of ependymal cells is present, bearing the same relation to the fibres of the optic nerve as the ependymal cells of the neural tube bear to the brain wall. This relationship emphasizes the fact that the retina is actually an evagination of the brain.

Elasmobranchs

The eye of the cartilaginous fishes are exceptionally large, those of the deep-sea dwelling holocephalians being largest of all in relation to body size. Eyelids consist of mere folds of skin which have but little power of movement. In some species, however, the lower lid is movable. The presence of an optic pedicel is characteristic, although lacking in a few forms. The eyeball is usually elliptical in shape with a short anteroposterior axis. Conspicuous features include the presence of cartilage in the sclera, the lack of intrinsic muscles in the ciliary body, and a large, permanently spherical lens which pushes the iris far forward so that it almost touches the cornea. The pupil is round in most forms but is slitlike in a few species.

Muscles in the iris are poorly developed and are altogether lacking in deep-sea forms. The presence of a core of ependymal cells in the optic nerves of a few elasmobranchs is indicative of a relationship to the condition found in the lamprey. In most elasmobranchs the surface of the choroid coat adjacent to the retina contains a layer of light-reflecting crystals composed of a chemical substance known as *guanin*. The layer is called the *tapetum lucidum*. Light rays striking the tapetum lucidum are reflected back, causing the eyes to "shine." This is, of course, most noticeable at night, and the presence of a tapetum in the many higher forms in which it occurs is generally restricted to those animals of nocturnal habit.

Visibility is thought to be enhanced under such conditions since light rays, instead of being absorbed by the pigment in the choroid coat, are reflected back to the sensitive retinal cells, thus conserving the light rays and making them more effective. The tapetum lucidum of elasmobranchs is exceptionally well developed. A thickened vascular portion of the choroid coat, called the *suprachoroid layer*, lies between the choroid and sclerotic coats on the medial aspect of the eyeball. It is present only in those elasmobranchs having an optic pedicel. Cones are absent from the elasmobranch retina except in a few species. It is probable that colour vision in lacking. Here, for the first time, an *area centralis* may be present, somewhat comparable to the *macula lutea* of primates except that it is devoid of pigment. No *fovea* is to be found. The suspensory ligament which holds the elasmobranch lens in place consists of a washer-shaped gelatinous membrane. The peripheral portion is fastened over the inner surface of the ciliary body and its centre border to a restricted, equatorially placed band on the surface of the lens. A portion of the ciliary body in the middorsal region extends downward and helps to support the lens.

The ventral side of the lens is supported by a papillalike projection of the ciliary body, which is attached anteriorly to the iris. A small muscle, the ***protractor lentis***, of ectodermal origin, is located on the lens side of the papilla. It is so arranged that contraction of its fibres causes the lens to swing forward, somewhat in the manner of a pendulum. The protractor lentis thus serves as a muscle of accommodation, moving the lens forward, rather than changing its shape, when focusing for near objects. The distance through which the lens can move is restricted by the shallowness of the space between iris and cornea, thus limiting accommodatory efforts to a rather narrow range.

Other Fishes

In *Latimeria* cones are extremely rare but rods are abundant. The eyes "shine" at night with a yellow-green colour like a cat's, but no really unusual features are to be found in the eyes of this ancient fish. A great many interesting variations are to be observed in the eyes of fishes. Only certain general features will be considered here. Typically there is a flattening of the anterior surface of the eyeball. Cartilage is present in the sclera, and in some forms bony plates may replace or supplement the cartilaginous elements. The cornea is composed of four layers, a *dermal layer* being interposed between conjunctival and scleral layers and a fibrous *autochthonous layer* lying inside the scleral portion. A *tapetum lucidum* is usually present in the choroid coat. The outer portion of the *uvea* in fishes is surrounded by a layer, the *argentea*, composed of silvery guanin crystals. This layer covers the dark pigment of the uvea and, by reflecting light, aids in rendering the almost transparent young fish inconspicuous. The argentea loses its significance during later life. A silvery *retinal tapetum* is present in a number of fishes. The light-reflecting crystals are located in the pigment layer of the retina. The visual cells of the retina consist of exceptionally long rods and two kinds of cones, single and "twin".

Among fishes the latter are to be found only in teleosts. Colour vision seems to be widespread among teleosts. An *area centralis* is present in the retinae of many bony fishes. A number of teleosts possess a *fovea*. Ciliary and functional iris muscles are lacking. Accommodation in the eyes of fishes is usually accomplished by changing the position of the spherical lens. In most cases a sickle-shaped *falciform process* is present in the form of a ridge of the choroid coat on the floor of the eyeball, projecting through the choroid fissure which, the teleosts, remains open. Beginning at the blind spot, its distal end extends forward and upward into the chamber of the vitreous humor. This end is attached to the lower border of the lens by a structure called the *companula Halleri*, or *retractor lentis*. Small muscle fibres of ectodermal (retinal) origin located in the retractor lentis pull upon the lens (which is suspended from above, like a pendulum), drawing it *backward*.

Deep-sea fishes which live in a realm of total darkness have relatively enormous eyes. Most of these forms possess luminescent photophores which furnish the only source of light that may affect their eyes. Recognition of enemies, of prey, of members of the same species, and of members of the opposite sex is thus possible in these abyssal regions. The eyes have undergone modifications all of which are conducive to increasing their sensitivity to light. Pupil and lens are extraordinarily large. Rod cells are extremely long and numerous, whereas cone cells are almost altogether lacking. In many forms the eyeball has assumed a tubular shape, its greatest diameter being in the region where the lens is situated. Surprisingly enough, relatively few deep-sea fishes are blind.

In some species, however, the eyes are degenerate. In others they are covered with a thick, opaque layer of skin. Only one fish (*Ipnops murrayi*) is known in which the eyes have disappeared altogether. This is the only vertebrate in which such a condition is known to exist. Better known blind fishes, such as those found in certain caves, have degenerate, microscopic, functionless eyes. There has been much speculation as to the reason for the degenerate eye condition is so many cave-dwelling forms. The consensus is that the fish did not become blind because they lived in darkness but rather than those with degenerate eyes found refuge in caves and became adapted to such an environment. In such fishes, eyes begin to develop during embryonic life but they fail to differentiate further after a certain developmental stage is reached.

In flatfishes, as in the flounder and halibut, both eyes in an adult are on the same side of the head. These fishes lie on the bottom of the ocean. The underside may be either the right or left, depending upon the species. Early in life the eye on the side on which the animal will lie migrates over or even through the head, bringing about the odd condition seen only in these fishes. The eyes of certain teleosts are adapted for vision in both air and water. Those of the mudskippers, *Periophthalmus* and *Boleophthalmus*, are set in the ends of retractable stalks. The so-called four-eyed fish, *Anableps*, of Central America, has two pupils in each eye. A horizontal partition, which appears in the iris during larval life divide the original pupil in two, the upper one being the larger. The upper pupil is usually out of water and is used in air vision. Two corneal and two retinal areas are also present.

Another fish, *Dialommus*, also appears to have two pupils in each eye, but these are actually circular, transparent areas in the otherwise pigmented opaque cornea. The use to which these areas are put is not understood. The capture off the Organ coast of a deep-sea fish with four eyes was reported in 1965. The fish, about 18 in long had two well-developed eyes with a secondary pair situated beneath them, enabling the fish to see light and movement of objects beneath it.

Ear or Auditory Organs

In its more typical condition each auditory organ consists of a membranous sac or vestibule, partially constricted into an upper portion or *utriculus* and a lower or *sacculus*. Three *semicircular canals* are connected with the utriculus, of which two are vertical and at right angles to one another, and the third is horizontal. One end of each canal is dilated into an *ampulla*. A slender tube, the *ductus endolymphaticus*, leaves the sacculus, and ends in a sac-like swelling, the *sinus endolymphaticus*, which apparently represents a portion of the embryonic epidermic involution from which the auditory organ is formed. A smaller sac-like outgrowth from the sacculus, the *lagena*, corresponds to the cochlea of the higher vertebrates.

The epidermic lining of this system of cavities is differentiated into patches or ridges of sense-cells (maculae or cristae), separated by supporting cells and innervated by the terminal branches of the auditory nerve. There is a crista acustica in each ampulla; and maculae acusticae are present in the utriculus, sacculus, and lagena. A fluid, the *endolymph*, fills all the cavities, and a similar fluid or *perilymph* occupies the spaces in the periotic capsule in which the various chambers are lodged. Among the more notable deviations from this type of auditory organ the cyclostome *Myxine*, apparently, has but a single semicircular canal with an ampulla at each end, and the vestibule is a single sac. *Petromyzon* has two canals, but lacs the horizontal

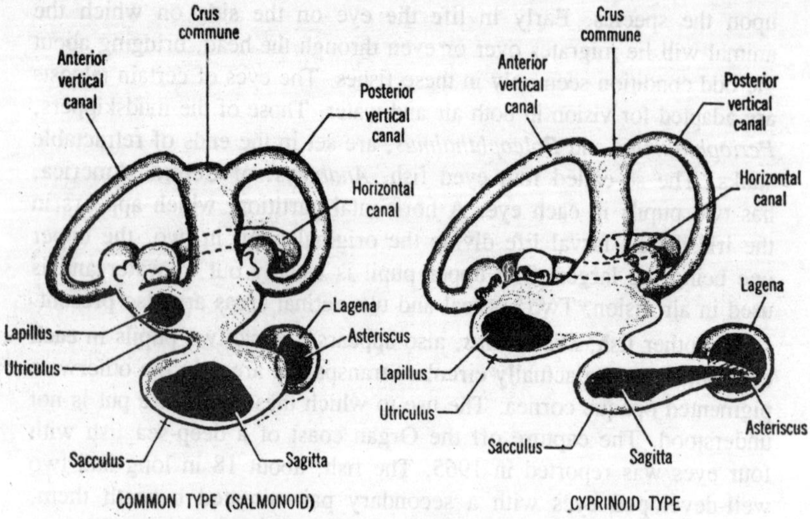

Fig. 9.10. Inner ear of fishes. Two principal types of membranous labyrinths as shown in view of organs of right side as seen from midline.

canal. In Elasmobranchs, including *Chimaera* (C), the ductus endolymphaticus retains its primitive connection with the exterior by means of a pore on the dorsal surface of the head.

In the Dipnoi (e.g., *Protopterus*) the paired endolymphatic sinuses divide into a number of caecal branches containing otoliths, which meet and interlace over the fourth ventricle. Otoliths, either in the form of fine, mucous-connected, calcareous particles, as in Elasmobranchs, or as massive solid concretions in Teleosts, are present in relation with the sensory areas of the utriculus sacculus, and lagena. In the few marine and a large number of freshwater Teleosts the auditory organ enters into a more or less intimate connection with the air-bladder by one of three different methods. The first and simplest is by the apposition of the extremities of a pair of caecal tubular prolongations from the air-bladder to the outer surfaces of the fibrous membranes which close a pair of vacuities in the outer bony walls of the periotic capsules, the inner surfaces being bathed by the perilymph surrounding the auditory organs. This method is characteristic of certain Serranidae, Berycidae, Sparidae, Gadidae, and Notopteridae, and probably in the Hyodontidae.

In the second method, of which several Clupeidae (*e.g.*, Herring, Pilchard, etc.) furnish examples, the periotic vacuities are open instead of closed, and the sac-like ends of the tubular extensions from the air-bladder are in actual contact with protruding outgrowths from the utriculus. The third method, by far the most elaborate, is by the intervention of a series of movably connected. "Weberian" ossicles, of which the most posterior on each side (the *tripus*) is inserted into the dorsal wall on the air-bladder, while the anterior one (*scaphium*) forms the outer walls of a median background prolongation (*sinus impar*) of the perilymph-containing spaces surrounding the two auditory organs. This is turn encloses a similar median prolongation (*sinus endolymphaticus*) from the two sub-cerebrally united endolymphatic ducts. The complex mechanism is present in the Cyprinidae, Siluridae, Characinidae, and Gymnotidae; and hence the term "Ostariophysi" as a collective name for these families. The physiological *raison d'etre* of the connection between the air-bladder and the auditory organ cannot yet be regarded as satisfactorily determined.

It is possible, as *Weber* thought, that it may be an auxiliary to the function of hearing by transmitting to the ear sound-waves impinging on the surface of the body and affecting the gases in the air-bladder. On the other hand, it may be urged with perhaps greater probability that the connection exists for the purpose of conveying to the ear stimuli due to the varying degrees of distension of the air-bladder, such as, it may be presumed, are naturally brought about by the variations of hydrostatic pressure which a Fish encounters in the course of its ascent or descent in the water. Whether regarded as an accessory to hearing, or as a means of regulating the distension of the air-bladder, the physiological value of the connection must be considerable, and on this point it is at least significant that the Weberian mechanism is characteristic of the dominant families of freshwater Teleosts at the present day.

10

MULTIPLICATION SYSTEM

Sexual reproduction is almost universal in vertebrates; only a very few vertebrates reproduce parthenogenetically i.e., without fertilization. A sexual mode of proliferation, which are widely found in most phyla of invertebrate animals and in plants are lacking. Among certain groups of invertebrates, the hermophrodite condition, as which the same individual may function as both male and female, is commonly found; in vertebrates the sexes are functionally separate in over whelming majority of cases.

Sexual reproduction in fishes, as in all organism, consists fundamentally of genetic recombination through the cellular processes of meiosis and fertilization. Accessory to these basic cellular phenomena are a vast array of anatomical, physiological and behavioural adaptations that function to bring together the sperm and eggs for fertilization and to support the development of the young. In fishes, the sexes are separate. Males develop sperms in their testes, and consist of spermatozoa capable of swimming and briefly resistant to the external environment. The female genital apparatus consists of an ovary where the eggs mature, but it differs according to the type of reproductive system adopted. Fishes can be oviparous, ovoviviparous or viviparous with many intermediate stages.

Development of Gonads

The male and female *gonads*, testes and ovaries, and derived from the coelomic epithelium near the inner or median aspect of the nephrotomes. Here the epithelium remains columnar, and soon projects into the ventral coelom as a continuous longitudinal ridge. It is probable that at first the modified epithelium is segmented as a series of "*gonotomes*", but if so, the latter must soon coalesce into a continuous ridge. Some of the epithelial cells enlarge to form the primitive sex-cells.

In the development of an ovary, portions of the epithelium sink inwards, carrying with them the primitive ova. Certain of the cells form the epithelial walls of a number of ovisacs, each of which encloses an ovum. As the ovisacs increase in number and size the germinal ridges project more and more into the coelom until, as ripe ovaries, they become suspended from its dorsal wall by a double peritoneal fold, the "*mesovarium*". The testes develop in a similar fashion except that the primitive sex-cells, which later give rise to spermatozoa, form the lining of a number of simple or ampulla-like tubules, the *seminiferous tubules*, and the suspensory fold is termed the "*mesorchium*".

Male Reproductive Organs

The male reproductive organs constitute the *testes*, *sperm ducts* and *capulatory* or *intromittent structures*. The tests is typically a compact organ, the shape of which shows much variation in members of different fishes. Histologically a testis is composed of numbers of *seminiferous tubules*. The number of the tubules varies from fish to fish. Each tubule is lined internally by given cells. The given cells undergo spermatogenesis and produce motile, flagellated sperms during breeding season. The several layers of cells that can be seen in a cross section of a fully developed seminiferous tubule represent different stages in

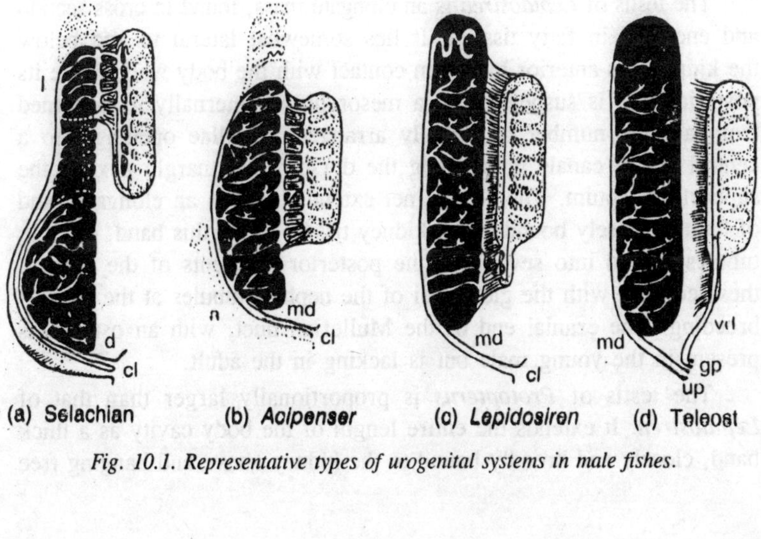

Fig. 10.1. Representative types of urogenital systems in male fishes.

the development of mature spermatozoa which finally are set free in the lumen of the tubule.

To collect sperms from various seminiferous tubules there are small ducts called *vasa efferentia*. These ducts unite together and form a common sperm carrying duct the *vas deferens*. There is a pair of vasa deferentia to carry sperms to outside the body mainly through the intromittent organs. In addition to their producing function, the testes of fishes are endocrine organs which elaborate the male hormone *testosterone*. The hormone producing cells which are probably *cells of Leydig* in fishes found between the seminiferous tubules along with blood vessels and connective tissues.

Male Reproductive Organs in Various Fishes

Choanates

Actinistian

In *Latimeria* the right testis is two or three times larger than the left but both produce sperms. The right testis extends through the middle quarter of the body cavity ventrolateral to the swim bladder. The epididymis is small and leads to the vas deferens, which enters the cloaca separate from the urinary sinus. The ovary of the female has a central cavity. The *Mullerian ducts* have not been described. In both sexes, masses of haemopoietic tissue are associated with the reproductive tract as well as with the viscera in general. In this respect the coelacanth resembles the other fishes.

Dipnoan

The testis of *Lepidosiren* is an elongate mass, found in cross section and enclosed in fatty tissues. It lies somewhat lateral to and below the kidney; its anterior half is in contact with the body wall, while its posterior half is suspended by a mesorchium. Internally it is formed from a large number of radially arranged ampullae opening into a central net of canals lying along the dorsomedial margin next to the supporting septum. The central net extends back as an elongate band of tubules closely bound to the kidney tissue. From this band, efferent tubules extend into several of the posterior segments of the kidney; these connect with the glomeruli of the nephric tubules at the time of breeding. The cranial end of the Mullerian duct, with an ostium, is present in the young male but is lacking in the adult.

The testis of *Protopterus* is proportionally larger than that of *Lepidosiren*. It extends the entire length of the body cavity as a thick band, closely and broadly bound to the kidney above and hanging free in the body cavity below. Anteriorly it is bound to the body wall and appears to be largely lymphoid in nature. The right gonad is attached anteriorly to the liver. The left gonad extends forward to the region of the *duct of Cuvier*. Posteriorly it is rounded off but has a short tubular extension. The seminiferous tubules are much longer than the ampullae of *Lepidosiren*. The tubules enter a central canal which extends along the dorsomedial margin of the organ and beyond the posterior end of the testis. The canalicular or rete extensions of right an left testis unite to form a median tube, which ends blindly just anterior to the urinary papilla. From this terminal pocket, efferent ducts enter the kidney to right and left and connect with a large number of renal capsules of the last two segments.

The cranial and caudal ends of the Mullerian duct remain; the remnants fuse at the midline but do not enter the urinary sinus. The testis of *Neoceratodus* is long and thick, tapered anteriorly and posteriorly, and triangular in section. The right testis is attached to the tip of the liver. The testis of either side is supported by a short mesorchium, its posterior half lying below the anterior half of the kidney. The seminiferous ampullae are long, radially arranged, and open into a central (dorsomedial) canal. Many efferent ducts extend from the central canal into the kidney and connect with capsules of the nephric tubules. The renal corpuscles of these tubules are retained and apparently remain functional as in the other lungfishes. The Mullerian duct is present, lying along the broad, dorsolateral surface of the testis. Posteriorly the Mullerian duct passes below, then lateral to the nephric duct, and from here to the midline where it fuses with that of the opposite side. The common sinus thus formed, ends blindly in front of the cloaca.

Actinopterygians

Teleost

The testes of the teleost are long and rounded in section; they extend nearly to the cloacal region. In texture they are finely granular as opposed to the coarsely granular or distinctly egg-laden ovary of the female. In fine structure there is usually a central cavity into which radially arranged short pouches (ampullae) or tubules open. The tubules may join together as they extend posteriorly, so that a central cavity as such is not present. In syngnathids the testis is a simple tube. The testis is not connected to kidney tubules, rather the central canal, of whatever form it takes, extends posteriorly lateral to the swim bladder and urinary bladder.

Eventually the ducts of either side join and open to the outside between the anus and excretory openings. In the cod the posterior ends of the testes are joined and a single duct extends to the genital opening. In some fishes the more posterior tubules of the testis are secretory and may be identified as *seminal vesicles*. The opening of the male genital sometimes differs from what has been described. It may open into the urinary bladder as in muraenids and *Anableps*; into the excretory, sinus in blennies, the salmon, perch and many others; into the anal margin in *Lota*; and into the joined anal and excretory opening in syngnathids (i.e., into a cloaca).

Polypterus

The male has an elongate testis closely connected with the kidney. The central canal begins anteriorly and in the thin posterior continuation there is an interconnected set channels. Posteriorly a single duct emerges and, following the wall of the nephric duct, opens in the urinary sinus. The male of *Polypterus* appears to be comparable to the male teleost in having the central canal extended posteriorly through the epigonal region, not to a separate pore, however, but to the cloaca. The posterior part of the male duct has been assumed to have arisen in a parovarial style.

Amia and Lepisosteus

In males of *Amia* and *Lepisosteus* the testis is long, band-like, and rounded at either end. It is attached by a broad mesorchium through which numerous efferent ducts pass into the kidney. The nephric duct shows no modification as a vas deferens.

Chondrichthyes

Shark

The testis is large and suspended by a broad mesorchium. The position in the body cavity varies from far anterior back to about the middle. The right is usually, larger than the left. In *Scyllium* there is a sequence of developmental stages of seminal ampullae. This sequence starts from the ventral free margin where new follicles are forming and progresses toward the dorsal attachment. At the time of breeding, a set of ripe ampullae connects through narrow *ductuli recti* with the central canal, which lies along the attached margin of the gonad. Many of the ampullae observed along the canal are empty; these are spent ones of previous reproductive periods.

The testes of most sharks are made up of typical ampullae, each connected with the central and each rise to repeated generations of

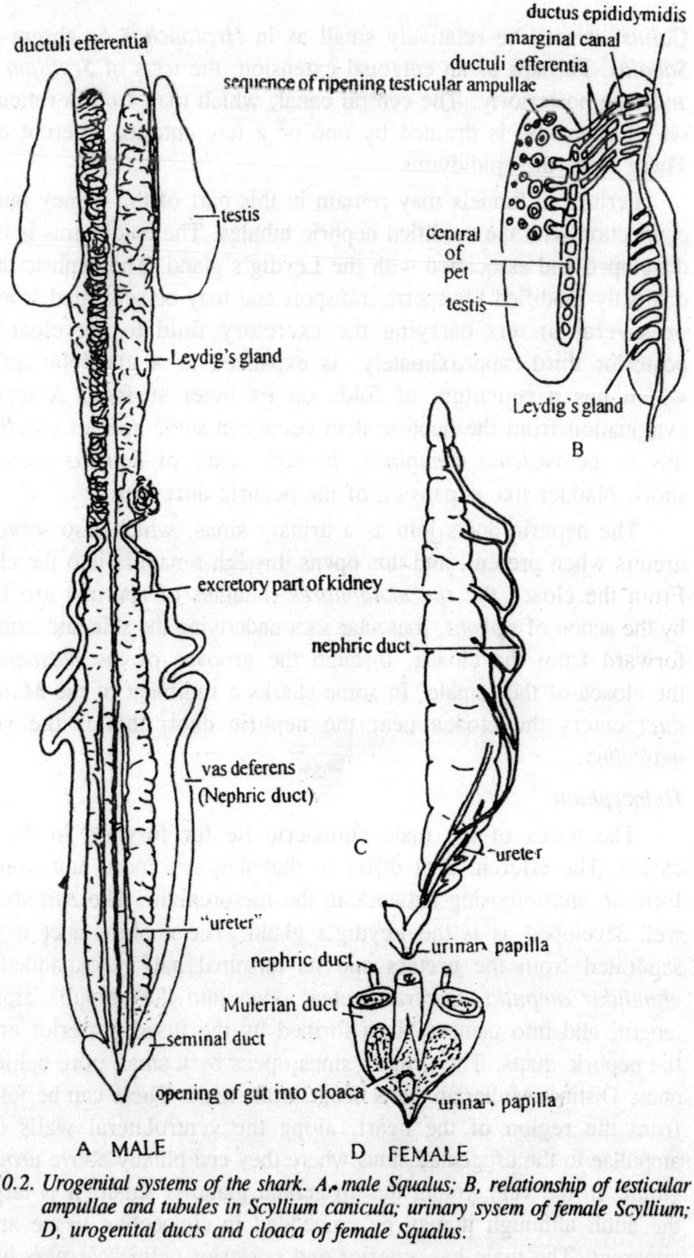

Fig. 10.2. Urogenital systems of the shark. A, male Squalus; B, relationship of testicular ampullae and tubules in Scyllium canicula; urinary sysem of female Scyllium; D, urogenital ducts and cloaca of female Squalus.

germ cells. In some sharks the testis is associated with an *epigonal organ* which is lymphoid in nature. This extends back to the cloaca in

Galius; it may be relatively small as in *Heptanchus* or absent as in *Squalus*. Perhaps as an epigonal extension, the tests of *Scyllium* meet and fuse posteriorly. The central canal, which may from a reticulated set of channels, is drained by one or a few anterior efferent ducts. These enter the epididymis.

Peritoneal funnels may remain in this part of the kidney but lose connection with the modified nephric tubules. The epididymis is highly developed and associated with the Leydig's gland. The nephric duct is distinctly modified for sperm transport and may be separated from one or several ureters carrying the excretory fluid to the cloaca. Its posterior third, approximately, is expanded as a glandular ampulla which has a reticulum of folds on its inner surface. A terminal evagination from the nephric duct occurs in some species (*Scyllium*); this is the *vesicula seminialis*. In both sexes of *Squalus* there is a short, bladder like expansion of the nephric duct here.

The nephric ducts join as a urinary sinus, which also serves the ureters when present, and this opens through a papilla into the cloaca. From the cloaca the *spermatophores* (bundles of sperms) are forced by the action of *siphons*, muscular sacs underlying the skin and extending forward from the cloaca, through the grooves of the *claspers* into the cloaca of the female. In some sharks a rudiment of the *Mullerian duct* enters the cloaca near the nephric duct; this is the *vagina masculina*.

Holocephian

The testes of the male chimaerid lie for forward in the body cavity. The efferent duct differ in that they are more numerous and form an anastomosing network in the mesorchium. The *epididymis* is well-developed as is the Leydig's gland. The nephric duct is partly separated from the ureters and its terminal half is expanded as a *glandular ampulla*. Several ureters open into the ampulla along its length, and into urinary sinus formed by the fused posterior ends of the nephric ducts. The urinary sinus opens by a small pore behind the anus. Distinct Mullerian ducts occur in the male. These can be followed from the region of the heart, along the ventrolateral walls of the ampullae to the urogenital sinus where they end blindly above *urogenital gland*. In the very young the urogenital gland is small; it is larger in the adult although it may be embedded in connective tissue and not apparent. The male has anterior and posterior pelvic *claspers* utilized in copulation.

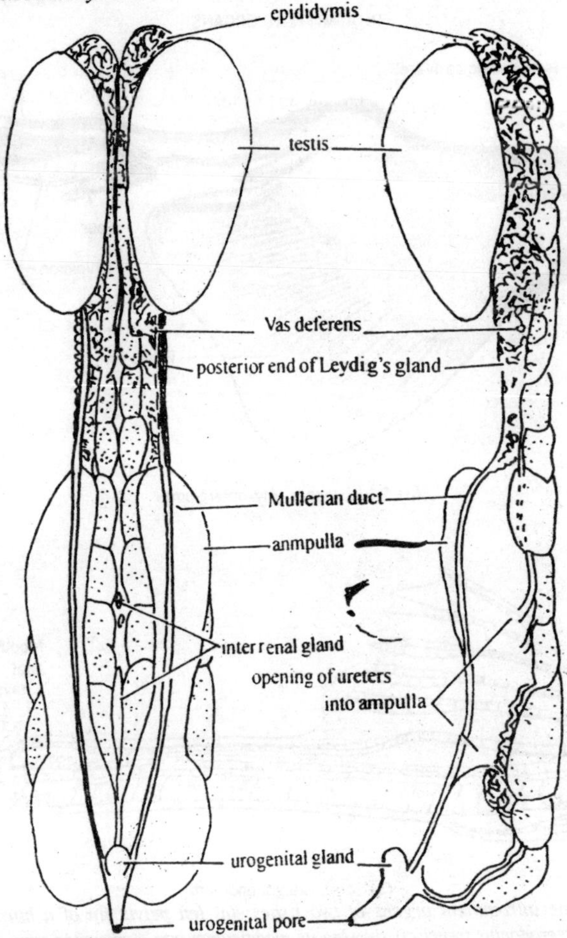

Fig. 10.3. Urogenital system of the chimaerid, Hydrologus. A, ventral view of male systems, showing testes lying in anterior part of body cavity; B, lateral view of male system.

Spermatozoa and their Formation

The process by which sperm cells are formed in the testes is called *spermatogenesis* or *spermiogenesis*. It is in this process that the materials that determine the characteristics which the offspring will inherit from its parents are readied for union with like materials in the egg cell. Not only do the spermatozoa or sperms of different fishes contain different hereditary materials, they differ in shape as well. It is also in this process that sperm cells gain a whip-like tail

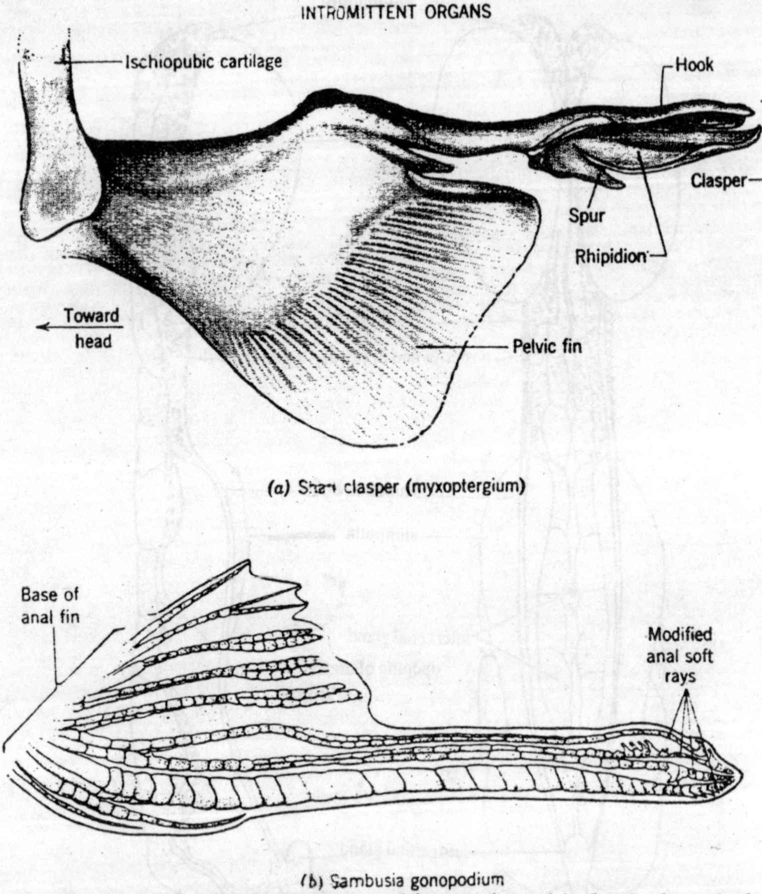

Fig. 10.4. Male intromittent organs of two fishes: (a) left pelvic fin of a horn shark (Heterodontus francisci) showing its modification into a myxopterygium, ventral view; (b) anal fin modified into a gonopodium in the mosquitofish (Gambusia affinis), lateral view.

which makes them motile and with which they make their way ultimately to and into eggs for fertilization.

Furthermore, during development spermatozoa attain the differences in shape which they exhibit as species characteristics. In order to insure fertilization, each male fish, like the male of other vertebrate species, produces huge numbers of sperm cells so tiny that a million of them can be held in a droplet. The spermatozoa, plus juicy contributions of the sperm ducts, compose the *milt* which the male fish exudes at *spawning* (like the semen of man). The sperms are inactive and immobile until the secretion of the sperm ducts has been encountered. Once they have become active and have started to move by the whipping motions of their tails they will soon perish unless an egg is encountered for fertilization.

The life span of sperms varies considerably according to species and according to substrate into which they are deposited. If the sperm cells are deposited in water, they live a much shorter time than if they are deposited in a female. If water has nearly the same salty content as the fish body fluid, the sperm cells live longer than if the medium is either higher or lower in effective salt content than body fluids. Viability of sperm cells is also affected by temperature; in general, they live longer in lower temperatures than in higher ones and quite as human sperms may be stored for *artificial insemination*, so it is possible to store the milt of fishes. Not all of the sperms manufactured in preparation for the reproductive period and passed from the body of the fish at the time of reproduction. Those that remain behind deteriorate in the gonads or the ducts and since they are composed largely of protein they appear to be effectively resorbed.

Female Reproductive Organs

In fishes the female reproductive organs are composed of a pair of ovaries, and oviducts. It is unusual to find unpaired ovaries in fishes. Even when this condition seems apparent, there is actually a fusion of the two gonads. Sometimes the ovaries are asymmetrical in position. In a number of elasmobranches only the right ovary becomes fully developed, the left ovary becoming atrophic during adult life.

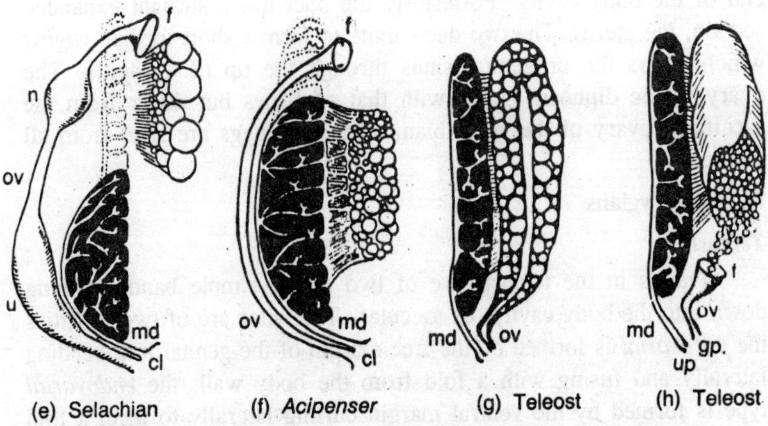

Fig. 10.5. Representative types of urogenital systems in female fishes.

Histologically the ovary is made up of germinal epithelium that undergoes oogenesis and forms ova that contain relatively enormous amount of yolk in oviparous fishes. The ova and then released out into the body cavity simple by rupturing the wall of ovary. This process is called *ovulation*. In fishes certain bodies called *corpora lutea*, may form from nurse cells of the ovarian follicles following *ovulation*. Presumably these structures have an endocrine function and may play an important role in the longer than usual retention of egg in the oviducts of oviparous species and in providing for the development and retention of the young in the uteri of ovo viviparous forms.

Female Reproductive Organs in Various Fishes

Choanates

Actinistion

The ovary of the female has a central cavity. The mullerian ducts have not been described. In both sexes, masses of haemopoietic tissue are associated with the reproductive tract as well as with the viscera in general. In respect the coelacanth resembles the other fishes.

Dipnoan

The ovary of *Protopterus* or *Neoceratodus* is long and hangs down from the wall as a band; the eggs are shed from its lateral aspect. There is usually a distinct epigonal extension. In *Neoceratodus* the ovary is bound anteriorly to the dorsal body wall; in *Protopterus* it is suspended by a short mesovarium. The convoluted *Mullerian duct* lies lateral to the ovary; its slit like fringed ostium opens into the anterior end of the body cavity. Posteriorly the duct has a straight expanded section, the uterus. The two ducts unite to form a short median *vagina* which enters the urogenital sinus through the tip of a papilla. The ovary in the dipnoan agrees with that of fishes but differ from the sacciform ovary of the amphibian, where the eggs are shed from all surfaces.

Actinopterygians

Teleosts

Ovaries in the teleosts are of two types: simple bands hanging down into the body cavity, or saccular. The latter are of two varieties the *parovarial* is formed by the free margin of the genital fold bending laterally and fusing with a fold from the body wall; the *endovarial* type is formed by the ventral margin curling laterally to meet a fold from the upper margin of the gonad. This folding of the genital ridge is observed of the salmon and in the region posterior to the ovary where a separate oviduct is produced in the salmon or *Osmerus*. The *oviduct* can be of the endovarial variety, attached to the lateral aspect of the swim bladder in front and the urinary bladder behind. But swim and urinary bladders are suspended in the medial septum. In either position, body wall or down on the swim bladder, the oviducts unite to form a genital sinus opening to the outside between anus and urinary apertures. In the cod the two ovaries join posteriorly and utilize a single oviduct. Not all females have saccular ovaries with oviducts.

In the salmon the ovary is a free band, or *parovarial*. The eggs pass from the body cavity into a funnel lying above the rectum and below the urinary bladder; this funnel opens between the anus and excretory pore. The funnel is formed by membranes extending out and upward from the terminal part of the intestine to the body wall; the dorsal mesentery of the gut is lacking in this region. In the muraenids the ovary is a simple fold; the eggs form on the lateral aspect and are shed into the coelom. Exit from the body cavity is by bilateral slit-like openings behind the anus. These opening are in the general position of the abdominal pores of other fishes.

Polypterus

The female has an elongate, band-like ovary hanging down into the body cavity from the mesovarium. There are numerous ridges on the other surface from which the eggs are shed into the body cavity. The oviduct lies lateral to the mesovarium, in the region of the posterior half of the ovary. It follows the course of the nephric duct to the posterior end of the body cavity, where it opens into the urinary sinus almost where that sinus opens into the cloaca. The oviduct is like that observed in the following actinopterygians.

Amia and Lepidosteus

The ovary is suspended by a broad mesovarium in which there are strands of tissue resembling the efferent ducts of the male. The ovary of *Amia* differs markedly from that of *Lepisosteus* in that it hangs free in the body cavity. From its lateral, ridged surface the ova are shed into the body cavity are carried to the cloaca through the large membranous funnel. The oviduct lies lateral to the mesovarium and below the nephric duct. The oviduct enters the nephric duct where the latter turns downward toward the urogenital sinus.

The ovary of *Amia* is like the testis in having a distinct vascular drainage which passes along the middle of the medial aspects and connects anteriorly with the main vascular channels. In *Lepisosteus* the ovary is more elongate than that of *Amia*. It is sacculate and attached in its middle part by a dorsally tapering, membranous funnel, in part formed from the mesovarium. This funnel leads into the oviduct lying along the body wall below the nephric duct, as in *Amia*. Posteriorly the oviduct enters the nephric duct in the region of its swollen bladder portion. The parovarial ovary of *Lepisosteus* is formed by folding of the genital ridge to meet a sheet extending down from the body wall. The ovary and its oviduct suggest the situation observed in teleosts but could be a *parallelism*.

Chondri Chthyes

Skarks

The ovary is generally shorter than the testis. Usually there are right and left ovaries, but in some sharks the left may be rudimentary (*Scyllium*, *Pristiophorus*, *Carcharias*, *Galeus*, *Musteus*, *Zygaena*). An *epigonal organ* may be associated with the ovary (*Scyllium*, *Heptanohus*). The eggs are shed into the body cavity and carried to the exterior through the *Mullerian ducts*. The common funnel of these ducts lies at the anterior end of the body cavity, below the oesophagus and just behind the heart. In some sharks there are paired ostia in the region of the pronephros. The Mullerian duct consists of the usual segments: funnel, tube, isthmus, uterus and vagina. In oviparous forms there are two specialized zones in the tube which secrete *albumin*, and a horny *shell* to enclose the egg; these zones form the *nidamental gland*. Fertilization apparently occurs in this area. *Squalus* is viviparous as are many sharks. In these the fertilized egg undergoes development in the uterus, most of the posterior half of the oviduct. The developing young may receive nourishment from the maternal tissues in some

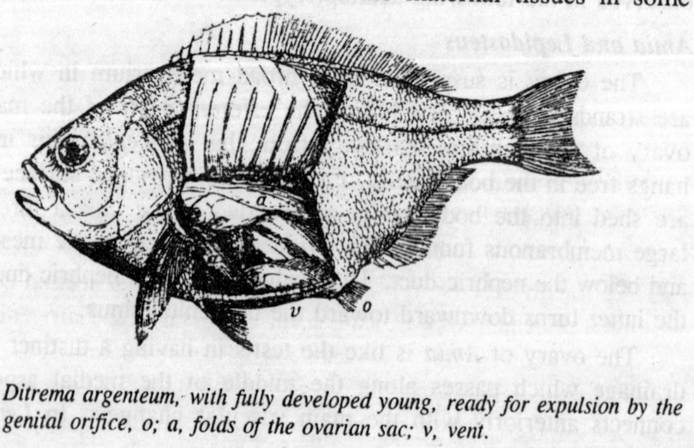

Fig. 10.6. Ditrema argenteum, with fully developed young, ready for expulsion by the genital orifice, o; a, folds of the ovarian sac; v, vent.

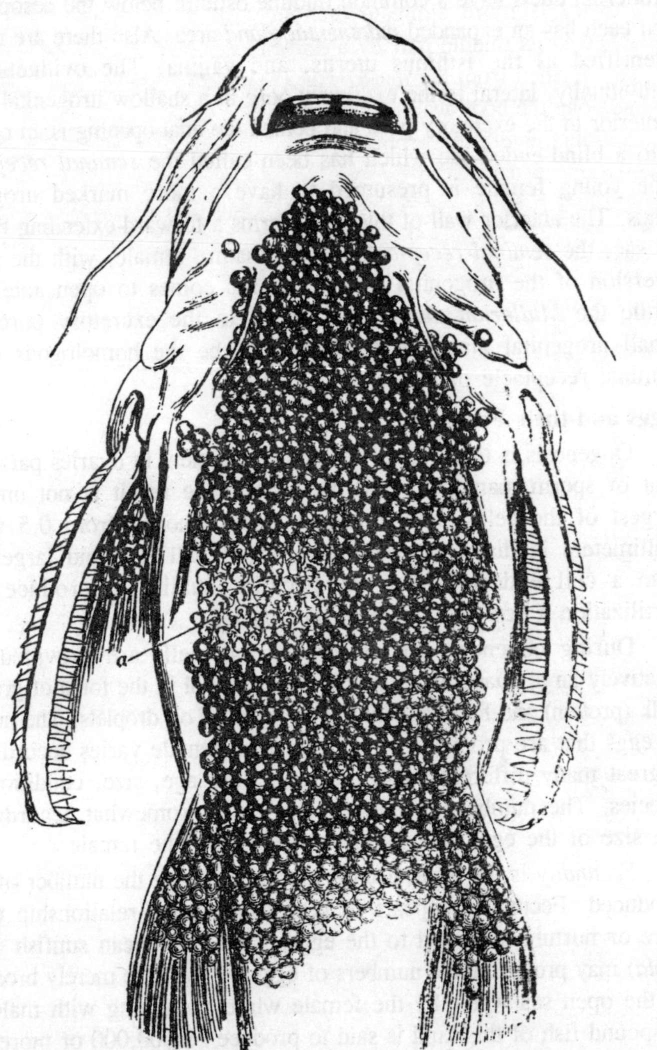

Fig. 10.7. Abdomen of Aspredo batrachus, with the ova attached; at a, the ova are removed, to show the spongy structure of the skin, and the processes filling the interspaces between the ova.

species (*metaviviparous*). The vagina is the terminal section connecting with the cloaca; it opens dorsolateral or dorsal to the urogenital sinus.

Holocephalian

The female is similar to the shark in the position of the ovary, the general nature of the oviduct, and the large shelled eggs. The

Mullerian ducts have a common midline ostium, below the oesophagus, and each has an expanded *nidamental gland* area. Also there are regions identified as the isthmus uterus, and vagina. The oviducts open individually, lateral to the excretory pore in a shallow urogenital sinus. Anterior to the excretory pore and behind the anal opening is an opening into a blind-ended sac which has been called the *seminal receptacle*. The young female is presumed to have a more marked urogenital sinus. The anterior wall of this sinus forms a forward-extending bladder or sac, the *seminal receptacle*. In the mature female, with the partial eversion of the urogenital cavity, the sac comes to open anteriorly, while the *Mullerian ducts* open lateral to the excretory pore. The small urogenital gland of the male may be the homologous of the seminal receptacle of the female.

Eggs and their Formation

Oogenosis is the process of egg development in ovaries parallel to that of sperm manufacture in the testes. The result is not only the largest of the cells to be found in fishes (mostly from 0.5 to 5.0 millimeters in diameter, with few to golf ball size and larger) but also a cell endowed with the peculiar quality to produce after fertilization a new individual of the species.

During oogenesis the developing egg cells are provided with relatively large quantities of stored food material in the form of granular yolk (protein) and fat commonly in the form of oil droplets. The number of eggs that are produced by an individual female varies according to a great many different factors including her age, size, condition and species. The number, of course, also varies somewhat according to the size of the eggs in relation to the size of the female.

Fecundity is the general term used to describe the number of eggs produced. Fecundity appears to bear some broad relationship to the care or nurture accorded to the eggs. Thus, the ocean sunfish (*Mola mola*) may produce huge numbers of eggs but they are merely broadcast in the open sea water by the female while swimming with males. A 54-pound fish of this kind is said to produce 28,000,000 or more eggs in a single season. The cod (*Gadus*), which is also a pelagic open-water spawner, produces as many as 9,000,000 eggs per female per season. Such species accord no care to their eggs, other than to emit them in the vicinity or in the company of males depositing sperms. Compare with this the sticklebacks (*Eucalia* and *Gasterosteus*) which make elaborate nests and then provide parental care to the developing fish. Here the usual number of eggs ranges between thirty and a hundred.

In guppies (*Lebistes*), which bear their young alive and thus accord an extreme amount of care (that is, internal incubation), the usual brood number is less than two dozen. Some fishes even develop at one time, only a single egg, or one in each ovary. The eggs of most fishes have *shells* secreted around them. The egg cell itself has only a tender membrane when ready to have the ovary. As the eggs of oviparous species pass through the oviducts, they go through glands that secrete the shells about them. In fishes that accord care to their young, either by having the parents stay on guard or by burying the eggs in gravel, the shells are often more tender and less limy or horny than they are in eggs exposed to waves currents, or other hazards. In many skates and some sharks, for example, the shells are horny-hard and almost finger-nail-like in strength. Such eggs resist damage, even drying, to such an extent that the eggs, when washed up on shore, may lie there for some time and still hatch when returned to the sea. An egg of the little skate (*Raja erinacea*) was once mailed without padding from the New Jersey coast to our Ann Arbor laboratories in a common letter envelope. It was out of the water in transit for several days and travelled some seven hundred miles. Yet it hatched successfully and became a pet for some months. Fish eggs have many other interesting characteristics than their covering. Some are buoyant (pelagic) and have a specific gravity about like that of fresh water.

Most of the marine food fishes are like this and the cod (*Gadus*) would be an outstanding example. Most stream fishes, on the contrary, have eggs which are heavy and sink (demersal), with a specific gravity which is greater than that of fresh water. Perhaps this is an adaptation that insures these eggs against being swept downstream at the time of disposition, which may take place in gravel beds in rather swift current in sites such as those chosen by stream chars (*Salvelinus*), the trouts and Atlantic salmon of the genus *Salmo*, and the species of Pacific salmons (*Oncorhynchus*). *Ora* also vary as to adhesive stickiness. Most buoyant eggs are *non-adhesive* and do not normally stick to anything for holdfast purposes. Most eggs that sink, however, are at least *temporarily* adhesive. Trout and salmon (Salmoninae) eggs are temporarily adhesive during the process of water hardening in the early minutes after deposition while they are absorbing water at a great rate. One can almost visualize the eggs as "sucking in" water at such a rate that they stick to things, things stick to them, or they adhere to one another.

In this process, the eggs of stream-spawning trouts and salmons pick up little particles of sand which help sink them to the bottom and anchor them there for covering and development. The stickiness of the eggs of other fishes such as the walleye (*Stizostedion*) serves the purpose of keeping the eggs from settling deep into bottom materials and becoming suffocated. Sticky, *demersal eggs* adhere to sticks and debris on the bottom without falling down into the deeper, softer materials which may be present. Adhesivencss can be due temporarily to the process of water hardening, or more permanently to a pasty, mucoid deposit on the egg which remains sticky throughout the whole developmental period. Eggs also vary as to shape, but those of most species of fish are spherical. The spherical shape seems to be the natural, basic contour of most animal eggs. In many anchovies of the herring-like family Engraulidae, the outline of the egg is elliptical, occasionally spherical.

But in the minnow genus *Rhodeus*, the European bitterling, and in various other fishes including the cichlids, the eggs are oval in outline (ovoid). In the gobies they range from ovoid to teardrop-shape. The eggs of fishes also vary considerably according to the appendages which they bear. In the hagfish (*Myxine*), for example, the egg in an elongate oviod, with a number of *tendrils* at each end. Apparently, these tendrils anchor the eggs to vegetation or other material. In the American smelt (*Osmerus mordax*), the egg has a low stalk which is adhesive and becomes attached to the stony bottoms of streams where the species most often spawns. In the brook silverside (*Labidesthes sicculus*), there is a single elongate filament that serves first for temporary *flotation* and then for *attachment*.

The eggs of some skates (Rajidae) and cat sharks (Scyliorhinidae) have tendrils that come out of each of the four corners of the more or less box-shaped case. Most flyingfishes (Exocoetidae) have eggs with numerous filaments, much like long hairs, that come out from the surface of the spherical egg. However, the eggs of a few flyingfishes have no filaments, some have the filaments confined to the two poles of the egg, and others, to one pole. Quite as spermatozoa exhibit different viability in different situations, so do eggs have various survival features. Interestingly, not all mature eggs are laid in any one period of spawning. Some remain in the ovary and the normally resorbed. It would appear, however, that when proportionately large quantities of eggs are not shed, the ovary is unable to resorb them. The individual may then become egg-bound, and the mouth of the oviduct plugged by a mass of fibrous tissue. Future passage of eggs may be barred.

11

EMBRYOLOGY

Fish eggs and larvae provide a relatively untapped source of biological material increased by the recent improvements in techniques for rearing marine species. Apart from their intrinsic interest, experimentally based information on these early stages is required for further progress in the advancing fields of fish culture and fisheries research. General text-books on ichthyology such as those of *Lagler* et al. (1962); *Nikolsky* (1963), *Norman* (1963), and *Marshall* (1965) and on reproduction in fish by *Breader* and *Rosen* (1966) provide both general and some detailed information.

Identifications of *Ehrenbaum* (1909), *D'Ancona* et al. (1931-1933), through the current series of plankton sheets issued by the International Council for the Exploration of the Sea, and with the help of the extensive bibliographies by *Dean* (1916) and *Mansueti* (1954). Most species of fish pass through a larval stage before assuming the adult form at metamorphosis. Sometimes the newly hatched fish is called a "prolarva" (or alevin in salmonids) until the yolk is resorbed, and then a "postlarva" (or fry). The term "larva" is used here for all stage to metamorphosis in marine fish, although alevin and fry may be used when referring to salmonids or other freshwater groups.

PARENTAL CONTRIBUTION

Apart from the more obvious genetical effects on differentiation, rate of development, body, form, size and behaviour, the parents, and especially the female, have an important influence on the viability of the offspring both on a species and individual level, in terms of (a) the condition for incubation, (b) fecundity and (c) egg size.

Conditions for Incubation

Differences of spawning seasons and time and of spawning sites and substrat mean that incubation can take place in a great variety of conditions which influence the early development and physiology of the offspring.

Eggs single, with no parental care

(i) Nonbuoyant, buried in sand or gravel—many salmonids, grunion *Leuresthes tenuis*, and lamprey *Petromyzon*; in peat or mud *Aphyosemion* and *Cynolebias* where the eggs undergo diapause during the dry season.

(ii) Buoyant, planktonic—most marine fish, e.g., gadids, clupeids, flatfish and deep-sea fish.

(iii) Nonbuoyant, loose or attached to substrata—a common fresh-water characteristic, e.g., cyprinids, pike Esox, or in littoral species, e.g., blenny *Blennius*, bullheads *Cottus*, and eels *Ammodytes*; also found in some marine species, e.g., herring *Clupea harengus*, capelin *Mallottus villosus*, catfish *Anarhichas*, and American flounder *Pseudopleuronectes americanus*. Tendrils for attachment: are found in many oviparous elasmobranchs, in the hagfish *Myxine*, smelt *Osmerus*, saury *Scomberesox*, and flying fish *Exocoetus*.

Eggs single, special environments

The bitterling *Rhodeus amarus* lays eggs in the gills of the fresh-water mussel and the lumpsucker *Careproctus* under the carapace of the Kamchatka crab.

Eggs single, with parental care

1. Parents carrying eggs—sea horses *Hippocampus* and pipefish *Syngnatinus* have brood pouches and the sheat fish *Platystacus* a specially modified area of "spongy" skin. Marine catfish, Ariidae, cardinal fish *Apogonidae* and *Tilapia* and mouth brooders, and *Tachysurus* incubates the eggs intestinally.
2. No nest, but eggs protected—found in many littoral forms, e.g., the bulheads Cottidae, blennies Blenniidae and gobies Gobiidae.
3. Ovovipiparity and viviparity—elasmobranchs include dogfish *Acanthias*, smooth hound *Mustelus vulgaris*, electric ray *Torpedo*, *stingray Trygon*, and the nurse hound *Mustelus laevis*. Teleosts include redfish *Sebastes*, *Heterandria*, *Anableps*, *poecinds* such as *Xiphophorus* and half beaks *Hemirhampus*.
4. Nests, often with parental protection and ventilation—also found in littoral species, e.g., blenny *lctalurus*, sticklebacks *Gasterosteus*,

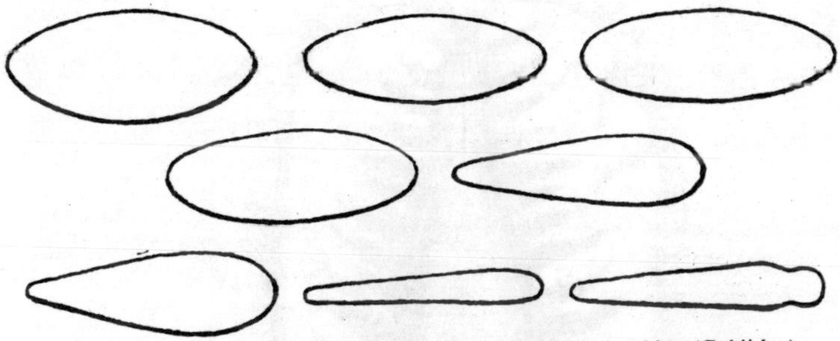

Fig. 11.1. Non-spherical fish eggs as shown in selected gobies (Gobiidae).

and in other freshwater species such as sunfish Centrarcidae, bowfin *Amia*, lungfish *Protopterus* and *Lepidosiren*, and in the Cichlidae. Bubble nests giving good aeration are found in tropical or swamp species, e.g., Siamese fighting fish *Betta splendens*.

Eggs massed

Angler fish *Lophius* and yellow perch Perca flavescens have massed but unprotected eggs; in the lumpsucker Cyclopterus and butterfish *Blennius pholis* the eggs are protected by the male.

Fecundity and Egg Size

In the higher latitudes the spawning season is often short and the eggs are liberated over a brief period of perhaps hours (clupeids) or over periods of some days, probably at certain times of the day or night (flatfish and gadids). Where the seasons are less marked spawning may occur over a much longer period or be intermittent throughout the year, especially where the time between generations is only a matter of weeks or months.

Fecundity may be considered as the number of eggs produced in one year by a female although this may be very difficult to determine where spawning is protracted. In general, fecundity is high where the eggs are liberated into open marine waters; it is lower in freshwater species and where there is parental care. There is also a strong tendency for fecundity and egg size to be inversely related. Apart from enormous interspecific differences, there are also considerable variations of fecundity within a species. Many authors have found that fecundity increases with length, weight, or age the relationship usually being of the form $F = aL^b$, where F is fecundity, L is length, and *a* and *b* are constants. Year-to-Year differences resulting almost certainly from environmental effects are also well established.

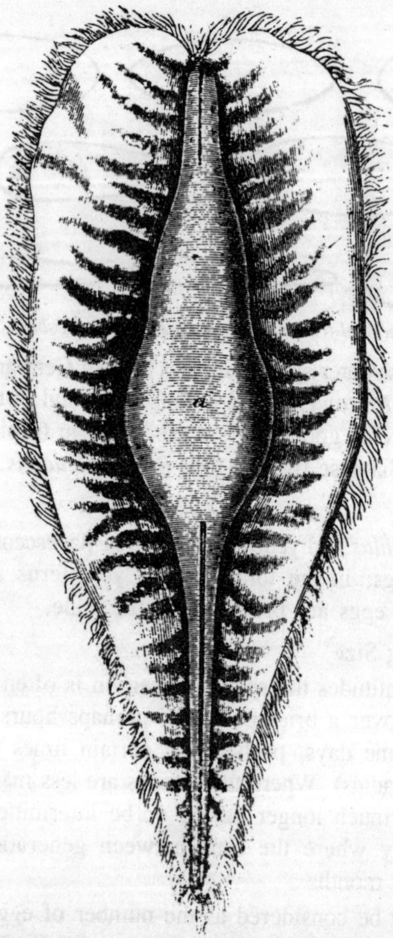

Fig. 11.2. Egg of Callorhynchus antarcticus. a, Cavity for the embryo.

For instance, sea temperature was correlated by *Rounsefell* (1957) with the fecundity of pink salmon, *Oncorhynchus gorbuscha*, higher temperatures apparently resulting in lower fecundities. (Here the effect of temperature on growth is a complication). *Bagenal* (1966) reported density-dependent factors operating in Scottish flatfish, high densities being correlated with low fecundity, *Anokhina* (1960) found that fecundity in Baltic herring could be related to feeding conditions, high fat content of the female being related to high fecundity.

Experimental studies *D.P. Scott* (1961) indicated in rainbows, trout, *Salmo gairdneri*, that an insufficient diet caused a reduction in egg

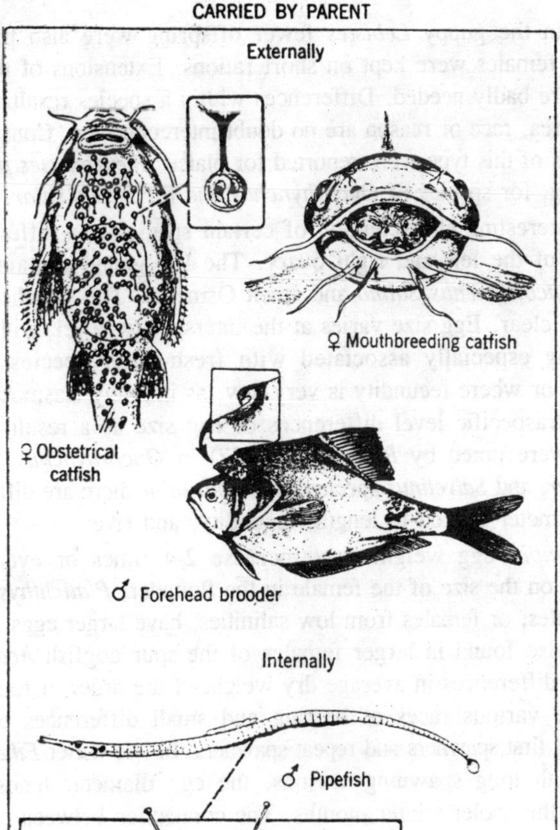

Fig. 11.3. Egg fate carried by parent.

number, in the guppy *Lebistes* fewer offspring were also produced when the females were kept on short rations. Extensions of this type of work are badly needed. Differences within a species resulting from latitude, area, race or reason are no doubt interconnected. Considerable differences of this type were reported for plaice, *Pleuronectes platessa*, for herring, for species of *Oncorhynchus* and for *Salmo salar*.

An interesting characteristic of certain species is a difference in fecundity of the left and right ovary. The left gonad contains more eggs in *Oncorhynchus Salmo* and smelt Osmerus. The significance of this is not clear. Egg size varies at the interspecific level, with larger eggs being especially associated with fresh-water species like to salmonids or where fecundity is very low, as in many elasmobranchs. As the intraspecific level differences of egg size as a result of area or river were noted by *Rounsefell* (1957) in *Oncorhynchus*, *Salmo*, *Cristivomer*, and *Salvelinus* species. In *Salmo salar* there are differences in egg diameter related to length, fecundity, and river.

In *Tilapia* egg weight may increase 2-4 times or even more depending on the size of the female in the flounder, *Platichthys flesus*, large females, or females from low salinities, have larger eggs. Larger eggs are also found in larger females of the spur dogfish *Acanthias*. There are differences in average dry weight of the order of four times among the various races of herring and small differences between very young first spawners and repeat spawners. In two darter *Etheostoma* species with long spawning seasons, the egg diameter tends to be greater in the cooler winter months. The connection between egg size and fecundity in related species may be correlated with the conditions for incubation. For example, *Tilapia tholloni*, a substrate brooder, has 500-3000 eggs depending on length. On the other hand, *T. mossambica* and *T. macrocephala*, which are mouth brooders, have less than 500 eggs that are considerably greater in weight (*Peters*, 1963). *Garnaud* reported two species of *Apogon*, one, *A imberbis*, with an egg diameter of 0.5 mm and fecundity of 22,000, and the other. *A conspersus*, with an egg diameter of 4.5 mm and fecundity of 150. The relationship between the egg size and egg weight within a species is shown for the different races of herring.

The winter-spring spawners have a low fecundity and large egg, an adaptation to poor food for the young, but a low predator population. In summer-autumn conditions fecundity is high and egg size law, presumably an adaptation to good food supplies and many predators. Intraspecific differences in fecundity and egg size deserve further study in terms of a link between ecological conditions and the physiology of maturation of the ovary. The chorion also hardens thus protecting the embryo in the early, vulnerable stages. Probably the inner layer of glycoprotein is mainly responsible for this and it has been suggested that the alveolar colloid, Ca ions, phospholipids, or hardening enzymes also play a part. In salmonids, *Zotin* (1958) reported hardening of the chorion because of an enzyme in the perivitelline fluid; Ca ions affect the enzyme rather than the chorion itself. *Ohtsuka* (1960) considered that a phospholipid was liberated from the cortex (not from the alveoli) in *Oryzias* eggs.

In *Fundulus* the chorion hardened with oxidizing agents but not with reducing agents. The soft chorion appeared to be impregnated with protein containing SH groups. Hardening resulted from oxidation of SH to SS groups by means of aldehydes produced from polysaccharides with x-glycol groups. *Zotin* distinguished between the initial enzyme action and subsequent hardening processes which last much longer and where the enzyme is no longer functionings. Thus, the initial enzyme reaction is blocked when Ca ions are bound by citrate or oxalate or by the use of NaCl or other chlorides. Later hardening is not susceptible to many of these factors.

The eggs of *Oryzias Gasterosteus*, and *Lampetra* can be activated by pricking. Other activating agents are surface-active chemicals and lipid solvents (perhaps emulsifying the cytoplasm at the animal pole) and thermal shock, electric fields, ultraviolet light, and high frequency vibrations. This type of artificial parthenogenesis usually leads to irregular cleavage, but stringent precautions are needed to prevent contamination with sperm in such experiments. Of interest is the ability of eggs and sperm to retain their fertilizability after leaving the parent.

According to *Yamamoto* (1961) fish eggs lose this capacity after a very short time, but this can be increased if they are retained in isotonic Ringer's solution. While this is true of some freshwater eggs, presumably as a result of water activation, in seawater the capacity for fertilization is retained for much longer—certainly for hours in the herring. *Nikolsky* (1963) states that sperm motility is short lived where spawning takes place in fast-flowing water, for example, 10-15 sec in *Oncorhynchus*. In slower flows, sturgeon sperm is motile for 230-290 sec, and in sea herring sperm may be motile for hours or days.

Observations of short-lived activity are difficult to make; thus, some of these figures must be considered approximate. Storage of gametes is useful technique in fish farming to allow controlled fertilizations in the laboratory and to obviate the need for transporting eggs in the susceptible pregastrulation stages. Salmonid gametes are best stored dry below 5°C; those of the herring may be held in buffered egg-yolk diluents, but are also best kept dry at about 4°C. While this permits storage for, at most, a few days long-term techniques are also possible. Herring sperm, but not eggs, were kept for some months in a diluent consisting of 12.5% glycerol in 3% salt solution (diluted seawater) at—79°C and crosses made successfully between spring and autumn spawning races. *Sned* and *Clemens* (1956) also succeeded in holding out-of-season carp sperm immotile for 30 days at 3°-5°C in frog Ringer. Carp sperm could also be frozen and stored at—73°C in isotonic Ringer containing 6-12% glycerol with some survival when thawed after 60 hss of storage.

More recently, *Truscott* et al. (1968) have shown salmon that sperm can be stored for 1-2 months at temperatures of—3°C to —4.5°C using diluents such as 5% ethylene glycol or 5% dimethyl sulfoxide, retaining 70-80% fertility. *Horton* et al. (1967) obtained alevins from salmon eggs fertilized with sperm frozen in liquid nitrogen with dimethyl sulfoxide as a protecting agent, but the fertility rate was low. *Hoyle* and *Idler* (1968) have also obtained fertile salmon sperm after storage in liquid nitrogen using ethylene glycol with added lactose or serine. Slow freezing produced the best results. *Mounib* et al. (1968) succeeded in storing cod sperm for up to 60 days using 17-24% glycerol and temperatures of —79° and —196°C. Initial experiments suggest that faster cooling rates gave the better results with cod sperm.

Incubation (Fertilization of Hatching)

The progress of cleavage, formation of layers, and morphogenesis have been described in a number of standard textbooks such as *Rudmick* (1955), *Waddington* (1956) and *Smith* (1957), with *Oppenheimer* (1947), and *Devillers* (1961) stressing structural changes from the viewpoint of experimental embryology. More detailed information is limited mainly to fresh-water species like the trout *Salmo trutta*, killifish *Fundulus*, medaka *Oryzias*, goldfish *Carassius*, and to the dogfish. Most fish eggs are round in the anchovy *Engraulis* and bitterling *Rhodeus* they are avoid, and in certain gobies pear-shaped.

Most species have telolecithal eggs with yolk more concentrated at the vegetative pole; some marine species have oil globules of varying size and number. Before fertilization the cytoplasm may be mixed with or separate from the yolk. The extent to which polarity exists at

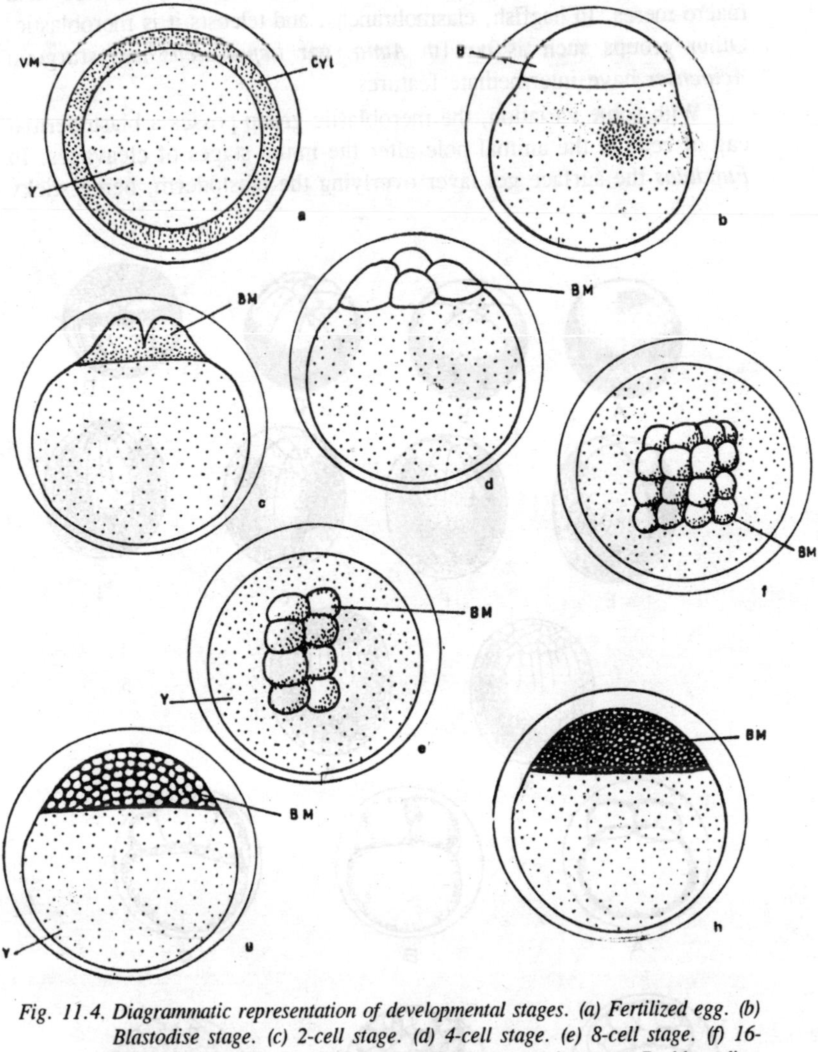

Fig. 11.4. Diagrammatic representation of developmental stages. (a) Fertilized egg. (b) Blastodise stage. (c) 2-cell stage. (d) 4-cell stage. (e) 8-cell stage. (f) 16-cell stage. (g) Early morula stage. (h) Late morula stage. B—blastodise; BM—blastomere; CYL—cytoplasmic layer; VM—vitelline membrane; Y—yolk.

this stage has not been described in many species. After fertilization (as the cortical alveoli release colloid, the perivitelline space develops and the chorion hardens) cytoplasm migrates to the future blastodermal region, most of it arriving by the first cleavage. The remaining cytoplasm forms a "halo" or periblast under the blastoderm. In lampreys cleavage is holoblastic but with the formation of micro- and macro-meres. In hagfish, elasmobranchs, and teleosts it is meroblastic. Other groups such as bowfin *Amia*, gar *Lepidosteus* and sturgeon *Acipenser* have intermediate features.

With some variation, the meroblastic group posses a blastodermal cap of cells at the animal pole after the initial stages of cleavages. In *Fundulus* the surface gel layer overlying the blastoderm, being sticky

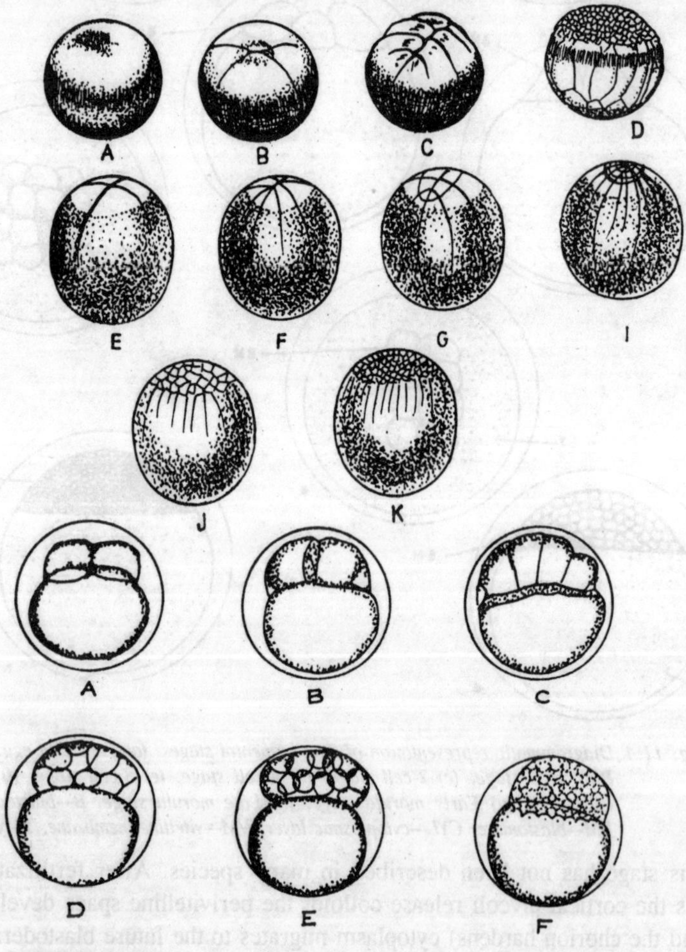

Fig. 11.5. Superficial view of the early developmental stages. (A-K) Acipencer. (A-F) Crenilabrus pavo.

on its inner surface, serves to hold the outer blastomeres together (*Triakaus*, 1951). Usually cleavage is not complete and in the deeper layers the periblast becomes syncytial and is involved in mobilizing the yolk reserves. There are substantial cohesive forces between the developing blastomeres and the surrounding periblast which are important in the subsequent morphogenetic movements. The blastoderm now commences to this and overgrow the yolk (epiboly) and at the same time invaginate at its periphery. The periblast seems closely connected with this spreading tendency of the blastoderm, which at the junction may be thickened to form a germ ring.

The syncytial periblast seems to have the property of autonomous spreading, and it is likely that cell proliferation in the blastoderm is relatively unimportant. *Devillers* (1961) suggests that the periblast acts as an intermediary between two "non-wettable" components—the blastoderm and yolk. As epiboly proceeds the blastopore contracts as does the surface gel layer over the yolk. In *Fundulus* this layer probably solates and passes inward. The form of the developing germ then seems to be controlled by a balance of the forces of adhesion of the deeper blastomeres with each other and the syncytium, of tension in the surface (yolk) gel layer and of contractility of the periblast.

The embryonic axis is laid down by a process of convergence and concentration in relation to the dorsal lip of the blastopore, the quantity of yolk present having some influence on the time at which this event occurs. Presumptive areas have been mapped in the early gastrulae of some species and show considerable variation. In earlier stags the eggs seem to be of the regulatory type. In *Fundulus* the 2-cell and 4-cell stage can withstand a loss of half the number of blastomeres.

The embryos of *Carassius* can be divided at the 8-cell stage, each part sometimes giving a normal embryo. Up to the 16-cell stage two embryo can be fused resulting either in twinning or an oversized single embryo. Removal of the yolk from the blastoderm before a critical stage is reached (8 cell stage in *Carassius*, 32 cell stage in *Fundulus*, blastula in *Salmo*) brings development to a halt. Incomplete removal of the yolk before the critical stage may, however, not prevent further development. It is likely that organizer substances rather than nutrient material diffusing from the yolk are more important. At later stages even the embryonic shield of *Fundulus* may be isolated and cultured to a fairly advanced stage, with the development of primitive axial organs, ears and eyes, and even with cardiac contractions and independent movement (*Oppenheimer*, 1964).

Hatching

The time of hatching is both a specifically and environmentally controlled character with temperature and oxygen supply exerting a considerable effect. Hatching results from a softening of the chorion because of enzymic or other chemical substances which are secreted from ectodermal glands usually on the anterior surface or from endodermal glands in the pharynx. In the sturgeon *Acipenser* the latter are innervated by the palatine nerve. The activity of the larvae, which may be enhanced by increase of temperature of light intensity or by reduction of oxygen tension, assists in breaking through the chorion. The biochemical aspects of hatching are dealt with by *Hayes* (1949),

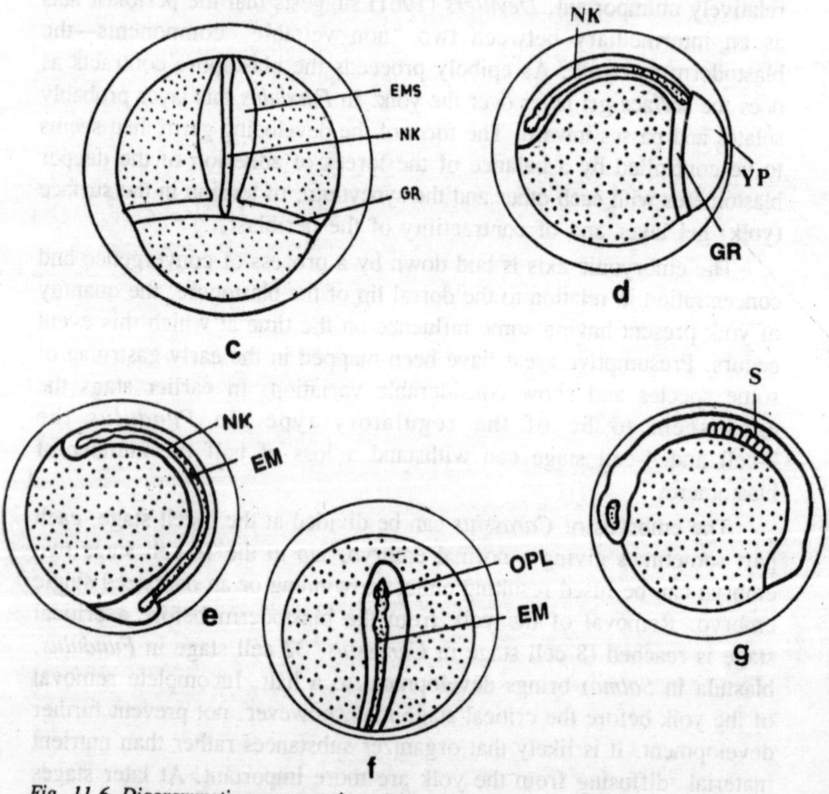

Fig. 11.6. Diagrammatic representation of developmental stage. (a) Three quarter of the egg surface covered. (b) Showing small yolk plug. (c) Showing closure of blastopore, side view. (d) Same, dorsal view. (e) Formation of somites. EM—embryo; EMS—embryonic shield; GR—germ ring; NK—neural keel; OPL—optic lobe; OPV—optic vesicle; S—somite; T—tail; Y—Yolk; YP—yolk plug.

Smith (1957), and *Deuchar* (1965). The enzymes have apparently been identified in a number of species, but certainly in salmonids Hayes' work shows there is doubt about their mode of action.

The chorion, which resists digestion by trypsin and pepsin, appears to be of "pseudo-keratin". *Kaighn* (1964) measured the amino acid and carbohydrate components in *Fundulus*. Cystine comprised only 1%, compared with 12% in keratin. It is therefore unlikely that disulfide links play a role in stabilizing chorionic protein as they probably do keratin. The hatching enzyme works best under alkaline conditions. pH 7.2.-9.6 and temperatures of 14°-20°C having been reported as optima. Very little hydrolysis takes place and Hayes speculates that the enzyme may be a reducing agent which liquefies the chorion. In *Oryzias* the enzyme is probably a *tryptase*. In *Fundulus*, Kaighn (1964) obtained purified chorionase and concluded that digestion of the chorion was mainly a proteolytic process, although he could not duplicate its action with other proteases. Whatever, the mode of action of the enzyme it is likely that a considerable part of the nutrient material in the chorion can be utilized by the embryo via the perivitelline fluid and the losses at hatching may not be too serious.

Larva

At hatching the larva is usually transparent with some pigment spots of unknown function. Notochord and myotomes are clear with usually little development of cartilage or ossification in the skeleton. A full complement of fins is rarely present, but a primordial fin fold is well-developed in the sagittal plane. The mouth and jaws may not yet have appeared, and the gut is a straight tube. Although the heart functions for a considerable period before hatching the blood is colorless in the majority of species and the circulation and respiratory systems poorly developed. The yolk sac is relatively enormous with, presumably, hydrodynamic disadvantages.

Pigmentation of the eyes is very variable, but where the eye is not functioning at hatching it very soon develops. The kidney is usually pronephric with very few glomeruli. Very little is known about the endocrine glands, gonads, and other organs of the body cavity as such an early stage. At the yolk is resorbed, the mouth begins to function, the gut and the eyes develop further, and the larva becomes fitted for transfer to sources of external food. One of the earlier systems to develop is that responsible for locomotion and support, the primordial fin being fairly soon replaced by medium fins and the skeleton laid down. This is one of the better known aspects of later development

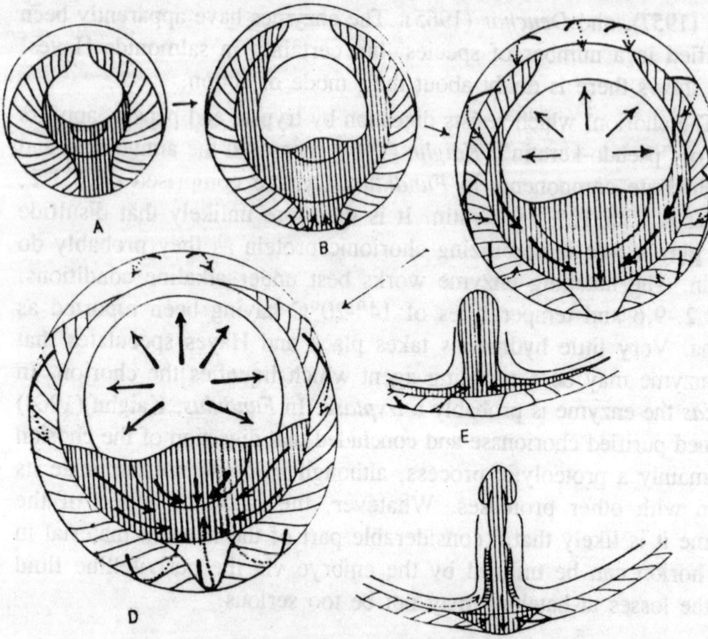

Fig. 11.7. Schematic representation of presumtive territories.

because it is a system less easily damaged in such delicate organisms and because of the importance of meristic characters in racial studies of fish. Branchial replaces cutaneous respiration as the gill arches and filaments appear. The swim bladder may or may not be present during the larval phase. It is possible that this and the eyes, which are potentially dangerous in making the transparent larva visible, and silvered in such a way as to render them inconspicuous.

Metamorphosis

A clear change or metamorphosis from the larval to adult form is to be found in many species. In others there may be a number of less marked metamorphoses, e.g., in salmonids and eels. The most obvious signs are the laying down of scales and other pigmentation and often the first appearance of haemoglobin in the circulation. The swim bladder and lateral line may also develop first at this stage. In flatfish there is rotation of the optic region of the skull and the changes in the normal orientation of the body so that they eventually come to lie on one side. There are often concomitant changes in distribution and behaviour such as schooling. Barrington (1961) gives detailed consideration to the physiological changes associated with metamorphosis

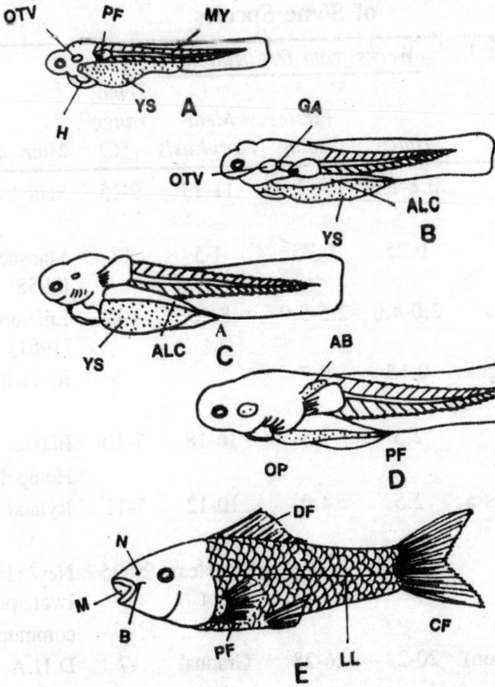

Fig. 11.8. Diagrammatic representation of developmental stages. (A) Freshly hatch larva. (B) One day old larva. (C) Two day old larva. (D) Four day old larva. (E) One month old larva. AB—air bladder; ALC—alimentary canal; CF—caudal fin; GA—gill arch; H—heart; LL—lateral line; M—mouth; My—myotome; N—nostril; OTV—otic vesicle; PF—pectoral fin, YS—yolk sac.

in salmonids, eels, and the lamprey, especially from the aspect of thyroid activity and osmoregulatory functions. The time to reach metamorphosis may be a matter of days in tropical species, a few weeks or months in the majority of fish from temperate latitudes, or periods of years in the sturgeon *Acipenser* and eel *Anguilla*. It is controlled not only genetically but also by temperature and food supply, which may affect the rate of growth, and possibly by social (hierarchical) factors as well.

Timing

To give some idea of the timing of the events just described, examples of the approximate duration of different stages under natural conditions. The modification of these times experimentally or by fluctuations in environmental conditions is discussed in the succeeding pages.

Table 11.1. Duration of Events in the Development of Some Species

Species	*Weeks from fertilization to* Hatch	First feeding	Meta-morphosis	Temp. range (°C)	Main author
Sconer scombrus (mickerel)	0.8-1.5	1.3-2.0	11-13	9-15	Sette (1943)
Roccus saxatilis (straped bass)	0.25	0.75	4-5	17	Mansueti (1958)
Osmeral eperlanus (smelt)	2.0-4.0	2.5-5.0	8-10	4-14	Lillelund (1961)
Acanturus triostegus (convict surgeon fish)	0.15	0.7	?	26	Randall (1961)
Cluped harengus (Clyde herring)	2.5	3.5	16-18	7-10	Blaxter and Hempel (1963)
Pleuronectes platessa (plaite)	2.5	4.0	10-12	7-11	Ryland (1966)
Oryzias latipes (madaka)	1.5-2.0	2.0-2.4	Not clear cut	20-25	New (1966); T. Iwai, personal communication
Salmo salar (salmon)	20-22	26-28	Gradual	-7	D.H.A. Marr, personal communication
Squalus santhias (spur dogfish)	ca.104	ca.104	< 104	4-12	Templeman (1944)
Scyliorhazus caniculus (spotted dogfish)	24-32	28-36	28-36	4-12	Amorosa (1961)

Metabolism and Growth

Rate of Development

Obvious specific differences in time to hatching may be masked by variations in ambient temperature, which is one of the most potent influences on rate of development. Detailed observations of temperature effects during development are scarce. *Fluchter* and *Rosenthal* (1965), however, showed between 3.5° and 9°C a doubling of heart rate in the embryos of the blue whiting, *Micromesistius poutassou*, and more rapid embryonic movements. Low temperatures retard hatching) and at a theoretical low temperature (the biological zero), the incubation period will be infinite. The product of incubation time (D) and temperature (T)—day-degrees—was originally thought to be constant i.e.

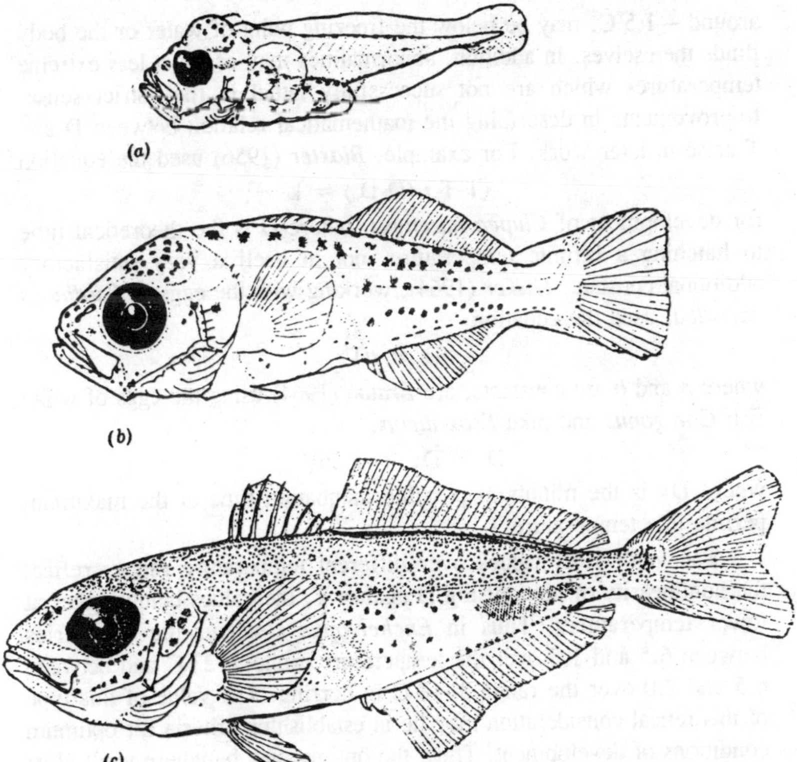

Fig. 11.9. *Post-hatching development of the bony fish, the jack mackerel (Trachurus symmetricus). (a) Larva of 4.9 mm. lacking pelvic fin and development of rays in dorsal and anal fins. (b) Larva of 10 mm. with pelvic fin just appearing but with rays established in other fins. (c) Juvenile of 28 mm. showing establishment of adult body from and further development of pigmentation over (a) and (b).*

$$TD = k$$

This was modified to use the temperature, not from zero, but from the biological zero (T_o), i.e.,

$$(T-T_o)D = k$$

There has been increasing criticism of the concept, for example, by *Kinne* and (1962) and *Garside* (1966), on various grounds. Plots of $1/D$ against T are *curvilinear* over wide ranges of temperature, simple linearity only applying over a narrow range. This invalidates the formulas given above. Furthermore, there may be infections even of the curvilinear relationship at extreme temperatures. The biological zero, which is usually given between 0° and —2°C and most often around —1.5°C, may be below the freezing point of water or the body fluids themselves. In addition, abnormalities may occur at less extreme temperatures which are not successfully lethal in the strict sense. Improvements in describing the mathematical relation between D and T arise in later work. For example, *Blaxter* (1956) used the equation

$$(T-T_o)\ (D-D_o) = k$$

for development of *Clupea harengus*, where *D* is the theoretical time to hatching a infinite temperature, not in itself a very satisfactory additional constant. *Lasker* (1964), working with the eggs of *Sardinops caerulea*, used the equation

$$D = aT^b$$

where *a* and *b* are constants, are *Braum* (1964) using the eggs of white fish *Coregonus* and pike *Esox lucius*,

$$D = Dx + 1.26^{Tx - T}$$

where Dx is the minimum possible incubation time at the maximum permissible temperature T.

The van't Hoff values over different temperature ranges reflect the nonlinearity of plots of log D against T, the values being higher at lower temperatures. Thus in *Enchelyopus cimbrius* the Q_{10} varies between 6.5 and 1.5 over the temperature range 5°-23°C and between 6.5 and 2.0 over the range 3°-18°C in herring. The value of this type of theoretical consideration may lie in establishing criteria for optimum conditions of development. Thus, the optima may be where van't Hoff values lie between certain limits. Certainly the day-degree concept is useful as an approximation for predicting events in normal hatchery practice. Other environmental factors influence the rate of development. Low salinities may accelerate or retard the time to hatching while oxygen lack has a retarding effect on development, especially at higher temperatures. *Laale* and *McCallion* (1968), found that the development of the zebra fish, *Brachydanio rerio*, could be arrested before gastrulation by the use of the supernatant of homogenates produced from other zebra fish embryos. This arrest, which could be reversed appeared to be an effect at the cellular level, the nuclei of the arrested embryos all being in interphase.

Yolk Utilization

The efficiency with which yolk is transformed to body tissue and the effect of the environment on utilization is important in that larger larvae may be expected to be stronger, better swimmers, less susceptible to damage, and less liable to predation. Efficiency at any time may be expressed as a percentage:

$$\frac{\text{dry weight increment of body}}{\text{dry weight decrement of yolk}} \times 100$$

More often of efficiency is measured from fertilization to final yolk resorption (or to maximum weight attained on the yolk reserves). This is goes efficiency, i.e.,

$$\frac{\text{dry weight of final body}}{\text{dry weight of original yolk}} \times 100$$

or from fertilization to intermediate stages as

$$\frac{\text{dry weight of body}}{\text{dry weight of original yolk - dry weight of remaining yolk}} \times 100$$

or more precisely

$$\frac{\text{dry weight of body}}{\text{dry weight of body +dry weight of yolk used for maintenance}} \times 100$$

The difficulties of measuring efficiency by dry weight lie in the need for taking samples of an egg population at different stages with the accompanying problems of initial differences in egg weight. Utilization of material from the chorion or losses of excretory products are also difficult to allow for, as are the possibilities of uptake of organic matter from the environment. Another serious problem when comparing, for example, environmental effects such as temperature on efficiency is the question of making dry weight measurements at "equivalent" stages. Both hatching and maximum weight (attained on the yolk) can be questioned for staging, hatching at different temperatures can result in larvae of quite different appearance, while full yolk utilization is often not complete when maximum weight is reached, some yolk remaining in the yolk sac.

Furthermore, the disappearance of the yolk sac is no certain indication that all the yolk has been used as it may be present in storage spaces within the larval body, for example in the subdermal spaces of cod and plaice larvae. *D.H.A. Marr* (1966) adopted the ratio as a criterion for *equivalent staging*, comparing in *S. salar* the efficiency

$$\frac{\text{dry weight of body}}{\text{dry weight of body +remaining yolk}} \times 100 \text{ (as percentage)}$$

of development at different temperatures between the 15% and 80% stages, *Ryland* and *Nichols* (1967) used the ratio

$$\frac{\text{rate of growth in length}}{\text{rate of yolk disappearance}} \times 100$$

for equivalent stages when comparing the efficiency of development at different temperatures during the yolk sac stages of *P. platessa*. The use of maximum weight as a stage for making comparisons still remains, however, a useful criterion and one with immediate meaning when deciding on the optimum conditions for hatchery practice. Efficiency over the whole process of yolk utilization is mainly between 40% and 70% although clearly cumulative efficiency must decrease as growth proceeds and the maintenance requirements increase.

Experiments with temperature indicate certain optima for maximum efficiency. Other influences on efficiency are the original egg weight at the intraspecific level in *C. herengus*, and light conditions, contour of the substrate, and turnover of water in the photonegative alevin of *S. salar* living with in the interstices of the spawning redd. Here highest efficiency is achieved under dark conditions, on a grooved substrate, with a rapid turnover of water. A word of caution is required where larval feeding may occur well before final yolk resorption. High efficiency may result from low activity, a high proportion of yolk being used for growth; if this is reflected in low feeding activity it could be a very undesirably trait.

Viviparity

Amoroso (1960), who gives a comprehensive review of this subject, out the rather indistinct barrier between ovoviviparity where the young develop, within the female only on their yolk reserves, and viviparity, where the nutrient requirements are obtained from the mother. In the first place, there are a wide variety of methods of obtaining these nutrients, by absorption through simple external surfaces, by swallowing, or by "placental" connections. Perhaps all these may be considered as viviparous. In the second place initial development may be ovoviviparous (on the yolk supply) with viviparity superimposed later. This is likely to be the case in the very early stages in all viviparous fish, but in the smooth dogfish, *Mustelus laevis*, and goodeid teleosts, for example, there is a change from one to the other rather later. The changes of weight found during the development of various species can be very striking.

Some of the ovoviviparous ones, where maternal nutrients are scarce, show decreases of organic matter between the fertilized egg and final embryo, for example, of 23-34% of *Torpedo* spp. This gives an efficiency of 66-74% which puts this species very much in the same category as oviparous forms although it is rarely certain what nutrient is obtained from the mother. The long gestation period of 4-6 months with this order of efficiency suggests some nutrients are being absorbed from the oviduct. In other "ovoviviparous" species there may be gains in organic matter, e.g., of 356% of *Mustelus vulgaris* and 1628% in *Trygon*. In *Mustelus laevis*, sometimes more strictly called viviparous, the gain is 1064%, although *Te Winkel* (1963) reported a fall in weight of organic matter early in the development of a close relative, *M. canis*.

Biochemical Aspects

Much work has been done in the past on the larger salmonid eggs which can provide a greater bulk for analysis or on whole ovaries. Some more recent work which include the use of isotopes, chromatography, and histochemistry is mentioned by *Deuchar* (1965) and *Williams* (1967).

Water Relations

The swelling of eggs, with the formation of the perivitelline space as a result of water uptake, is a general phenomenon signifying fertilization. The chorion at fertilization is permeable to water and also to urea, glucose, salts and certain dyes. It seems like that the colloidal material liberated from the cortical alveoli cannot escape through the chorion and creates an osmotic pressure which draws in water. This effect can be inhibited by high osmotic pressure in the outside medium. The chorion then hardens, a process taking a matter of a few hours and the osmotic forces become matched by the resistance of the chorion. In hypertonic solutions the eggs of salmonids and of *Mullus barbatus* lose water only from the perivitelline space and not from the embryo. Some loss does, however, occur in acipenserids. The use of D_2O on water-activated and early fertilized eggs of *Oncorhynchus tshawytscha* seems to confirm the view that only the perivitelline space is penetrable by water. Subsequent use of ^3H_2O, $^{22}NaCl$, and Na ^{31}I on the water-activated eggs of *Salmo gairdneri* has shown a definite but limited permeability of the vitelline membrane to anions, cations and water. Unfortunately, this was not done on fertilized eggs.

Recently, however, *Potts* and *Rudy* (1969) have confirmed with ^3H_2O that the vitelline membrane of fertilized eggs of *S. salar* has a high permeability before laying and during water hardening. Subsequently permeability is low until the eyed stage. The use of ^{24}Na showed that sodium exchange is confined initially to the perivitelline fluid but accumulation within the embryo occurs during the eyed stage.

Terner (1968) reported that during the eyed stage *Salmo gairdneri* external substrates such as ^{14}C-labeled pyruvate and acetate were apparently taken up and metabolized as judged by the presence of $^{14}CO_2$ in the respiratory CO_2. *Wedemeyer* (1968) also found that ^{65}Zn was taken up by developing eggs of coho salmon *O. kisutch*. Almost all was bound to the chorion, but 26% was found in the perivitelline fluid, 2% in the yolk, and 4% in the embryo, *Mounib* and *Eisan* (1969) found that both ^{14}C-labeled pyruvate and glyoxylate could be utilized in the form of an exogenous substrate by salmon (*S. salar*) eggs. Lactate was produced and any carbon atom in these compounds could be incorporated by the eggs into organic acids, lipids, nucleic acids or proteins.

The formation of ^{14}C- amino acids indicated the presence of an active transminase system. The use of freezing measurements on the yolk of herring and plaice eggs gives and alternative picture. In the herring which has a demersal egg, the yolk is not regulated osmotically until after gastrulation when it is covered by cells. However, in the pelagic floating egg of the plate, regulation occurs from fertilization, indicating the ability of the vitelline membrane to osmoregulate. This seems necessary from the buoyancy aspect. If Gray's (1926) measurements of wet and dry weights of the embryo and yolk of trout are generally true of salmonids, then the yolk as a high density nutrient (41% dry weight) requires considerable quantities of water for transformation to the relatively watery embryo (16% dry weight). Obtaining this water may be a problem if the permeability of the vitelline membrane remains limited throughout much of development. *Smith* (1957) suggests that growth may be retarded until hatching occurs and water becomes more readily available. It is likely that highly desiccated yolk is only feasible in the demersal egg; in *Sardinops caerulae* the water content of the larval yolk is about 91% (Lasker, 1962).

Chemical Composition

The chemical composition of the eggs of two species are given in Table 12.2. Further data are given by *Phillips* and *Dumas* (1959), who showed in particular that there was no difference in the constituents of different sized eggs of *Salmo trutta*. There is considerable difficulty in deciding on the sequence of utilization of various materials for energy production. Analysis of the change in chemical components may be unreliable where substances like carbo-hydrates are being synthesized.

Table 11.2. Analyses of Eggs

Species	*Dry* weight %	*Percent of wet weight* *Protein*	Total *fat*	*oil*	Phos- *pholipid*	Carbo- *hydrate*	*Ash*
S. gairdneri (irideus) (rainbow trout)	33.8	20.2	—	3.6	3.8	0.2	1.3
Sardinops caerulea (California sardine)	29.3	21.0	3.8	—	3.2	<0.3	2.1

Heat production and respiratory quotients are difficult to determine accurately in small eggs, and CO_2 liberation may be masked by a buffered external medium. Amberson and *Armstrong* (1963) found the RQ's of *Fundulus* eggs over the first 6 days of development were 0.90, 0.78, 0.77, 0.76, 0.72 and 0.72 suggesting very early carbohydrate metabolism. In *Oryzias*, *Hishida* and *Nakano* (1954) found 0.75 after fertilization, 0.92 at gastrulation, and 0.70 later. Another problem to overcome is the retention of nitrogenous excretory matter within the egg which gives values of protein metabolism which are too low. The original concept of *Needham* (1931) that materials were used up for energy in the sequence

carbohydrate—protein—fat

has to some extent been supported by *Devillers* (1965). However, *Hayes* (1949) was of the opinion that the sequence in salmon eggs was

hatch
↓
fat—protein—fat—protein

and Smith (1957, 1958) that it was

hatch
phospholipid—protein + carbohydrate—protein—phospholipid—
triglyceride fat

Further work is required in this field, especially of a comparative nature.

Carbohydrate Metabolism

Since carbohydrate is present in small quantities and synthesis is continuously occurring, its role is particularly difficult to ascertain by bulk analysis. *Hayes* (1949) reported a gradual increase in carbohydrate level during salmon development with a temporary fall in glucose level at hatching. Glycogen is probably synthesized near hatching and is present exclusively in the embryo, being stored in the liver near the end of the yolk sac stage. Glucose, in aqueous phase, is present in relatively greater quantities in the embryo than in the yolk due to its higher water content. *Smith* (1957) described falls in total carbohydrate after gastrulation (linked with the establishment of the circulatory system and therefore a higher potential for metabolism) and a further fall in glucose level at hatching, perhaps because of an interruption of glycogen synthesis. He did not report the general increase found by Hayes.

Terner (1968) has more recently found that the ova of *Salmo gairdneri* have a free store of glucose which increases during development and only falls at hatching. Carbohydrate is certainly being synthesized in development for use in the final stage of energy production and for mechanical and osmoregulatory work.

Protein

Protein is the dominant raw material in the yolk and the main source for tissue formation. The use of labeled amino acids in *Oryzias* shows a slow passage from the yolk to a low molecular weight pool, a more rapid uptake occurring when the vitelline circulation is established. The proportion of yolk used for energy requirements is not easy to determine since the egg retains some nitrogenous metabolities (shown by an increase of non-protein nitrogen with age) and only small amounts of ammonia are excreted. Of particular interest is the finding by *Read* (1948) of ornithine carbomoyltransferase and arginase in the embryos of *Squalus suckelyi* and *Raja binoculata*, suggesting an early functioning ornithine urea cycle with the retention of urea for osmoregulatory purposes.

The use of protein for energy seems to decline after hatching, while the amount used over the whole of early development depends on whether calculations are made when some yolk remains or whether, at a later stage, the larva starts to consume its own tissues. While it has been suggested that 40% of the original protein may be metabolized more recent detailed studies using isotopes on Oryzias indicate that all protein being resorbed from the yolk in the 9 days subsequent to gastrulation was being transformed into new tissue and none lost in combustion. Regarding different types of protein it appears that in *Oryzias* 40% of phosphorus is incorporated as phospho-protein: it falls after gastrulation and the loss is very rapid at hatching, especially in *S. gairdneri* (*irideus*). *Deuchar* (1965), reviewing analyses of amino acids present in the same species, reports that aspartic and glutamic acid are always present, valine and leucine appearing after gastrulation.

Fat

There is some ambiguity about the role of fat in energy production during different phases of development. Fat is undoubtedly used as a fuel, possibly 70-90% being consumed over the whole period of development. *Hayes* (1949) reported the main loss of fat to be in the fourth week after hatching in the salmon. It seems that the triglyceride fats are the last to be utilized-before food is required from external sources. The energy requirements before hatching may also be met by fat, according to Hayes, or by fats in combination with protein. According to *Phillips* and *Dumas* (1959) fat is synthesized from protein in the eyed stage of the eggs of brown trout *Salmo trutta*. *Terner* et al. (1968) found that C-labeled acetate in the incubation medium of the eggs of *S. gairdneri*, *S. trutta*, and *Salvelinus fontinalis* were incorporated into free fatty acids and other lipids of the egg, which were probably the substrates for endogenous respiration. In addition, labeled acetate was found in the egg phospholipids, presumably as an intermediate stage in the synthesis of complex fats. These authors suggest that the lipids of the embryo are not transferred directly from the yolk but are resynthesized by the embryo after the breakdown of triglyceride fats in the lipid pool of the yolk.

Respiration

In the sense of obtaining oxygen, respiratory problems vary with the environment of the egg and larva, but they are probably rather limiting with pelagic eggs. Eggs deposited on or in a substrate may act as an oxygen "sink" the oxygen level at the egg surface always being less than that of the surroundings even in high velocities of current. Hayes *et al.* (1951) calculated a flow of 1.2×10 $\mu l/cm/cm^2$ thickness/min through the chorion of salmon eggs. Sensitivity to low oxygen tensions or anoxia varies with both species and stage of development. In salmonids, cleavage may continue in anoxic conditions, perhaps this is possible because requirements are low and oxygen is stored within the egg, especially in the perivitelline fluid. Gastrulation is more easily blocked by lack of oxygen or by respiratory poisons such as cyanide or azide.

Development can also be retarded by blocking oxidative phosphorylation, with dinitrophenol. During oxygen lack there are often increases in lactic acid production but, in the short-term, retardation of development to anoxia is reversible *De Ciechomski* (1965) succeeded in keeping the demersal eggs of *Austroatherina* for 17-18 days in "Vaseline oil". Development was at first normal, but no heart beat or movement was observed. Transfer to water resulted in the heart starting to beat and some movements occurred, but abnormal pigmentation developed, and no embryos hatched. Eggs of three pelagic species were killed by immersion in the oil. Various adaptations are found which assist in obtaining oxygen. The spawning redd of the salmonid fish has a rapid current of water passing through it; parents guarding their eggs may ventilate them (e.g. the stickleback, *Gasterosteus aculeatus*) while the lungfish *Lepidosiren* has special pelvic "gills" for oxygenating the eggs in its nest; viviparous species obtain oxygen as well as nutrients from the female. After hatching functional gill filaments are often lacking; in herring they do not appear for some weeks at a length of about 20 mm. Oxygen is obtained over the body surface and almost certainly via internal body surfaces such as the pharynx and gut.

The ultramicroscopic corrugations described by *Jones* et al. (1966) and Lasker and *Threadgold* (1968) on the epidermis might possibly be directing the flow of water over the body surface. In less favourable environments external gills may be found, for example, in the bichir *Polypterus*, in the lungfish *Protopterus* and *Lepidosiren*, in the loach *Misgurnus fossilis*, and in *Gymnarchus niloticus*. In newly hatched trout, *Salmo trutta*, respiratory currents are produced by the pectoral fins which also keep the body clear of silt. After a short time mouth and gill movements replace those of the fins and then clogging of the gills by silt is prevented by a coughing reflex or aggregating the silt with mucus. Intake of oxygen has been measured in a number of species. In the egg stage this may be very low and variable before fertilization. In *Oryzias latipes* there is no sharp rise until a few hours after fertilization, although it seems likely that the respiratory rate becomes steadier at this stage (*Nakano*, 1953). Before any movements occur within the egg, relative values of oxygen uptake between different stages and species are quite valid, but subsequent to movement being possible the oxygen consumed can be dominated by bursts of activity.

Oxygen uptake may be used to express metabolic functions, total or general metabolism consisting of two components, active metabolism (owing to activity) and basal metabolism (owing to maintenance functions). Standard metabolism applies to practically motionless fish under experimental conditions; in the fry of *Salmo salar*, for example, the basal metabolism was 64-69% the standard rate. For active fish metabolism is proportional to KV^n, where V is velocity and K and n are constants. The function KV^n is called the "scope or activity" and has been estimated for a number of species. Some values of oxygen uptake expressed as Qo_2 (ml/mg dry weight/hr) are given in Table 12.3 based where possible on subjective criteria of activity and inactivity. Without the type of precise control, these are the best available date at the present time. The standard Qo_2's show some similarity considering the variety of the material used. Active Qo_2's will naturally fluctuate widely, depending on the type and extent of activity permitted. The use of certain anesthetics to control activity may also help in establishing at least some of the factors controlling oxygen uptake.

Table 11.3. Scope for Activity

Species	*Stage*	*Range of* increase of Qo_2 *owing to activity*	*Author*
Salmo salar (salmon)	Eggs	3	Hayes *et al.* (1951)
Salmo salar (salmon)	Fry	7-14	Ivlev (1960a)
Sardinops caerlea (California sardine)	Larvae	3	Lasker and Theilacker (1962)
Clupea harengua (herring)	Larvae	9-10	Holliday *et al.* (1964)

Cumulative oxygen uptake has been measured over short periods as well as fairly long period of development. In *Fundulus heteroclitus*, the killifish, it was 80 μl/egg from fertilization to hatching (Scott and Kellicott, 1916). In salmon eggs (Hayes *et al.*, 1951) the uptake was about 0.2 ml/egg/hr at fertilization rising to 3.4 ml/egg/hr at hatching, the cumulative total between these stages amounting to 1400 ml/egg. Privolniev, quoted by Devillers (1965), found only 850 ml/egg from fertilization to an age of 120 days. Such values may be compared with the short yolk sac period of the California sardine, *Sardinops caerulea*, where only 12 ml/larva were used from hatching to 180 hr. subsequently Lasker and Theilaekar, (1962). Obviously cumulative uptake and uptake per unit time will vary with size, age, and temperature. *Devillers* (1965) considers that increases in oxygen uptake with age may be shown by the exponential equation

$$Q = ae^{kt}$$

where *t* is time and a and k are constants. Qo_2's will also vary with environmental conditions. In herring eggs and larvae the Q_{10} for oxygen uptake (Q_{02}) was 2.0 between 55° and 14°C. Q_{02} also increased with a salinity shock but not with constant rearing salinities. There were only slight effects because of light intensity and starvation. In salmonids oxygen uptake falls at low ambient oxygen tensions, but the young stages of fish will resist oxygen lack quite well for a time. This raises the problem of oxygen debt in such experiments and indeed in any respiration measurements where activity is intense. The accumulation of such a debt can be tested by continuing respiration measurements after activity or anoxia.

Hayes et al. (1951) found no oxygen debt after periods at reduced oxygen tensions in salmon eggs, suggesting there had been a reduction in metabolism under these conditions. Salmon fry however, an oxygen debt amounting to 48% of general metabolism after a period of activity. Fry of bleak, *Alburnus alburnus*, had a negligible debt when stemming a water velocity of 2 cm/sec. but one of about 15% in 6.42 cm/sec.

An overall review of respiration rates, including that of larval fish, allowed *Winberg* (1960) to deduce that the relationship

$$Q_s = 0.3 \quad \omega^{0.8}$$

represented the best fit for the data on standard metabolism in microliter per hour at 20°C for organisms of wet weight W mg. He provides conversion factors for other temperatures but stresses that such relationships are designed to evaluate experimental results rather than replace them. *Blaxter* (1966), recalculating the data of *Holliday* et al. (1964) for herring, concluded that in larvae of mean wet weight 1.53 mg respiration amounted to 17μlO_2/day at 8°C. Using Winberg's relationship a value of 23 μl O_2/day is arrived at, fair agreement considering his relationship is operating at its limit.

Deductions can be made about the type of metabolism involved in respiration. The consumption of 1000 1 O_2· corresponds to the utilization of different materials as follows:

$$1000 \text{ ml } O_2 = \begin{cases} 1.05 \text{ mg protein} \\ \mu 1.23 \text{ mg carbohydrate} \\ 0.50 \text{ mg fat} \end{cases}$$

1 mg of protein yields 4.25 cal, 1 mg of carbohydrate yields 4.15 cal, and 1 mg of fat yields 9.45 cal;

therefore,

$$1000 \text{ 1 } O_2 = \begin{cases} 4.5 \text{ cal from protein} \\ 5.1 \text{ cal from carbohydrate} \\ \underline{4.7 \text{ cal from fat}} \\ 4.77 \end{cases}$$

The value of 4.77 cal/1000 μl O_2 is sometimes called the oxycaloric coefficient. This makes it possible to calculate the requirements for growth and metabolism from measurements of growth rate and respiration. On the basis of 1000 μl O_2 being equivalent to 4.77 cal, and 1 mg wet weight of growth corresponding to 1 cal, *Ivlev* (1960b) concluded that young Baltic herring of wet weight 66.5 mg required 16.2 cal/day for growth and maintenance.

Table 11.4. Calorific Intake in Some Species

Species	*Wet* weight *(mg)*	*Daily* intake *(cal)*	Temp. *(°C)*	*Author*
Engraulis japonicus (anchovy)	1500-6300	78-426	14-20	Takahashi and Hatanaka (1960)
Alburnus alburnus (bleak)	250	38.3	19.22	Ivlev (1960b)
Clupea harengua (herring)	66.5	16.2	(?) 15-17	Ivlev (1960b)
C. harengua (herring)	1.53	0.18	8	Blaxter (1966)

Smith (1957, 1958) reviews the work on growth, heat production, and respiration. High respiration and high rates of growth appear to be correlated, at least at some phases of development. In particular, specific growth rate and heat produced show very similar trends at various stages of development of the rainbow trout, *Salmo gairdneri* (*irideus*). Up to hatching, plots of heat production on specific growth lie on a straight line with positive slope passing through the origin; after hatching the line passes through an intercept on the ordinate. There are two components in heat production—one resulting from growth and the other from maintenance. Temperature may affect the relative requirements of growth and maintenance thus giving differences in efficiency of development with temperature optima.

Growth

Growth is influenced in the early stages by the ratio between embryo and yolk weight, by the problems of efficient yolk utilization and the need to mobilize the nutrient reserves. Higher maintenance requirements per unit weight in small organisms, surface-volume ratios, change of diet with age, feeding and searching potential are also important. Growth has often been shown in terms of length but there is a need for uniformity, i.e., whether the caudal fin or fin fold is included and length gives no information on the condition (fatness) of the organism or of the changes in body proportions with age; this is especially true in the yolk sac stage and at metamorphosis.

Similarly, wet weights introduce the error of varying water content; thus, dry weight is the most satisfactory measure. Example of the relative quantities of yolk and embryo in a large and a small egg at different stages of development have already been given. The increase of weight of the embryo and decrease of the yolk tend to be logarithmic in nature. It is clear that in the earlier stages relatively less yolk is being used for maintenance than for growth, giving a higher efficiency of utilization. Without external feeding the larval weight begins to fall before all the yolk is resorbed. *Farris* (1959) divided the growth in length of four species of pelagic larvae after hatching into three phases: an early rapid phase after hatching, a slow phase near the completion of resorption, and a subsequent negative phase if no food was available. This shrinkage in terms of length was also described by *Lasker* (1964) in the larvae of *Sardinops caerulea*. Growth of salmonid and other fresh-water fry in hatcheries has received considerable attention from an economic point of view, to provide maximum growth at minimum cost.

Growth in marine conditions, where sampling of the same population is difficult owing to mortality, migration, or changes in net avoidance with age, is less adequately known. Some estimates of survival of fish larvae depend on a knowledge of growth rate and this is often inadequate or based on flimsy experimental data. Comparisons from tank experiments are difficult owing to mortalities, size-hierarchy effects, or other possible aquarium artifacts. Examples of size hierarchies in rearing experiments are given, showing the enormous range of size which may be found from an initially fairly uniform stock. Thus, *Shelbourne* (1964), after a series of rearing experiments on plaice, found the length ranged from 7.5 and 37.5 mm.

In dense populations there were also relatively larger numbers of small larvae. *Magnuson* (1962) tested the effect of food supply on *Oryzias* kept at densities. As long a food was adequate, growth was not retarded, social hierarchies did not develop did not develop and aggressive behaviour was not manifested. Territorial behaviour only occurred when food was limited spatially. Whether size hierarchies occur in natural conditions is not known. It seems they must stem from inherently different growth rates or from some types of dominance hierarchy—the first could occur in the wild, but the second is more likely to occur in crowded tank conditions.

Food supply and temperature are the most potent environmental factors in controlling growth in the later stages and there are few data available on this except for salmonids. In the egg stage low temperatures often produce longer larvae at hatching although *Lasker* (1964) found the maximum length of *Sardinops caerulea* larvae at intermediate temperatures. There are then probably temperature optima for yolk utilization which partially give rise to these effects, but often the range of temperature used may be insufficient to show them. On the assumption that high temperatures are nonoptimal other conditions producing smaller larvae, such as low oxygen tensions and current velocities for salmonids as well as high salinities in the desert minnow, *Cyprinodon macularius*, shown by *Sweet* and *Kine* (1964) and in cod, *Gadus macrocephalus*, by *Forrester* and *Alderdice* (1966), may also be considered nonoptimal. Combinations of temperature and salinity suggest that the optimum conditions for growth in length over the ranges used were 26°C for 70% salinity, 28°C for 35% and 33°C for freshwater in *Cyprinodon*, and 6°C and 19% for *Gadus macrocephalus*. The mortality, length, and maximum weight of salmon alevins (on the yolk supply) was greatest when they were reared in the dark, on a grooved surface, where presumably activity was minimal.

Endocrines, Growth and Metamorphosis

Russian, Canadian, and other work on the role of the endocrine system in fish larvae is fully reviewed by *Pickford* and *Atz* (1957) and the role of the thyroid in metamorphosis by *Barrington* (1961). The main techniques used were histological, histochemical, or immersion in dilute solutions of thyroid hormone or antithyroid drugs. Evidence from a number of sources suggests that the pituitary in inactive until metamorphosis, for example, in herring *Clupea harengus*, eels *Anguilla anguilla*, bream *Abramis brama*, milkfish *Chanos chanos*, and sturgeon *Acipenser stellatus*. The thyroid varies in its historical "activity" in the early larval stages but at metamorphosis in sturgeon, flatfish, eels, herring, pilchard *Sardina pichardus*, and bone fish *Albula vulpes* there are signs of secretion of colloid into the lumen of the thyroid follicles. Whether this is storage, or release into the circulation is taking place, can be satisfactorily tested only by measuring hormone levels in the blood.

Earlier, thyrozine may retard growth and accelerate morphogenesis (e.g., in salmon, sturgeon, *Misgurnus fossilis* and *Lebistes*) at a concentration of 1 ppm or less. Rate of oxygen consumption may also be increased. In the developing eggs of *Scyliorhinus canicula*, the functioning of the thyroid seems highly dependent on temperature. At 8°C use of ^{131}I showed the thyroid concentrated and bound iodine, but there was no evidence of thyroid hormone formation. At room temperature thyroid hormone reached a measurable level. Thiourea at concentrations of about 100 ppm may delay hatching, retard yolk resorption, and inhibit growth in salmonids, inhibit morphogenesis in sturgeon larvae, and decrease oxygen consumption is salmon and sturgeon larvae by as much as 15%. Thiourea seems to improve the ability to utilize oxygen at low tensions, although it is not clear whether this is because of its antithyroid action. Thiourea might well improve survival in suboptimal conditions of oxygen concentration that might occur, for example during transportation. The effect of thyroxine on promoting metamorphosis was tested on the eel (*Vilter*, 1946).

There was an accelerating, but not a sudden, effect with an indication that an increasing threshold controls the sequence of events in metamorphosis. In other words, the later stages of metamorphosis require higher concentrations of thyroxine. Other experiments on lampreys showed no effect of thyroxine, thyroid extract, iodide, or iodine on metamorphosis (see *Barrington*, 1961). It may be concluded that there is a suggestion of the thyroid playing a role in growth and differentiation in the larval stages but further work is required, especially at the important metamorphosis stage. If the pituitary is not functioning until metamorphosis the thyroid must be relatively independent of pituitary control in the early stages of life.

Feeding, Digestion and Starvation

Often the mouth is not completely formed at hatching, but rapid development in many marine fish larvae leads to the possibility of taking external food before the yolk is finally resorbed. Without success in feeding there is eventually self-metabolism and loss of weight. Yolk may be transported and stored within the body, especially in the base of the primordial fin and other subdermal spaces, and fat may be stored in the mesenteries of the larval gut. Unlike demersal fresh-water species there is no vitelline circulation in most pelagic larvae although *Scomberesox*, *Trachypterus*, and a few other species are known to be exceptions.

Usually there is a yolk sac sinus in connection with the heart and with lateral branches to the subdermal spaces; in the species with a vascularized yolk sac these spaces are much less inflated. After final resorption of the yolk the larvae retain their potential to feed for some days depending on species, egg size and temperature. The concept of a "point of no return" was introduced by *Blaxter* and *Hempel* (1963) for herring larvae. This point, after which the larvae are still living but too weak to feed, may vary from 15 days after fertilization for summer spawners in warmer water to 45 days for winter spawners. In the larval cisco, *Lauchichthys artedi*, it was 27-36 days after hatching at 3°-4°C.

The activity of larval *Sola solea* drops from about 70% of the time active at first feeding to 20% some 5-6 days later when, if feeding is unsuccessful, inanition occurs. Information on the time to point of no return in relation to spawning and food supply may help to predict the probability of survival in different broods. Most fish larvae are predatory with a large mouth and well-developed eyes. In herring the gape of the jaw increases by 50% during the yolk sac stage: the gap of 0.3-0.4 mm at first feeding depends on the length of the larvae and, therefore to some extent, on the original egg size. An elastic ligament at the articulation makes it possible for even larger organisms to be taken. Some species have very small gapes which is partly a function of their small size (e.g., lemon sole *Microstomus kitt*, sprat *Sprattus sprattus*, pilchard *Sardina pilchardus*, sardine *Sardinops caerulea*, and many gadoids). These must require very small food organisms for their early survival, a factor which is of considerably less importance for many fresh-water fish.

The forward darting movements of fish larvae in order to engulf the prey are described in a number of species, for instances, in *Coregonus* and *Esox* by *Braum* (1964) and in sole, *Solea solea*, by *Rosenthal* (1966). *Iwai* (1964) found three types of feeding in the larvae of *Plecoglossus altivelis*: a predatory snapping, a respiratory current with a sieving action by the gill rakers, and a ciliary current associated with the olfactory pits. The extent of which fish larvae are herbivorous and might make use of such currents is not clear. Green food in the gut may be incidental to the swallowing of water or due to the fecal material from prey species, and its presence is no evidence of utilization. In the anchovy, *Engraulis anchoita*, however, phytoplankton in the guts of young stages can be correlated with the formation of gill rakers which presumably act as a sieve.

The utilization of dissolved organic matter has also never been adequately tested. *Morris* (1955), in a reappraisal of the theory, concluded from circumstantial evidence of a low incidence of feeding in larvae collected from nature, that it was likely they were utilizing at least fine particulate matter, which could be collected by mucous cells within the buccal cavity or during swallowing of seawater for osmoregulatory purposes. Morris estimated that the larvae of the night smelt, *Spirinchus starksi*, might pass about 1.15 ml of water per day across these mucous surfaces, possibly containing 0.011 mg of organic matter, a very small quantity even for a larva of 0.9 mg wet weight. The recent work of *Terner* (1968), mentioned in Section IV, D, showing that the ova of *Salmo gairdneri* can probably incorporate exogenous substrates into their endogenous reserves, may also be significant.

The guts of many pelagic marine larvae are often empty, a puzzling feature when one considers the fairly high food requirements in the early stages. Apart from a lack of suitable food this may result from the capture by net of the weaker feeders only, rapid digestion, defecation during capture, or diurnal feeding rhythms. The role of vision in feeding has been tested experimentally and thresholds of light intensity measured. The fall off in rate of feeding corresponds with the dusk and dawn periods. Some species can take food in the dark, e.g., cisco, especially when it is present in high concentrations. Some, like sole, take food in the dark for most of the larval life and others, like plaice, only at later stages around metamorphosis. When vision is important the daily feeding period must vary considerably with season, latitude, clarity of water, and even average cloud over.

Ivlev (1960b) calculated, from the food requirements of young Baltic herring, their rate of feeding and the density of food, that they needed to fed 15 hr/day in August, in other words nearly all the hours of daylight. *Blaxter* (1966) estimated that much younger stages of herring had 10 hr/day to feed in the southern winter spawning groups and up to 24 hr/day in the more northern summer spawners. Variations in feeding time affect searching power, and therefore survival and growth, and these may be further influenced by the temperatures prevailing in different seasons. Success in early feeding, feeding drives, and learning factors have been studied.

At very early stages *Braum* (1964) found only 3-8% of feeding movements in *coregonus* were successful, but 30% in *Esox*. In the yolk sac stage, herring larvae take food successfully in 3-10% of their feeding movements but later in 80-90%. In plaice, feeding is 60-80% successful throughout larval development. In herring larvae from 20 to 40 mm in length the feeding drive depends on factors such as satiation and especially on the activity of the prey. Live organisms, although more difficult to catch, seem to increase the feeding drive and are probably very important in the young stages, where the main visual process may centre round movement perception. Larval cisco, however, showed a much higher rate of feeding on dead than live *Cyclops*.

The gut is usually a straight or simple tube at first feeding and the food is often digested near the anus. This is an interesting reverse of the normal situation in adult vertebrates. It seems likely that the gut is unspecialized at this stage and digestive enzymes are secreted along its length, although no critical work has been done on this. *Harder* (1960) gives a detailed account of the changes in the gut during the early growth of clupeids and engraulids; *Ryland* (1966) gives a rather simpler account for the plaice. Movement along the gut is mainly by peristalsis, although *Iwai* (1964) found cilia in the gut of ayu, *Plecoglossus altivelis*, especially in positions posterior to the liver. Backwardly directed ciliary currents and forward peristalis seemed to cause a circulation of the gut contents perhaps to aid digestion.

In further reports use of the electron microscope showed intermingling of ciliated cells and columnar cells with microvilli in both *Plecoglossus* and *Hypomesus olidus*. Reduction of cilia in later stages suggested a transition from cilia to microvilli. Whether cilia are to be found in many species is still open to investigation. Rates of digestion have been measured by rate of disappearance of the gut contents (the gut wall and body being transparent) or by interposing differently coloured food in the normal diet and observing the first signs of coloured feces. *Kurata* (1959), using herring 12 days old, found that the gut clearance rate ranged from 12 to 19 hrs. or more at 9°C depending on the extent of food intake. The rate of digestion varies from 9 hrs. at 7°C to 4 hrs. at 15°C judged by transparency of the gut contents in this species. Reports of larvae in poor conditions or dead in net hauls may be due to selective capture or poor washing and are not certain evidence for mortality. *Shelbourne* (957) estimated the condition of plaice larvae in the North Sea and found more signs of inanition early in the year.

Measurements of condition factor [weight/length^3 x 1000] in herring larvae compared with starving larvae in tanks suggested that larvae in the sea were often very near the point of starvation although, over a number of years, it was not possible to correlate high condition factors with a good food supply. *Ivlev* (1961) looked at various aspects of starvation in young catfish, *Silurus glanis* and bream, *Abramis brama*. With complete starvation the survival time was 34-36 days corresponding to a weight loss of 33-39%. With rations below the maintenance requirement survival was prolonged, but even on a full maintenance diet, the weight remaining constant, the fry died after 126-151 days, indicating that a "need" for growth may be characteristic of these young stages and that merely to maintain uniform weight is physiologically or developmentally inadequate.

The current velocity stemmed by young carp *Cyprinus carpio* and roach *Rutilus rutilus* for 5 min was reduced by 10 times, from about 100 to 10 cm/sec, over 52 days of starvation. Age had a considerable influence on survival time without food. In catfish, roach, and bleak *Alburnus alburnus*, it ranged from 3 to 6 days at yolk resorption and from 109 to more than 180 days when 100 days old. Losses of weight of 30-35% were possible before death, with survival times of 30-80 days. Starving fish were much more susceptible to unfavourable conditions and also to infection and predation.

Sense Organs

Vision

The newly hatched larvae of many species have unpigmented, and presumably nonfunctioning, eyes (e.g., sole *Solea solea*, mackerel *Scomber scombrus*, whiting *Merlangius merlangus*, pilchard *Sardina pilchardus*, and sardine *Sardinops caerulea*); other have pigmented eyes (e.g., plaice *Pleuronectes platessa*, cod *Gadus morhua*, herring *Clupea harengus*, and salmonids). *Schwassmann* (1965) examined sections of the eye and brain of larval *Sardinops caerulea* after hatching. On the first day there was no pigment and little differentiation, but by the second day pigment with a stratified retina and visual cells had developed. At 5 days the optic nerve was myelinated and the optic chiasma showed some interdigitation. Considering there are only about 40 individual fibres in the optic nerve when feeding commences at 3 days, it seems probable that the eye at first feeding is only capable of coarse movement perception. The larval eye is also limited in other respects. In newly hatched and *Oncorhynchus* and in *Clupea harengus* and *Pleuronectes platessa* and in a number of other species up to metamorphosis, there is a pure-cone retina, although the adult has both cones and rods. *Lyall* (1957a, b) examined the developing eye of the trout, *Salmo trutta*, and concluded that the rods might develop from single cones or by an outward migration of bipolar cells, the latter view being supported by *Blaxter* and *Jones* (1967).

Retinal pigment migration and other retinomotor responses associated with dark and light adaptation only develop when the rods appear and are therefore absent in herring and plaice until the larvae metamorphose. This, may be true of many other marine species. Thus, the larvae are, visually, poorly equipped, with a single type of visual cell, no ability to dark or light adapt, and with a coarse retinal mosaic. *Oncorhynchus* species acquire rods and retinomotor responses very rapidly after hatching compared with the marine fish which have been examined. The range of light intensity over which light and dark adaptation takes place is then 10^1-10^{-1} meter-candles (mc), which corresponds fairly well with the thresholds for feeding.

The acuity of the larvae is poor compared with the adult. Although the retinal cells are small and fairly closely packed, the eye is also small as is the focal length of the lens. *Baerends* et al. (1960) found that young cichlids, *Aequidens portalegrensis*, could be trained to distinguish stripes 1.5 mm apart at 3 cm body length; this improved to 0.3 mm at 11 cm body length. *Blaxter* and *Jones* (1967) calculated the acuity of the eyes of larval herring (minimum separable angle) to be 200 minutes at 1 cm body length improving to 50 minutes at metamorphosis.

Thresholds and spectral, sensitivity were measured by behaviour techniques in herring, plaice, and sole larvae. Using a negative phototaxis as criterion the visual thresholds were about 10^{-5}-10^{-6} mc, the spectral sensitivity curves being plateaulike with peaks which correspond to different cone populations. Distances of perception and visual fields have been measured by a number of workers.

Neuromast Organs

It has probably not been fully appreciated that many species of fish larvae have free neuromast organs. Iwai reviews earlier reports and his own work on *Tribolodon hakonesis*, *Tridentiger trigonocephalus*, and *Oryzias latipes* (1967a). These organs are usually situated in lateral rows along the body with the small hump of sensory cells being the main element visible especially in sea-caught or fixed specimens. Rearing, with the use of phase contrast microcopy, makes it possible to retain and see the cupulae which are relatively enormous—0.05 mm long in *Oryzias* but 0.1-0.4 mm long in some other species. In the adult it is likely that they are more often sunk into pits. The ability of larvae to avoid capture before the eyes fully develop is almost certainly a result of these neuromast organs.

Activity and Distribution

Phototaxis and Activity

Much of the published work on phototaxis in fish larvae in aneodotal. Generalizations that marine fish larvae are photopositive and demersal fresh-water larvae photonegative do not stand up to a detailed investigation using different temperatures, light intensities, species, or ages. Changes of the sign of the phototaxis may be related to temperature, especially where a sudden change is applied, e.g., in Coregonus. Herring larvae show a positive phototaxis at high light intensities which becomes negative below a certain threshold.

Activity levels are frequently associated with diurnal light changes and phototaxis. *Hoar* (1956, 1958), *Ali* (1959), and *Heard* (1964) review much work on activity, rheotaxis, schooling and migration of young salmonids. Emergence of the alevins of sockeye salmon. *Oncorhynchus nerka*, seems to be controlled by incident light on the gravel and can be delayed by artificial light at night. The younger fry of pink salmon, *Oncorhynchus gorbuscha*, for example, are negatively phototactic and hide among the stones of their home stream by day. They become active at night any may rise to the surface and swim downstream or get displaced by the current. After schooling both pink and chum salmon *O. keta*, prefer lighted conditions and presumably substitute schooling for hiding as their protection against predators. They show then a positive phototaxis, stem the current by day displaying a cover reaction only with abrupt changes of light, and again become displaced downstream at night when the visually controlled rheotaxis phases out. Coho salmon fry, *O. kisutch*, are rather different in behaviour; they are strongly territorial and generally more active, but their responses to light and their diurnal changes in activity are much less marked. The behaviour of the fry of the different species of *Oncorhynchus* can, in fact, be related rather generally to their migratory habits. Pink and chum fry migrate to the sea quite soon after hatching, whereas sockeye fry tend to move into lakes, and coho remain in the home stream for a year or so. Loch trout, *Salmo trutta*, also show incipient territorial behaviour at an early age. *Woodhead* (1957), in a laboratory study of the larvae of *Salmo* species, related changes in photokinetic activity levels to ambient light intensity and age. Between 0.005 and 100 mc the photokinetic activity of *S. trutta* gradually increased above a basal dark level, and this species and *S. gairdneri* (iridesu) and *S. salar* all showed increased activity with age. A positive photokinesis, together with a negative phototaxis, would work together to keep the larvae within the stones of their home steam. *Woodhead* and *Woodhead* (1955) found a similar increase in the activity of herring larvae with light intensity above 0.3 mc, and there was a basal level of activity in very low illumination as there was in the salmonids.

Vertical Distribution

Diurnal variations in distribution are linked to activity, phototaxis, and brightness discrimination. Some earlier work on changes of vertical distribution did not take into account the need for opening and closing the plankton nets at defined depths and for preventing differential not avoidance by day and night. A general lack of larvae by day, especially larger specimens, partly results from this latter factor, as emphasized by *Bridger* (1956). With slow two nets it can be difficult to estimate the relative effects of net avoidance and vertical distribution when comparing day and night catches. *Strasburg* (1960) used slow two nets to catch tuna larvae. Some species such as skipjack, *Katsuwonus pelamis*, were rarely caught by day at all while catches of all species were always much higher by night. He concluded that there was some net avoidance, but the main diurnal difference was owing to vertical migration. Stevenson (1962) used high speed nets for larvae of Pacific herring, *Clupea pallasii*. The catches at night were far higher and the average size of the larvae was greater, from which he concluded that the larvae were concentrated at the surface by night and spread over a fairly wide stratum by day. There seemed to be net avoidance, even of high speed nets, by the larger larvae in the daytime. *Ryland* (1964) found a reverse distribution of larval plaice and sand eels *Ammodytes*, which were spread over a wide depth range from the surface to about 35 m by night and much more concentrated between 5 and 10 m during the day. The use of high speed nets in this study and in that of *Colton* (1965) on haddock larvae, *Melanogrammus aeglefinus*, prevented diurnal variation in size and presumably overcome the net avoidance factor.

Buoyancy and Pressure

There are a number of devices which provide buoyancy in pelagic fish eggs and larvae. In marine species, found in high salinities, maintenance of the body fluid concentration below that of the environment acts as a buoyancy mechanism. In the early stages lack of a skeleton also keeps the specific gravity fairly low. In fresh-water the eggs are often demersal so that the much greater problem of maintaining buoyancy does not exist. In fact, in salmonid eggs the specific gravity may be very high in the early stages because of low water content of the yolk. It is sometimes said that a large perivitelline space is a mechanism for buoyancy. However, this normally contains water of the same concentration as the environment and is thus much more likely to act as a shock absorber, particularly in pelagic fish eggs subject to wave action. The embryo and yolk may regulate the body fluids from fertilization (e.g., in pelagic plaice eggs: by means of the vitelline membrane. In the demersal herring egg there is no regulation until the yolk is covered by cellular tissue at gastrulation. Many species such as pilchard *Sardina pilchardus*, mackerel *Scomber scombrus*, hake *Merluccius merluccius* and sole *Solea solea* have oil droplets, but these are not typical of pelagic eggs for they are absent in the sprat *Sprattus sprattus* and in *Gadus* and *Pleuronectes* species. Tendril like outgrowth found, for example, in *Exocoetus* and Scomberesox eggs are as likely for attachment as for buoyancy.

There may be a mechanism to adjust specific gravity to the spawning medium. *Solemdal* (1967) found that variations in the osmotic pressure of the blood of female flounders. *Pleuronectes flesus*, at the spawning season depended on the salinity. This, in turn, caused changes in the specific gravity of the eggs when spawned. In water of low salinity the eggs were larger and their osmotic pressure lower. In the low saline water of the Baltic (6.5%), fertilization and development were possible but the eggs rested on the bottom. In the sprat there is a correlation between low salinity and larger eggs and also in herring and plaice.

In the larval stages subdermal spaces as in the plaice and cod reduce specific gravity as long as the body fluids and maintained at an osmotic pressure below that of the environment. Other species have long processes to the dorsal of lateral fins, e.g., the angler fish *Lophius* and deal fish *Trachypterus*. It may be significant that *Orton* (1957) quotes the latter species as having only moderate fin folds and subdermal spaces, but whether the increased surface area assists buoyancy is open to question; these may be organs to reduce predation. In fresh water, the larvae of salmonids can afford to retain dense yolk reserves as they rest on the bottom. Others like carp and bream have cement organs for attaching themselves to weed.

The first, filling of the swim bladder is well known in fresh-water fish. In the physostomatous salmonids like *S. trutta*, *S. gairdneri*, *Cristivomer namaycush*, and *Coregonus clupeaformis* the swim bladder is filled 1-3 months after hatching, depending on temperature or feeding. Air is taken in from the surface via the pneumatic duct and without access to the surface the swim bladder remains empty. The pneumatic duct remains open under such conditions and functions later. Lake trout, *Cristivomer namaycush*, were able to swim considerable vertical distances (280 m) with unfilled bladders suggesting that development in deep water presented no problems for filling the swim bladder. Physoclists like *Gasterosteus*, *Lebistes* and *Hippocampus* also require access to air to fill to swim bladder but if deprived of this at the critical time the pneumatic duct closes and the swim bladder cannot be filled later.

Pressure effects on fish larvae were tested briefly by *Bishai* (1961). Newly hatched herring survived compression to and decompression from 4 atm and young plaice 37-50 mm long to and from 2 atm. Young salmonids could live at 5 atm to the end of the yolk sac stage, sometimes showing increased activity during compression or decompression. Older fry, about 20 days after yolk resorption, could withstand compression to 2 atm but had difficulty in dealing with air bubbles inside the body and in the swim bladder during decompression. Older fry still, about 60-140 days after yolk resorption, could not even withstand compression to 1.2 atm. At later stages, the ability to withstand compression increased but decompression was lethal. *Qasim* et al. (1963) observed changes in the vertical distribution of larval teleosts subjected to pressure changes. Young plaice larvae tended to swim upwards when the pressure was increased from 1 atm to 2 atm and sank when the pressure was returned to normal. Metamorphosing plaice did not show this response. Larvae of the blenny *Centronotus gunnellus* responded in a similar way to pressure changes equivalent to 25 cm of seawater and larvae of another blenny *Blennius pholis* to only 5 cm of seawater. Only the last of these species has a swim bladder. It is possible in such experiments that the larvae could have been responding to water currents produced by the pressure changes. In this is not so, one may ask how larvae without swim bladders can respond to pressure.

Locomotion and Schooling

Most species show sings of activity within the egg, and movement is often an important part of the hatching process. The relatively large yolk sac at hatching must be a hydrodynamic embarrassment to the larvae, but in salmonids, where it is especially large, there is little movement as the larvae grow in a rather passive way within the stream bed. The buoyant oil globule found in many marine pelagic larvae may also make equilibrium difficult. Two sorts of movement are found at yolk resorption—a serpentine eel-like mode of swimming in the long thin bodied clupeoid type and a more maneuverable movement allowing "backing-up" with the use of the pectoral fin and marginal fin folds in the shorter flatfish type of larvae.

Swimming performance in terms of burst speeds maintained for a few seconds, where the theoretical 10 body lengths/sec seems to fit quite well. Improvements in performance can be correlated with development and upturning of the caudal fin.

If schooling occurs it usually starts at metamorphosis. Presumably the larvae, being transparent, do not provide very good mutual stimuli for keeping schools together visually. *Show* (1960, 1961) analyzed the development of schooling in the silverside *Menidia* and found responses initiated between two fry at a body length of 8-9 mm with increasing participation as they grew. By 11-12 mm as many as 10 could be seen in parallel orientation. Fish reared in isolation in paraffin-coated bowls took a few hours to become integrated into existing school. Generally speaking there was a considerable latency in schooling when "isolates" and "quasi-isolates" (isolated 5-7 days after hatching) were brought together at a size where schooling was normally well-developed. Surprisingly, this initial latency was less among "isolates" than "quasi-isolates". In the newly emerged fry of *Oncorhynchus* species, schooling ceased in the dark and following subjecting to light they needed 15 min or more to complete the school again, although in all species except coho this time became reduced with age. In herring schooling started about 2 months after hatching at a body length of 25-30 mm. Starving larvae tended to school more positively, but the school rapidly broke up when food was offered.

Rheotropic (optomotor) responses (orientation to currents or moving backgrounds) are often well-developed in the larval stage as shown in some experiments on swimming performance and from the many observations on salmonid larvae by day and night. This responses seems to be maintained by either visual or tactile cues, and the sign may change under shock treatment. In young salmonids a temperature rise of 12°C changed the rheotaxis from positive to negative, presumably a defensive response.

Searching Ability

The ability to search for prey or, in the case of visual feeders, "optically filter" the environment, depends on cruising speed and the distance at which food is perceived. Perception distances depend on the movement of the prey, its orientation in relation to the eye, contrast with the background, and illumination. Activity in terms of cruising speed may well be reduced a period of feeding and is certainly influenced by feeding drives in older larvae. Starvation may also cause a reduction in activity. The volume searched per hour has been estimated in whitefish and herring larvae. When multiplied up on a daily volume searched becomes very dependent on day length, that is, on latitudinal and seasonal effects.

Mortality, Tolerance and Optima

From hatchery, rearing, and experimental studies it is well-known that fish eggs and larvae go through periods of mortality. Susceptibility to mechanical shock or nonoptimal temperatures and salinities may vary with age. Marine fish eggs seem especially delicate until the completion of gastrulation although this has not been systematically tested. Perhaps sensitive morphogenetic processes in the early stage or failure to osmoregulate are the cause. Similarly there are stages, such as the eyed stage in salmonids, where eggs may be particularly resistant to damage. After hatching, further stages of mortality may be experienced which can sometimes by connected with the further development of certain organs and with such physiological changes as the transition from cutaneous to gill respiration or the development of the swim bladder. Rearing studies on marine fish show a high mortality at the end of the yolk sac stage in species such as cod *G. morhua*, haddock *Melanogrammus aeglfinus*, pilchard *Sardina pilchardus*, herring *Clupea harengus*, and lemon sole *Microstomus kitt*. These "critical phases" are probably in part the result of unsuitable food being supplied, especially from the aspect of size. The question of whether critical phases at yolk resorption occur under natural conditions is much vexed question, examined especially by fishery biologists looking for the factors controlling brood survival *J.C. Marr* (1966), in reviewing this hypothesis, concluded in two species (mackerel *Scomber scombrus*, and Pacific sardine *Sardinops caerulea*) that mortality was more likely to be constant or steadily decreasing without the catastrophic periods advocated by earlier workers. Undoubtedly some of the five species mentioned have small mouths and the finding of suitable food in the sea at the end of the yolk sac stage before the point of no return must be a serious problem.

The lethal levels of various environmental factors such as salinity, temperature, oxygen, tension, pH, radiation and mechanical stimuli have been assessed for fish eggs and larvae. Salinity tolerance is surprisingly wide, ranging in marine species which have been examined from about 5%° up to 45-50%° with even a wider range over short periods of exposure. Predictably, the lethal temperature varies with level of adaptation. A wide variety of criteria have, unfortunately, been used for defining lethal temperature, ranging from 50% mortality after one day (or longer) to the number of days to 50% mortality. Temperature fluctuations are much damped down in the sea and may be negligible in deep water. In the tropics the seasonal fluctuations may also be slight, and this may lead to a narrow range of tolerance. In 10 species of fish larvae from the Indian Ocean, *Kuthalingam* (1959) found the range from lower to upper lethal temperature was only 3°-4°C (about 28°-32°C). Temperature tolerance may also vary with age. *Combs* (1965) measured long-term temperature thresholds over the period from fertilization to hatching in *Oncorhynchus* species. In *O. nerka*, for example, the low temperature threshold was about 5°C and the high temperature threshold about 13.5°C. At the 128-cell stage and later both *O. nerka* and *Chinook*, *O. tshawytshca* had a greater range of tolerances. In fresh-water or high latitude resistance to cold must be adequate or else adaptations are developed which prevent contact with extremely cold water. This flounder, *Pseudopleuronectes americanus*, is one of the few species of flatfish with a demersal egg and it spawns inshore in very cold water. Temperature optima may be judged by hatching success. For instance in *Coregonus clupeaformis* the highest percentage hatching was at 0.5°C with a very rapid fall off above 6°C. In herring temperature optima vary with the race. In medaka *Oryzias* the best survival was obtained when the rearing temperature was alternated between 22° and 30°C every 12 hrs. Generally, outside the optima not only does the percentage hatch fall but there are more abnormalities.

Oxygen is apparently only limiting at its low level. However, super-saturated solutions may be harmful if air bubbles are swallowed by the larvae. This "gas disease" can be fatal if the larvae cannot eliminate the bubbles from the gut as their buoyancy is disturbed. The eggs of the salmonids *S. salar*, rainbow trout, *S. gairdneri*, Pacific salmon *O. tshawytscha* and *O. kisutch*, brook trout *Salvelinus fontinalis*, and lake trout *Salvelinus* namaycush, have been tested at low oxygen tensions. Most were able to resist levels of 2.4 ppm or even less (25% saturation) although development was retarded and the larvae smaller at hatching. Reductions in current flow had a similar effect. *Alderdice* et al. (1958), using chum salmon eggs, measured the retardation in hatching resulting from low dissolved oxygen levels at different developmental stages. The intermediate stages, about one-third of the way from fertilization to hatching, were most susceptible. *Hayes* et al. defined the limiting tension not from its lethal aspect, but the tension at which normal oxygen consumption started to drop. In *S. salar* eggs this was 3.0 ppm at 20 days, 7.5 ppm at 45 days, and

4.7 ppm at hatching. Other salmonid data on limiting tensions are summarized by *Wicket* (1954). In the egg of chum salmon, *O. keta*, the calculated limiting tension varied from 0.72 to 3.7 ppm increasing with age (temperatures from 0.1° to 8.2°C). In the eggs of pike, *Esox lucius*, it varied from 1.2 to 4.1 ppm depending on age and temperature. This type of finding is of value in assessing the survival chances of fish eggs, buried in gravel. *Wicket* (1954), for example, has shown that the flow rates and oxygen content of the water may be quite inadequate for chum salmon eggs, being even as low as 2 mm/hr and 0.2 ppm.

Marine and other fish larvae were studied by *Bashai* (1960a). Herring larvae at hatching were "restricted" at 27.32% saturation and at 55-64% saturation a few days later. Larvae of the lumpsucker, *Cyclopterus lumpus*, just tolerated 43% saturation 3 weeks after hatching and young plaice 4-5 cm in length just tolerated 14% saturation. He found that the larvae of the salmonids, *S. salar* and *S. trutta*, could survive very low levels of saturation, 3-10%, after hatching, but this ability decreased with age, the minimum level tolerated being 16-28%. Again, criteria for defining lethal levels vary, and it is difficult to compare the results of different workers. To some extent acclimation to low oxygen increases resistance to hypoxia and there are also specific differences in tolerance.

Bishai (1960c) has also summarized work on the effects of pH on larval fish. Compared with seawater (pH about 8.0) which is fairly well buffered fresh-water has a pH which is more variable and more susceptible to change by effluents or run-off. Bishai found that herring larvae could survive values between 6.5 and 8.5. Lumpsucker larvae had a low tolerance level of about 6.9, plaice (4-5 cm) 6.2-6.5, while salmonid larvae had a range of tolerance from about 5.8 to 6.2 up to 9.0, or even 10.0 in sea trout alevins. To some extent the levels of tolerance depend on the substances used to change the pH. It is possible that CO_2 for reducing pH has an additional toxic or respiratory effect compared with the use of HCl. Certainly *Alderdice* and *Wickett* (1958) found high levels of CO_2 inhibited oxygen consumption in the eggs of *O. keta*.

Radiation effects on salmonid eggs and larvae are reviewed by *Hamdorf* (1960) and *Eisler* (1961). Strong visible light may cause early hatching mortality, poor growth, and greater pigmentation. Exposure to light is less serious when the intensity is low or in the later stages, especially after the onset of pigmentation. It is, perhaps, not surprising that light is harmful to eggs and larvae which normally develop in complete or semidarkness among gravel or stones; increased activity probably contributes to its deleterious effect. Other species which develop in sand as the grunion, *Leuresthes tenuis*, or near the sea bed such as herring may also show a poorer rate of hatching under lighted conditions. Careful controls are required in experiments since the light may cause other changes such as raising the temperature on increasing growth of phytoplankton in the water. The main effect on many other species is the accelerate hatching. Ultraviolet light has been tested on *Fundulus* and causes abnormalities of the skeleton, cylopia and twinning. In sockeye salmon, *O. nerka*, it causes premature hatching, vertebral abnormalities, high mortality and delays pigmentation.

Marinaro and *Bernard* (1966) exposed the eggs of pelagic fish such as pilchard *Sardina pilchardus* mullet *Mullus*, horse mackerel *Trachurus* and "sargue" *Diplodus annularis* to sunlight with and without ultraviolet filters, finding a lower rate of hatching with ultraviolet. The transparency of many fish eggs and larvae may be an adaptation to prevent absorption of light as well as acting as a means of camouflage. X-ray have been tested, for example, by Solberg (1938) on *Fundulus* embryos. They were most sensitive after fertilization with a decrease of 10 times in sensitivity between early cleavage and 4-5 days later. Other work is reviewed by Eisler (1961) especially on salmonids. X-rays have a number of effects such as higher mortality, decrease of growth, deformities, decrease in erythrocytes, and destruction of hematopoietic tissue and increased pigmentation.

Generally, susceptibility decreases with age and in optimal conditions of temperature. *Neyfakh* (1959) used X-rays for inactivating the nuclei in the developing eggs of the loach, *Misgurnus fossilis*. Irradiation during certain phases of development was followed by death at a very specific time, suggesting a periodical functioning of the nuclei. Fish eggs in almost any environment are subject to mechanical stimuli which are potentially harmful—movement of gravel within a spawning redd, current borne objects on the sea bed, or wave action at the surface. The resistance of the egg may be measured by the maximum lead it will take before the chorion bursts. They indicate an increase in the resistance of the chorion as it hardens subsequent to fertilization and a decrease in the strength of the chorion prior to hatching, presumably resulting from hatching enzymes.

Changes in the chorion may in part account for periods of varying resistance to external conditions. Other harmful influences may be hatching enzymes, especially for embryos which may be somewhat unhealthy in other respects, or where eggs are very crowded and the enzymes are concentrated in the water after hatching. *Stuart* (1953) observed sediments attached to the chorion of loch trout. These might well reduce oxygen intake. The alevins, however, seemed to have the ability to disperse such sediments by means of respiratory currents passing over the body surface. With gill respiration sediments were aggregated by mucus and rendered less harmful. A recent trend in research on tolerance is to use combinations of factors such as temperature, salinity, and oxygen concentration to see how they interact and to find out the optima.

Kinne and *Kinne* (1962) used such combinations during the development of the desert minnow, *Cyprinodon macularius* (i.e., temperature between 10° and 37°C, salinities from 0 to 85%, and 70%, 100% and 300% saturation of air in water). Mortality was least at near lethal temperatures when the salinity was held between one-half seawater and 35%. The lethal temperature was lower in hypoxial conditions. It seemed that of the salinities used 35% was the optimum for incubation. Lewis (1966), using menhaden larvae, *Brevoortia tyrannus*, found that temperature tolerance was greatest at salinities of 10-15%, suggesting that isotomic conditions were optimum for survival. *Forrester* and *Alderdice* (1966) calculated from experimental data that the optimum salinity and temperature for hatching eggs of cod, *Gadus macrocephalus*, was 19.4% and 5.3°C.

Meristic Characters

Counts of vertebrae, myotomes, scales, gill rakers or fin rays have been used in racial studies. The variability in number and liability under different environmental conditions are striking examples of rather flexible raw material for evolutionary forces. Usually mean counts from samples of fish are required, the differences from race to race being, inadequate for identification of individuals. While correlations between the counts in adults or young and the environmental conditions on the spawning ground first suggested that meristic characters were liable, confirmation has been obtained by experimental studies. They show that factors such as temperature, salinity and oxygen content superimpose their effect on the range of meristic counts determined by the genotype.

There is also an individual variation even in fish from the same parents reared under the same conditions. Earlier work, mainly on salmonid and inshore or estuarine species, has been reviewed by Taning

(1952) and Barlow (1961). Since it is necessary to keep the young until the meristic characters form there has been less work on species which require to be fed and reared for a considerable time after hatching. Most work has been done on the effect of temperature on vertebral and fin ray counts. Generally, a V-type relationship is found where the average vertebral count is minimum at an intermediate temperature. For example, in sea trout, *Salmo trutta*, Taning found this was at about 6°C; in plaice it is at 8°C; at lower and higher temperatures the average counts were higher. With fin rays the V may tend to be asymmetrically inverted with counts highest at intermediate temperatures.

In other instances the mean vertebral counts are inversely related to temperature, but the lack of an inflection may result from an insufficiently wide range of experimental temperatures being used. Apart from the factors, both intensity and duration of light and CO_2 may have an effect. Higher intensities in the grunion, *Leuresthes tenuis*, reduce the number of vertebrate. A 16-hr day caused lower caudal vertebrae and possibly anal fin ray counts in *Oncorhynchus nerka*, while in sea trout, *Salmo trutta*, increasing tension of CO_2 produced lower number of vertebrae. The work of *Heuts* (1949) and *Lindsey* (1962) using different temperature-salinity combinations in *Gasterosteus aculeatus* were somewhat inconclusive resulting from the problem of getting sufficient numbers of survivors in an extensive series of experiments, some of which were in unfavourable conditions.

Certainly, it was shown that temperature effects vary with the salinity level. Using a fresh-water and brackish water race, Heuts found the greatest effect of temperature on fin ray counts at the salinity of adaptation. Differential mortality and the unsolved problem of controlling, for example, oxygen and temperature independently have often led to difficulties in interpreting results. It is the underlying reasons for meristic lability which is, perhaps, of most interest to the physiologist. Sensitive periods have been found during which lability of vertebral counts is as its greatest. In sea trout it is around gastrulation, with a further period when the last vertebrae are being performed. It is also before hatching in herring and killifish. However, in plaice and paradise fish meristic characters are still susceptible to modification after hatching.

Generally speaking, fin ray counts are determined later than those of vertebrae. Later studies on showed that transfer to a higher temperature 2 days after fertilization produced fewer vertebrae, whereas transfer after 6 days resulted in a higher vertebral count. Regular diurnal alternations of temperature between 22° and 30°C gave vertebral counts intermediate between those resulting from sustained temperatures of 22° and 30°C. Of the various hypotheses put forward to explain meristic lability those of *Gabriel* (1944) and *Barlow* (1961) fit the facts best. Often, but not always, environmental factors which delay hatching produce higher counts. Alternatively, higher counts may be considered to come from nonoptimum conditions. It is likely that the environmental factors change the relationship between the growth and differentiation. *Hayes* et al. (1953), in their studies of morphogenesis in salmon at temperatures from 1° to 15°C, certainly found that the relative appearance of certain anatomical characters varied with temperature.

If differentiation is late more tissue is available to be differentiated leading to a higher count. Thus one may consider an interaction of the Q_{10} for growth and the Q_{10} for differentiation of the character; as the temperature is varied so the relationship between growth and differentiation varies. Where V-shaped curves are obtained there are then points of infection for the temperature coefficients. *Garside* and *Fry* (1959), using normal and reciprocal hybrid fry of speckled trout, *Salvelinus fontinalis*, and lake trout, *S. namaycush*, found that the mean myomere count was lower where the fish developed on the speckled trout yolks, which were smaller. There was also an inverse relationship between myomere count and the degree of twinning in Siamese twins of speckled trout, the count being lower where was less shared tissue.

Rearing And Farming

Techniques

Three lines of approach are of interest at the present time:

1. Tropical and temperate fish culture for food
2. Rearing of salmonids and sturgeon to maintain population size after intensive fishing or hydroelectric schemes
3. Temperate marine fish rearing and farming.

A good general review on fish farming by Iversen (1968) is available and on tropical species by *Hickling* (1962). Some of the most popular species are the common carp, *Cyprinus carpio*, mullets, *Tilapia*, milkfish *Chanos chanos*, the carp *Putinus javanicus* and grass carp *Ctenopharyngodon idella*. Many of the species are reared in still water fertilized artificially or by sewage. The last four species mentioned are herbivorous which gives maximum efficiency of food utilization.

Some of the species, e.g., *Tilapia* and *Chanos*, can be kept in salt or brackish water. In the case of *Tilapia* and the Israeli carp, breeding takes place in the holding ponds; in the European carp special breeding ponds are used. On the other hand, the fry of *Puntius*, *Chanos*, *Ctenopharyngodon*, and mullet must be caught in the wild and transferred to rearing ponds.

Generally in tropical culture there are no serious problems of feeding the young which are of sufficient size to take the food available. The extensive of salmonids presents no problem of food supply for the young, which are relatively large when the plentiful yolk has been resorbed. Rearing is usually in running fresh-water although acclimation of the young of *Salmo* species to marine conditions, is now being developed, with improved growth and freedom from disease. The question of dietetics is well advanced in the group. Marine fish rearing farming, although attempted from time to time over the last hundred years, have only recently made significant advances. Starting with the liberation of millions of plaice and cod eggs or yolk sac larvae in coastal areas, only the later experiments by *Dannevig* (1963) seem convincingly exploitable.

Very recently, *Mugil* and *Solea* have been established as self-perpetuating species in some of the Mediterranean coastal lagoons. Following the earlier rearing in aquaria of some species to an advanced stage by highly empirical methods, a better experimental approach has now been adopted. Success in rearing *Pleuronectes platessa*, *Solea solea*, *Microstomus kitt*, *Clupea harengus*, *Sardinops caerulea*, Engraulis and *Scomber* and *Sardina pilchardus* has depended on some or all of the following factors:

1. Black-walled tanks and uniform overhead illumination have helped feeding success by providing contrast of the food against the background and even distribution of food in the tank.
2. Tank hygiene.
3. *Food supply*. The use of *Artemia* and *Balanus* nauplii, *Mytilus* trochophores, young *Tigriopus*, small nematodes, rotifers and oligochaetes and secondary foods like Dunalielia has improved survival rates, so also has the extensive collection and selection of natural plankton.
4. *Antibiotic*. Particularly with flatfish, dosages of mixed sodium penicillin (50 IU/ml) and streptomycin sulfate (0.05 mg/ml) have improved survival in the egg stage.

There are interesting physiological and behavioural aspects of rearing where the fish may be artificially crowded and conditions of stress set up. Size hierarchy effects prevail but growth seems to be density dependent only from the beginning of metamorphosis. Bitten fins are found more often in the smaller members of a tank community but are not always dependent on density. Perhaps more serious is the incidence of abnormal (albino or semi-albino) pigmentation in flatfish. It is generally worse in smaller fish and in very crowded tanks but may be reduced by using *spawning* stock acclimated to hatchery conditions. Survival to metamorphosis is also better in larvae from acclimated spawning stock.

Sensory Deprivation

Hatchery-reared flatfish do not survive well when transferred to the sea, presumably because of high predation, hatchery-reared salmonids usually have a poorer swimming performance. In fact, the hatchery environment by preventing contact from predation, by often rather uniform physical and chemical conditions and lack of shelter may prevent the development of appropriate defensive responses and muscular system. The need for high density may also encourage undesirable intraspecific relationships at the expense of the more desirable interspecific responses. The more obvious shortcomings of the aquarium tank may be accompanied by sensory deprivation in a more general and less easily defined way.

The uniformity of the aquarium environment may lead to insufficient sensory input. For instance, *Qasim* (1959), when rearing the young of *Blennius pholis*, obtained better survival with a day-night regime than with continuous light. *Blaxter* (1968a) rearing herring, found a hint of better survival when the light conditions over the tank were continually changed and when air jets playing on the surface kept up a continuous series of ripples. *Hoar* (1942) found that young salmon would feed on chopped earthworm in darkness during the day, but not at night, perhaps because of an activity (or retinomotor) rhythm. Such rhythms might eventually be suppressed by continuous light. Activity rhythms with periods of rest might play as important a part in the development of the sense organs and nervous system as sufficient sensory input during the active phases.

12

HORMONAL GLANDS

The vertebrate endocrine system is present in it essence already in Lampreys and Hagfishes, further similarities with higher vertebrates appear in both the shark and bony fishes. Fishes possess a well-developed endocrine system comparable to that of other vertebrate. The recent demonstration of tissue with parathyroid like function (*Rasquin* and *Rosenbloom* 1954) has filled a gap formerly thought to exist in the list of fish endocrine organs. Major contributions in the fish endocrine have also been made in studies of physiology of migration (*Fontain*, 1954, *Oliverean* 1954) and reproduction (*Bretsch-neider*, *Wit* 1947, *Docid* 1955, *Hoar*, 1957).

Important Endocrine Glands in Fishes

Following are important endocrine glands found in fishes:

1. Pituitary gland.
2. Thyroid gland.
3. Chromaffin tissue (supra renal body, medullary tissue represent the adrenal).
4. Interrenal tissue (Adrenocortical tissue).
5. Gonads.
6. Ultimobranchial tissue.
7. Intestinal mucosa.

Following are main tissue with possible endocrine function:

1. Carpuscles of stanius.
2. Pineal organ.
3. Thymus
4. Urohypophysis
5. Pseudobranches.

Pituitary Gland

The arrangement and organisations of the component parts of pituitary gland varies greatly in different species. *Olivereace*, 1954, has recently discussed the terminology and the homologus of the different

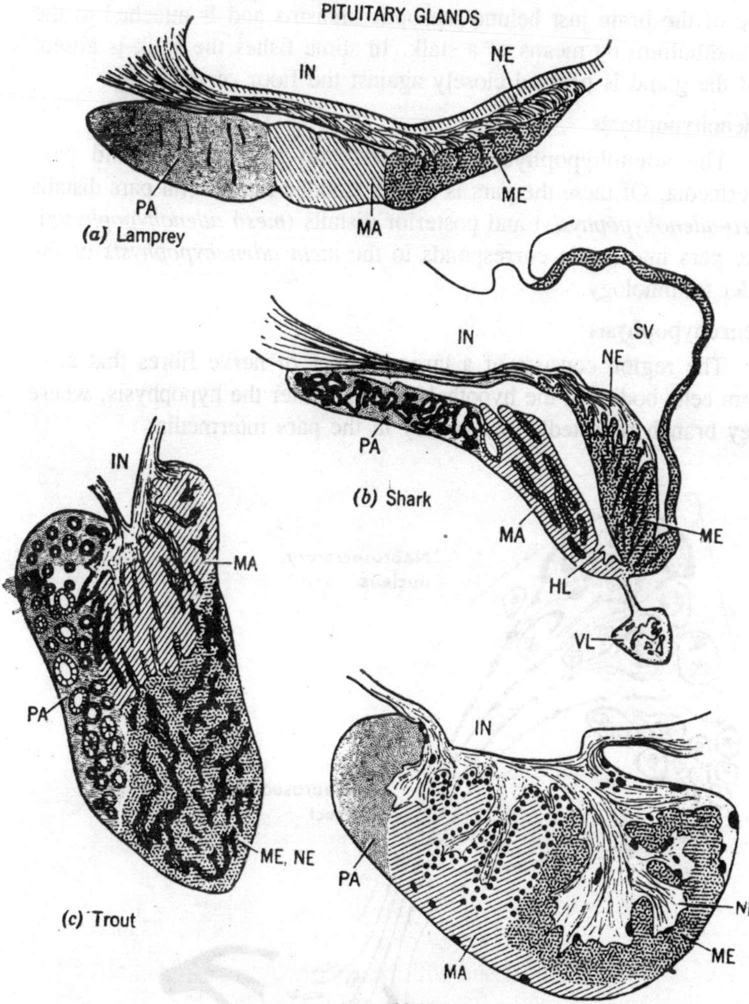

Fig. 12.1. Diagrams of the pituitary gland of fishes: (a) lamprey (Petromyzon); (b) dogfish shark (Squalus); (c) trout (Salmo); (d) perch (Perca). HL–lumen of hypophysis; IN–infundibulum; MA–meso-adenohypophysis; ME–meta-adenohypophysis; NE–neurohypophysis; PA–pro-adenohypophysis; SV–saccus vasculosus; VL–ventral lobe.

area. The pituitary gland or the hypophysis develops from two part as follows: (*a*) Neurohypophysis derived from the brain. (*b*) Adenohypophysis which develops as an ectodermal up-growtu the roof of the buccal cavity (Rathke's pouch). The hypophysis of the ventral side of the brain just behind the optic chiasma and is attached to the infundibulium by means of a stalk. In some fishes the stalk is absent and the gland is pressed closely against the floor of the brain.

Adenohypophysis

The adenohypophysis is divided into pars distalis and pars intermedia. Of these the pars is further divided into rostral pars distalis (*Pro-adenohypophysis*) and posterior distails (*meso adenohypophysis*). The pars intermedia corresponds to the *meta adenohypophysis* of the older terminology.

Neurohypophysis

The region consists of a large number of nerve fibres that arise from cells bodies in the hypothalamus and enter the hypophysis, where they branch repeatedly and ramify in the pars intermedia.

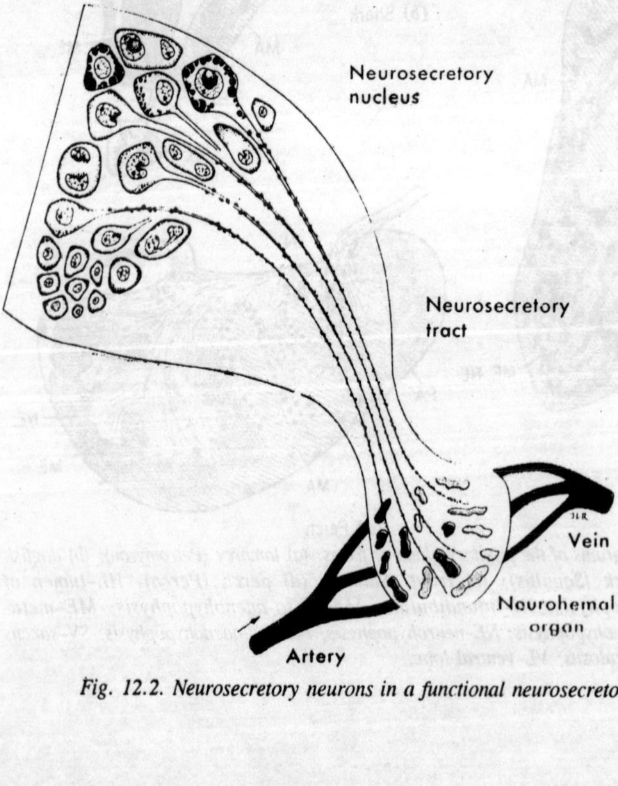

Fig. 12.2. Neurosecretory neurons in a functional neurosecretory unit.

Pituitary in Elasmobranch

The neurohypophysis of this group is diffused and intermingled with the pars intermedia. The two parts are often been collectively referred to as neurointermediate lobe. Neurosecretory axons arose from the paraoptic nuclei and lateral hypothalamus terminate in the neurohypophysis but, these are absent from the anterior influndibulum floor. A few such fibres termed the *saccus vasculosus* are folded and highly vascularized structure laying posterior to the neurohypophysis. The pars distails lies below the infundibulun and is divisible in the procentral and rostral zone. The ventral lobe is a peculiar feature of the Elasmobranches adenohypophysis and varies greatly in size and shape among different species. The presence of an anatomically differentiated media eminence is questionable.

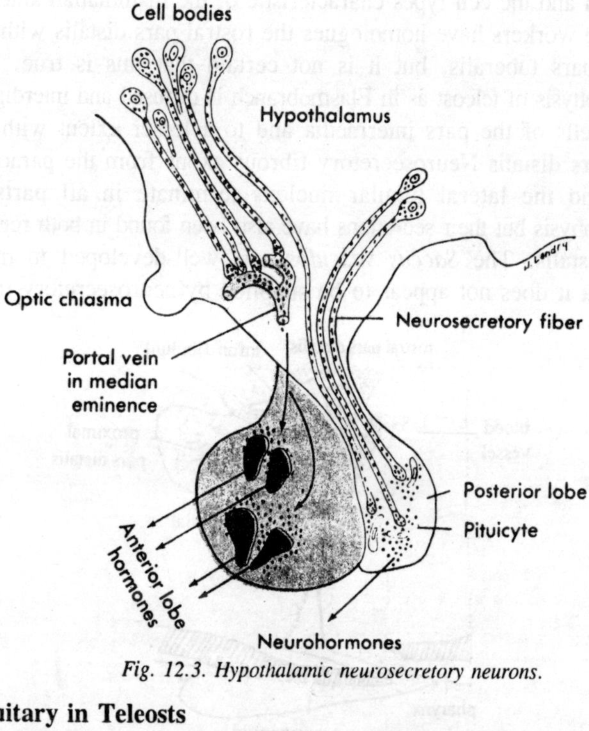

Fig. 12.3. Hypothalamic neurosecretory neurons.

Pituitary in Teleosts

In teleosts the adrenohypophysis consists of a pars intermedia and an intermediate lobe which is sometimes divisible topographically into a rostal and procentral region. This region of pars destalis appears to

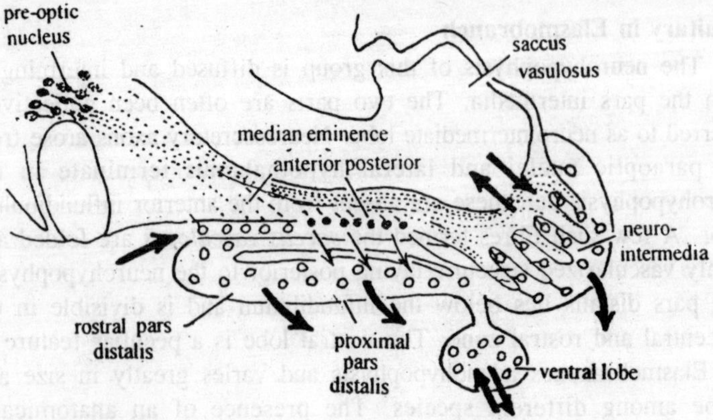

Fig. 12.4. Selacinian pituitary. Solid arrows, veins; broken arrows, arteries; thin arrows, portal veins.

contain cell and the cell types characteristic of the mammalian anterior lobe. Some workers have homologues the rostral pars distalis with the tetrapods pars tuberalis, but it is not certain that this is true. The neurohypophysis of teleost as in Elasmobranch is diffused and interdigites with the cells of the pars intermedia and to a lesser extent with the cells of pars distalis Neurosecretory fibrous axons from the paraoptic nucleus and the lateral tubular nucleus terminate in all parts of neurohypophysis but their secretions have also been found in both regions of pars distalis. The *Saccus vasculosus* is well-developed in many teleosts but it does not appear to be supplied by neurosecretory tract.

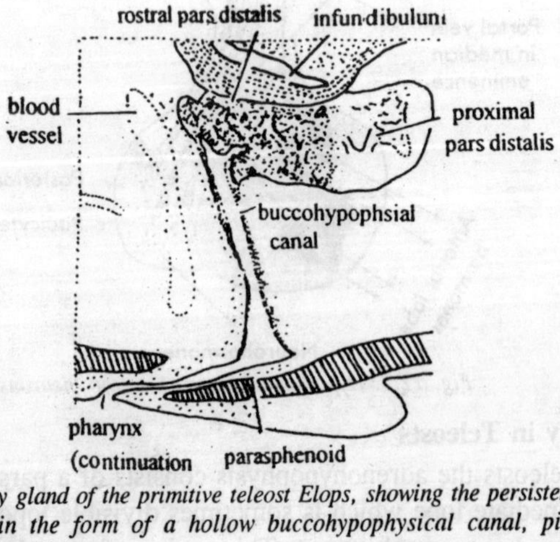

Fig. 12.5. Pituitary gland of the primitive teleost Elops, showing the persistent Rathke's pouch in the form of a hollow buccohypophysical canal, piercing the parasphenoid bone.

Pituitary in Dipnoi

Among lung fishes the neurophypophysis is separated from the pass intermedia and a pars tuberalis absent. A cleft is typically present between the pars destalis and pars intermedia and the saccus vasculoses does not developed. A primitive system of portal veins convergs blood from median eminence to the pars distalis. Except for the pars tuberalis, the Dipnoi pituitary is automatically comparable to the pituitary of tetrapod vertebrates.

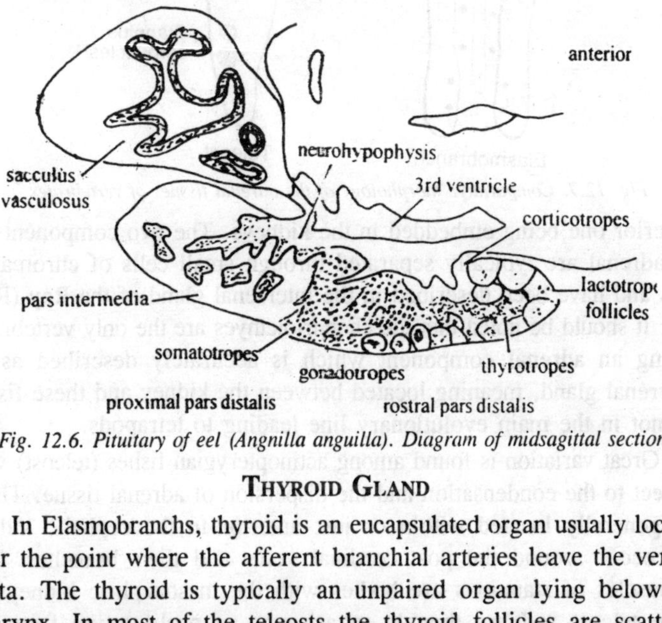

Fig. 12.6. Pituitary of eel (Angnilla anguilla). Diagram of midsagittal section.

THYROID GLAND

In Elasmobranchs, thyroid is an eucapsulated organ usually located near the point where the afferent branchial arteries leave the ventral aorta. The thyroid is typically an unpaired organ lying below the pharynx. In most of the teleosts the thyroid follicles are scattered along the ventral aorta. Thyroid follicles are often found in the head kidneys of the platy fish (Platypolcitus). The number increasing with the age. In a few teleosts such as 'Bermuda' Parrot fish, the follicles are arranged to form a compact unpaired gland.

CHROMAFFIN TISSUE

Adrenal tissue are present in all vertebrates from cyclostomata to mammals but profound differences are encounted in the arrangement of the functional components i.e., steroid producing cells and catecholamine producing cells. In *Elasmobranches* the stereo genetic tissue is composed into several well form bodies lying between the caudal ends are kindeys. These are the *interrenal glands*. Paired aggregation of chromaffin pairs are present between kindeys. The more

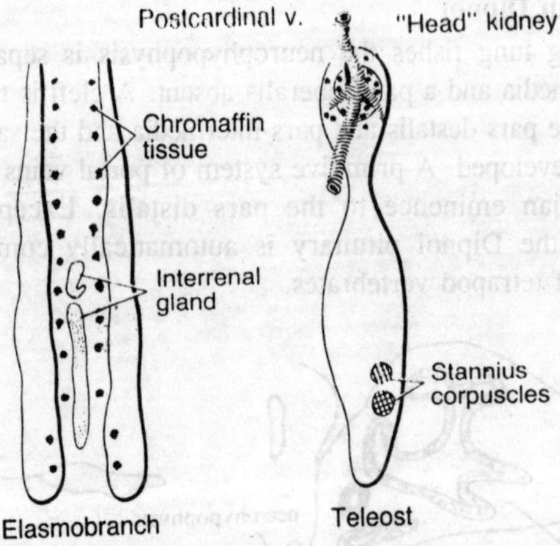

Fig. 12.7. Comparative morphology of the adrenal tissues of vertebrates.

posterior one being embedded in the kidneys. The two components of the adrenal are typically separated through small cells of chromaffin cells and have been described in the interrenal gland of the Ray (Raja sp.). It should be noted that the chondrioicthyes are the only vertebrates having an adrenal component which is accurately described as an interrenal gland, meaning located between the kidney and these fishes are not in the main evolutionary line leading to tetrapods.

Great variation is found among actinopterygian fishes (teleost) with respect to the condensation and the dispersion of adrenal tissue. These are generally located within or just anterior to the cephalic kidney and occur around the post cordinal veins and their branches. The corpuscles of stannious are bodies with the mesonephric kidneys of some teleost and are thought to arise as an proliferation from the urinary duct. Some workers regarded it as *holocrine gland* comparable to adrenal cortices of mammal due to presence of adreno cortical in the corpuscles.

Interrenal Tissue

'*Baecker*'—1928 and '*Dettus*'—1940 have described the interrenal tissue of many fishes. *Abion*-1940 has made an extensive review of the literature.

In Elasmobranchs

In elasmobranchs interrenal is located posteriorly in one or more compact masses between the opisthonephric and is discreate enough to permit surgical removal.

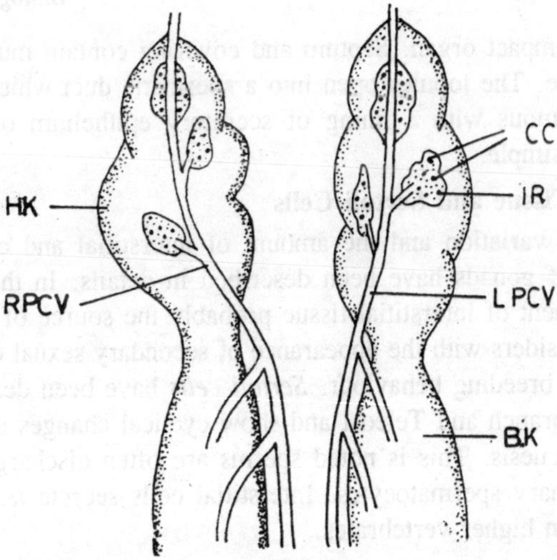

Fig. 12.8. Diagram to show location of interrenal glands in fishes. BC–body kidney; CC–chromaffin cells; HK–head kidney; IR–interrenal tissue; LPVC–left post cardinal view; RPCV–right post cardinal view.

In Teleost

In teleosts the interrenal gland is located historically. Here it is confined to the cranial regions of the head kidney and usually forms a sheath of one or more layers of acidophillic cells around the posterior cardinal veins and their branches.

In Dipnoi

In dipnoi (*Protopterus*) the interrenal and chromaffin cells intermingled and are located along large veins capillaries ventral to the kidney.

GONADS

In fishes, the maturation and possibly also the elaboration of gametes requires the presence of sex hormones in addition to secondary sex characteristics such as coloration, breeding tubercells and the maturation of gonopodium depends upon the pressure of endocrine substances from the sex glands.

Testis

Testis is made up of lobular units. In teleost lobular unit may be short and long tubules and internally divided space with their apex at the centre of the organ and their broader ends directed towards the periphery. In most of fishes however the complex network of elongated and intergreatly divided lobules, and lobular spaces are bound together forming a compact organ. Septum and covering contain much elastic but no muscle. The lobules open into a spermatic duct which may be long and tortuous with a lining of secretory epithelium or may be straight and simple.

Interstitial Tissue and Sertoli Cells

Seasonal variation and the amount of interstitial and connective tissue cells of gonads have been described in details. In this species the development of interstitial tissue probably the source of testicular androgen considers with the appearance of secondary sexual characters and complex breeding behaviour. *Sertoli cells* have been described in both Elasmobranch and Teleost and show cyclical changes in relation to spermatogenesis. This is noted sperms are often discharged at the stage of primary spermatocytes. Interstitial cells secrete *testosterone* hormone as in higher vertebrates.

Ovary

The ovary is made up of numerous ovarian follicles embeded in connective tissue. The arrangement of ovarian follicles within the supporting tissue varies greatly in different groups. New batches of ovarian follicles arise periodically from the germinal epithelium. At first each contains a small yolk egg which is deeply basophillic in fixed and stained material. The follicular cells which surround the developing ova are arranged in several well marked layers. In *Rhinobalus granulutus* (granulosum) an intermediate membranous covering and outer-theca consisting of a compact interna and a closely organized externa. The follicular cells have interesting and varied history. In all cases even a detailed examination has been made. Some of the follicles tend to produce mature ova and the all other become atretic or hypertropic and develop into an enlarged glandular mass of cells which has been termed as a *preovulation corpus luteum*. This development has been carefully studied in cyclostomes and teleosts.

Islets of Langerhans

Within the pancreas of higher forms and sharks, islets of Langerhans develop from the tubules of the otherwise called digestive gland. Islet tissue in the actinopterygian fishes tend to be concentrated. There is usually one islet or sometimes two large islets in the region of the bile duct. In the agnathan fishes pancreatic islets have been observed in the lamprey but not in the hagfish.

In the lamprey there are three masses of cells in the wall of the gut just behind the junction of the pharynx and the midgut. These masses have been shown experimentally to affect the sugar concentration of the blood. Whether or not they actually produce insulin and glucagon hormones. In teleost small masses of islets tissue are located into the region of the gall bladder, spleen, pyloric caeca and small intestine. This may or may not be anatomically related to the zymogenous tissue of different pancreas. There is usually one principal islets mass and in some fishes, the islets tissue is concentration in a single compact organ either caping the gall bladder or located near to it.

Thymus

In the actinopterygian fishes a rather massive thymus is observed in the medial wall of the branchial cavity. This thymns apparently arises from contributions of several of the branchial pouches. The thymus of *Latimeria* is comparable to that of the actinopterygian but is lobular. In the cyclostomes all of the gill pouches give rise to thymus tissue dorsally. The anterior pouches also have ventral anlagen. In the adult only the dorsal parts remain.

Parathyroid

There are no parathyroid glands in the sharks and in the actinopterygian fishes. Unquestionable parathyroids appear first in the amphibians, and this may be correlated with the loss of gills. It should be remarked here that a secretory or endocrine function is assumed for the pseudobranch of fishes, especially those buried in the tissue. In addition the gills of marine teleosts have "chloride secreting" cells which help in maintaining the internal osmotic state. Such secretory cells could be the foreunners of endocrines.

Ultimobranchial Body

An ultimobranchial body is said to be present in the sharks, holocephalans, and some ray-finned fishes. In some of the teleosts, there is a flat discoidal mass lying in the connective tissue between the floor of the esophagus and the sinus venosus. Experimental work with extracts of this tissue suggests that it is parathyroid in function. It has also been viewed as an accessory thyroid. The embryological source of this tissue is in question but has been attributed to the last gill pouches or the wall of the gut in the region of this pair of pouches. In a few teleosts (characids), there are accessory branchial organs, diverticula of the fifth, branchial pouch, which appears to be endocrine in function. The gland does not seems to have been located in cyclostomas.

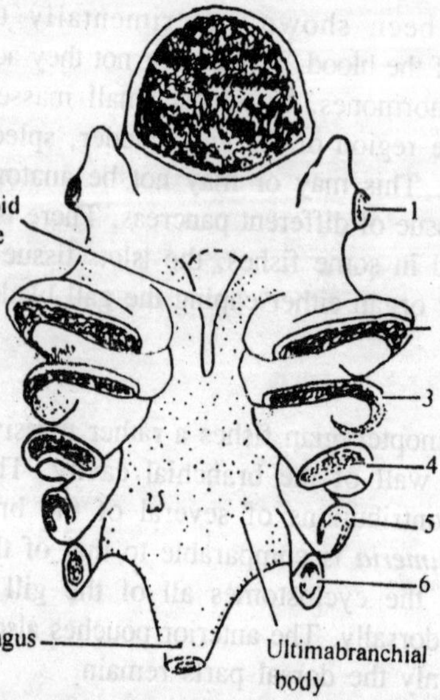

Fig. 12.9. Pharynx of shark embryo, viewed from below. 1. Spiracle; 2. to 6, gill slits.

Intestinal Mucosa

The intestinal mucosa being producing hormones which join with nervous controls to regulate the pancreatic secretions. The intestinal hormones are *secretin* and *pancreozymin*. Secretin produces flow of the enzyme carrying liquids from the pancreas and pancreozymin enhances flow of the zymogens. Both hormones are found in the anterior portion of the small intestine and other tissue with possible endocrine function.

Corpuscles of Stannius

The corpuscles of Stannius, which occur in the kidney of many actinopterygian fishes, have often been viewed as potentially endocrine. These arise as many (40 or more) diverticulae from the nephric ducts in the anterior part of the opisthonephros (segments 9 to 12) of *Amia* and *Lepisosteus* or, in teleosts, as a few or even a single pair of diverticulae in the posterior region of the kidney. These corpuscles have not been observed on other vertebrates and are thus assumed to have evolved within the actinopterygian group. They are not observe in the chondrosteans (*Acipenser*). It has been suggested that they are homologous with the *Mullerian ducts*, but such an homology is unlikely. Furthermore, these corpuscles are not a part of the adrenal system. Experimental work has revealed no function for them. They appear to be only modified kidney tubules, perhaps related to the Leydig's glands of chondrichthians. The number of these appears to have undergone reduction in the teleosts. There are none is some.

Pineal Organ

The pineal organ, outgrowth from the roof of the brain, is recognized as a *photoreceptor* in fishes (it is vestigial in the hagfish), but its glandular structure suggests endocrine activity. Removal of the pineal organ of the guppy (*Lebistes*) is followed by reduced growth rate, skeletal abnormalities, are marked stimulation of both pituitary and thyroid glands. Increased activity was also noted in the corpuscles of Stannius and in remnants of the pronephric tissue. It is suggested that the action of the pineal is secretory and its action is mediated through the pituitary and thyroid glands. The presence of a pineal nerve also indicates a more direct (nervous) association.

Table 12.1. Endocrine Glands in Fishes

Name of the gland	*Secretion*	*Action*
1. *Pituitary*		
A. Posterior lobe (Neurohypophysis)	Vassopressin, oxytocin and antidiuretic hormone.	Cell permeability (?) (?)
B. Intermediate zone or lobe. [Meta adenohypophysis (GH)] Growth hormone	Intermediate (MSH)	Malanogenesis. malaim disperson. Size increase and metabolic effect (?)
C. Transitional lobe or zone (Mesoadenohypophysis)	Corticotropin (ACTH)	Controls crotin secretion (melanogenesis)
D. Ant. lobe or zone (Proadenohypophysis)	(I) Thyrotropin (TH)	Controls thyroid secretion
	(II) Gonadotropin (LH, FSH)	Controls secretion of gonadal hormones
	(III) Prolactin	Melanogenesis together with MSH
	(IV) Hypothalamus producing substance (?) Malanophore concentrating hormone (?)	Balancing through muscular contractions.

2. *Thyroid gland*	Thyroid hormone (S)	Inhance maturity, controls metabolism but apparently has no direct control of iodine metabolism, involved in metamorphic changes and possible in osmoregulation.
3. *Ultimobranchial*	Unknown	Ca^{++} metabolism (Parathyroid function)
4. *Supra-renals chromaffin (along kidneys)*	Epinephrine	Concentrate pigment granules in melanophore, controls B.P. rise and affects pulse rates etc. (Effects comparative to those of sympathetic nervous system)
5. *Adrenal cortical tissue*	Adrenal caortical steroides	Promote utilization of stored fat, carbohydrate, metabolism, Protein catabolism. No relation and electrolite metabolism.
6. *Pancreatic islats*	Insulin	Carbohydrate metabolism.
7. Sex glands	Androgen (S) (?) in male.	Development of secondary sexual characters, reproductive behaviour, maturation of gametes.
	Estrogen (S) (?) in female).	Development of secondary sexual characters reproduction behaviur, maturation of follicles.
8. Intestinal mucosa	Secretin (Pancriozymin)	Stimulates pancreatic secretion.

13

SUPPORTIVE SYSTEM

The term *skeleton* refers to the framework of the animal body. In vertebrates it is composed of cartilage, bone, or a combination of the two and serves for support, attachment of muscles, and the protection of certain delicate vital organs which it more or less completely surrounds. Furthermore, it maintains the definite form of the animal and is provided with joints so that movement is made possible. The word *endoskeleton* is used to denote internal skeletal structures. The presence of an endoskeleton is one of the distinguishing characteristics of chordates. Although in most chordates the integument is soft and contains no hard skeletal parts, there are many members of the phylum in which bony elements are present in the skin, being derived from the dermis and providing the animals with a protective armor. Dermal scales are present in most fishes living today, in a few amphibians, in crocodilians, turtles, and even armadillos among mammals.

The term *dermal skeleton* is used in referring to such structures and their derivatives. Sometimes the dermal skeleton is spoken of as the *exoskeleton*, but the latter term is more properly used in connection with skeleton of invertebrates. It is difficult to decide whether the endoskeleton or the dermal skeleton is phylogenetically older. The presence of bony dermal plates in the fossil remains of extinct ostracoderms and placoderms indicates that they appeared very early in the evolutionary history of vertebrates and should be considered to be primitive structures. However, during the embryonic development of animals possessing both an endoskeleton and dermal skeletal structures, the endoskeleton appears much earlier than the dermal skeleton. This would lead one to infer that the endoskeleton is phylogenetically older. Although many modern vertebrates at first sight appear to lack a dermal skeleton the, membrane bones of their skulls, as well as their teeth, are believed to represent dermal skeletal elements.

Axial Skeleton

Vertebral Column

The vertebrae of fishes exhibit a broad spectrum of morphological from species to species. This is attributable to the varying degrees to which scleroblasts invade the notochord sheath and the notochord proper, to the nature and extent of the contribution of the sheath and notochord

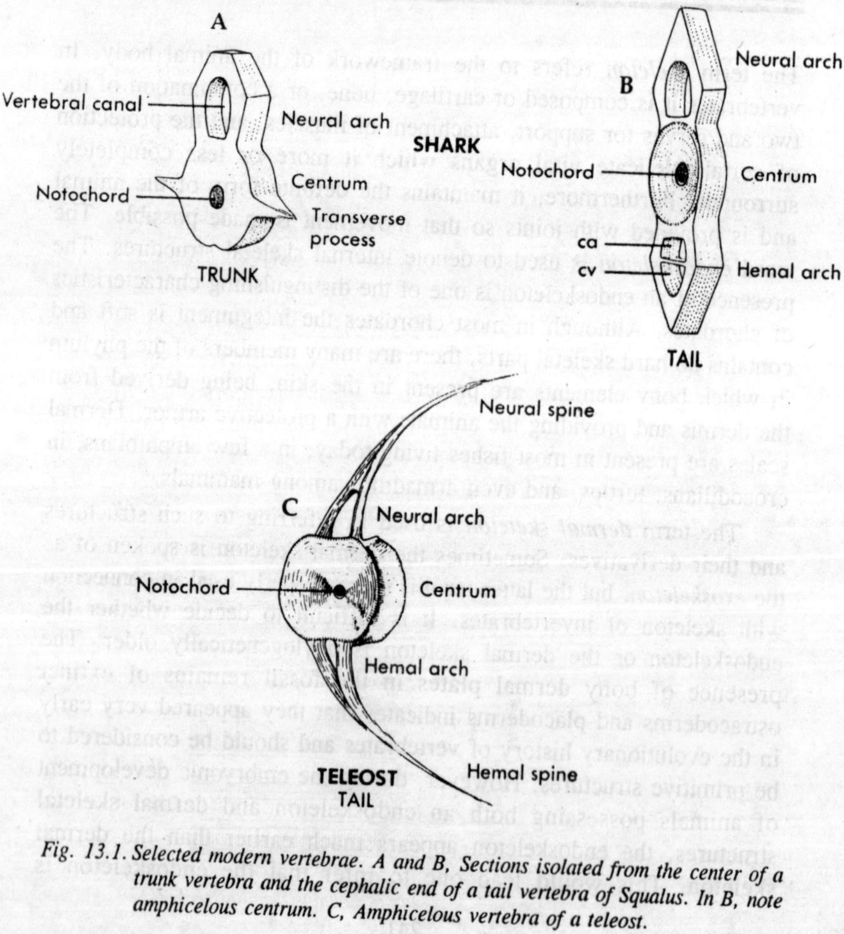

Fig. 13.1. Selected modern vertebrae. A and B, Sections isolated from the center of a trunk vertebra and the cephalic end of a tail vertebra of Squalus. In B, note amphicelous centrum. C, Amphicelous vertebra of a teleost.

to the adult centra, to the number of chondrification and ossification centres that develop in the perichordal blastemas, and to the extent to which separate foel of skeletogenic activity remain independent or unite during ontogeny.

Although a fish has only two varieties of vertebrae, *trunk* (dorsal) and *coudal*, both may differ gradually along with length of the column from lead head to vent, and from vent to the tip of the nail. Agnathans have a strange vertebral column if, indeed, they may be said to have one. The only skeletal elements are *lateral neural cartilages*, one or two pairs per body segment depending on the species. Caudally, the lateral neutral cartilages fuse to form a single dorsolateral cartilaginous plate perforated by foramina for spinal nerves. In hagfishes, lateral neural cartilage, are limited to the tail. These lateral neural cartilages may be vestigial vertebrae, primitive vertebrae, or they may bear no phylogenetic relationship to vertebrae. Whether you call them vertebrae depends on how you wish to define "vertebra". How would you define the term?

In sharks, the notochord is present throughout the length of the vertebral column and is considered within each centrum. As a result, the centra composed of chordal and perichordal cartilage, are concave at each end. A centrum that is concave at each end is said to be *amphicoelous*. The spinal cord is enclosed in a continuous tunnel, the *vertebral* (*neural*) *canal*, consisting of paired *dorsal plates* that constitute a neural arch, paired *dorsal intercalary* plates between the arches, and, in some species, paired wedge-shaped *supradorsal cartilages*. *Squalus acanthias* lacks supradorsals. The neural arches and intercalary

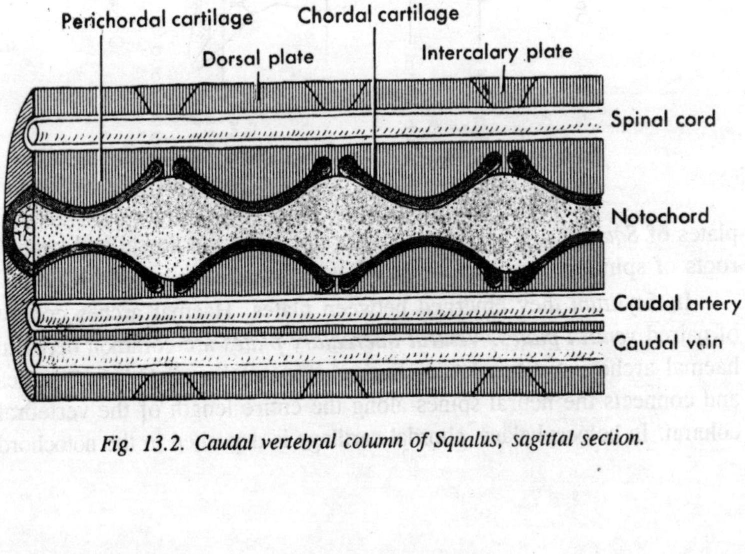

Fig. 13.2. Caudal vertebral column of Squalus, sagittal section.

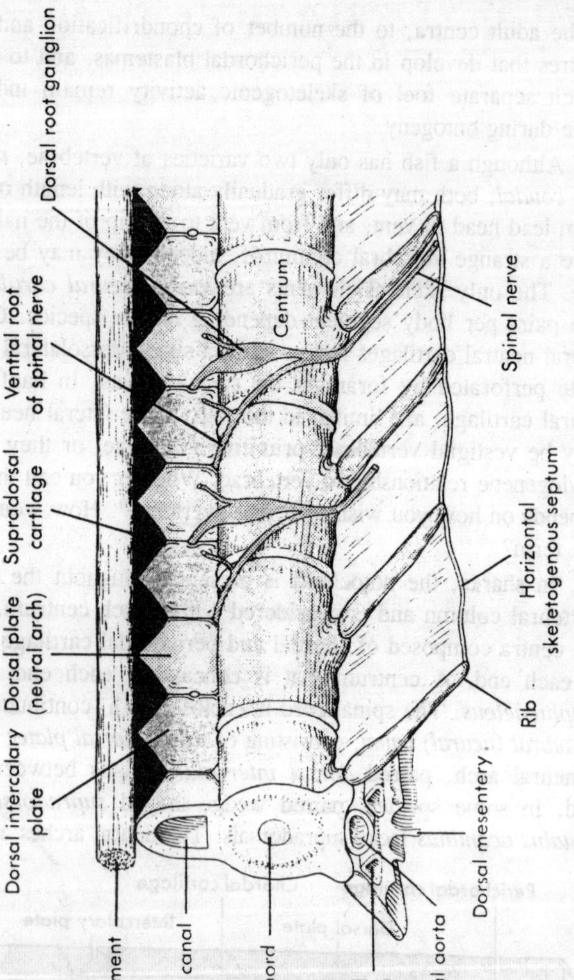

Fig. 13.3. Vertebral column and associated structures of trunk of Scyllium, a shark.

plates of *Squalus* are perforated by foramina for the dorsal and ventral roots of spinal nerves and for blood vessels.

In *Scyllium* they emerged between plates. Haemal arches consist of paired *ventral plates*. *Ventral intercalary plates* are common between haemal arches. A fibroelastic, ligament present in all fishes, overlies and connects the neural spines along the entire length of the vertebral column. In holocephalans, chordal cartilage is deposited in the notochord sheath, which becomes greatly thickened. Thereafter, calcification may convert the sheath into calcified cartilaginous rings that are much more numerous than the body segments. The rings constitute the major component of the column in holocephalans. Any minor components resulting from perichordal chondrification differ along the length of the column and from species to species. Modern dipnoans, chondrosteans other than *polypterus*, and the cociacanth *Latimeria* have no centra. The notochord is present and uncostricted throughout its length, and its thick fibrous sheath contains little chordal cartilage or bone.

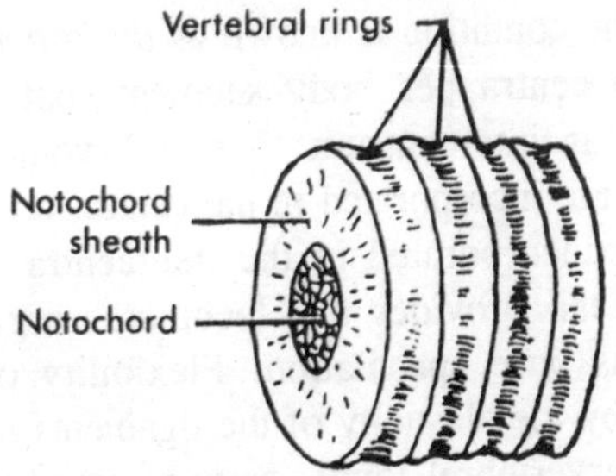

Fig. 13.4. Calcified vertebral rings from an extinct holocephalan.

Associated with the notochord in each body segment are paired *basidorsal*, *basiventral*, *interdorsal* and *interventral* cartilages, which are products of isolated skeletogenic loci in the embryonic perichordal blastema. The chondrostean *Polypterus* and living holosteans have well-ossified complete centra. Teleosts have well-ossified amphicoelous vertebrae. The core of each centrum is a dumbbell-shaped vacuole where the notochord previously existed, and the space between successive vertebrae is occupied by a porous cartilage-like material that probably includes notochordal tissue. The centra and neural arches of successive vertebrae are interconnected by a complex system of collagenous and elastic ligaments that facilitate lateral undulation of the body for locomotion. In the posterior trunk and tail these ligaments are frequently ossified to form a long, delicate, bony rod. The neural spines are often very tall, and in lower teleosts they are sometimes surmounted by *supraneural bones*.

A variety of processes, mostly unlike those of tetrapods, protrude from the arches and centra. Elasmobranchs and many bony fishes have two centra and two sets of arch in each metamere of the tail, and a few have duplication of centra and arches in the trunk also. Consequently, the number of centra in those regions is double the number of myomeres

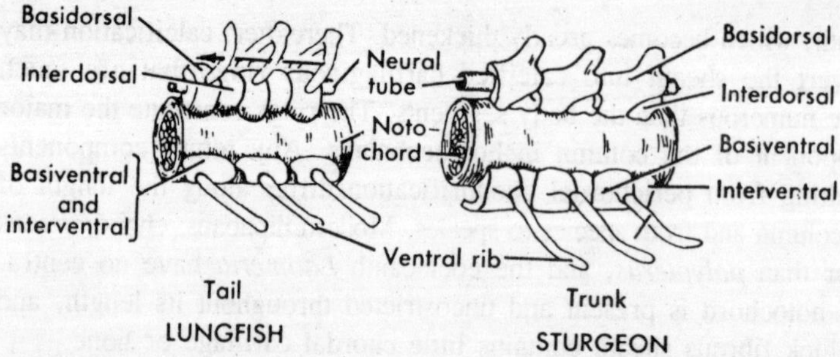

Fig. 13.5. Vertebral components in adult lungfish (Neoceratodus) and sturgeon.

and spinal nerves. The condition is known as *diplospondyly*. *Amia* has what looks like two centra per body segment, but only one bears arches. The other is an intercentrum. The embryonic basidorsal and basiventral cartilages are incorporated in the centra, and the inerdorsals and interventrals are incorporated in the intercentra. It may be that diplospondyly in the tail provides this locomotor organ with greater flexibility, but this is pure speculation. Flexibility of the vertebral column is facilitated by the elasticity of the ligaments that interconnect vertebrae, not by intervertebral joints, as in tetrapods. No movement is possible between the first vertebrae and the skull. They are united by cartilage or nonelastic connective tissue. The terminal vertebrae of fishes are part of the caudal fin.

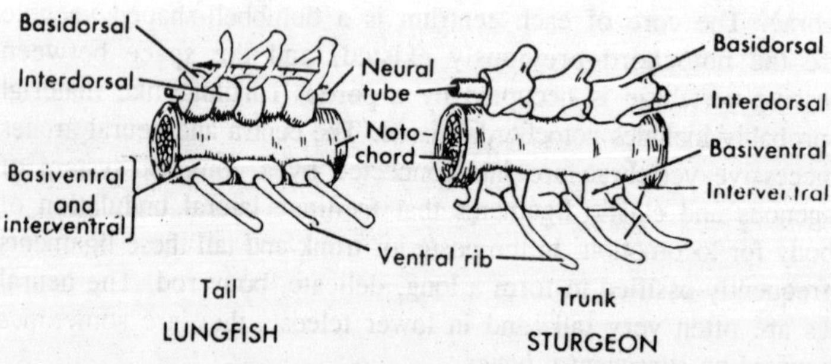

Fig. 13.6. Tail vertebrae of Amia.

Median Fins

Median fins include the *dorsals*, the *caudal*, and the *anal*–these are structures limited to fishes or at least specialized aquatic forms. The tetrapod groups are not involved in this discussion although it can be noted that whales have dorsal fins and *caudal flukes*. Similar fins were present in the ichthyosaur reptiles (also mosasaurs and others) with the difference that the caudal flukes of the whale are fleshy transverse planes, while the tail of the aquatic reptiles was like that of fish.

Dorsal and Anal Fins

In actinopterygian fishes there is great variety of fin structures and fin arrangements. The dorsal fin may be preceded by a series of *spines*, as in the perch; this fin may be divided into two parts or it may be continuous with the caudal fin. The salmon has a small "*adipose*" fin between dorsal and caudal. This small fin seems to be a new structure peculiar to some of the teleosts. The skeletal support of the dorsal fin or its parts is fairly simple. *Radials* extend upward from the spinal column toward the fin, and on the ends of these there are one or two *basal elements* which support the *fin rays* or spines. The anal fin is similarly constructed. In the more primitive (fossil) actinopterygians, there was a single dorsal fin, an arrangement retained in *Acipenser* or *Lepisosteus*. This style is obscured in *Amia* where the dorsal extends much of the length of the body, and in *Polypterus* where many *finlets* replace the usual dorsal fin. These finlets occupy much the same space as the dorsal fin of *Amia*, but there is no gap between them and the caudal fin.

Most actinopterygians have a simple anal fin, but sometimes it becomes continuous with the caudal or is subdivided into finlets. In the choanate fishes, there were two dorsals, the more posterior one generally the larger. There was a single anal in the same position as that of the actinopterygian. *Latimeria* has brought this pattern down to the present, differing in that the anterior dorsal is the larger. These fins in the rhipidistian or actinistian have large, plate-like basal structures, with or without radials, in lobe-like extensions of the body resembling those of the lateral fins The dipnoans retain the simplest relationship between basal, radical, and fin ray.

The Chondrichthyes, generally have two dorsals but the most primitive living genera (*Ohlamydoselachus*, *Heptanchus*, *Hexanchus*) have only one. This exceptional condition may be due to secondary loss of one fin, or evidence of both patterns of dorsal fins in sharks. The known Devonian species of sharks had two fins and this suggests that loss has occurred in this primitive living genera. The Holocephalans have two dorsals; the second is long and low in *Hydrolagus*. The skeletons of these fins are complex in the shark, resembling those of the lateral fins or being simply a jumble of closely packed cartilaginous elements. The anterior dorsal of *Hydrolagus* has a massive basal plate with a large anterior spine. The posterior dorsal has simple basals with no apparent radials. In Hydrolagus only the anterior dorsal has a spine and this is very large.

Caudal Fin

The caudal fin is an important contributor to forward thrust in swimming that is why it is well-developed in most fishes. The caudal fins may be divided into several categories on the basis of the position of the vertebral column in caudal region and the fin lobes (Already given in the text). Each median fin is supported by a series of parallel bony rods called *radials* and the dermal fin rays or the *dermotrichia*. A typical radial consists of three pieces, of which the proximal segment is large and roughly dagger shaped. It is called the interspinous bone or the *axonost*, as it has between two neural spines. The middle piece is short while the distal pieces consists of very small, double bony nodule. The three segments of each radial are disposed in an obliquely backward direction. The dermotrichia in *Labeo* are jointed and branched structures having a bony texture. Besides these a second set of horny rays called the *active trichia* are also present to support the free margin of the fin.

Ribs

Ribs consist of a series of cartilaginous or bony elongated structures attached at their proximal ends to vertebrae. In the primitive condition there is pair of ribs for each vertebra. In some vertebrates the ribs consist of little more than cartilaginous tips on the transverse processes of the vertebrae. They may be fused permanently with the transverse process *pleurapophyses*) or else a joint may appear between the two. In higher forms they may be stout, strongly arched structures surrounding the thoracic cavity and uniting ventrally with a median bony structure, the *sternum*, or *breastbone*. The ribs in these cases protect such vital organs as the heart and lungs by enclosing them with a basketlike framework. Ribs first appear in fishes and are of two kinds, dorsal and ventral. Many fishes possess either one type or the other, but in *Polypterus* and many teleosts, both types are present.

Dorsal Ribs

The upper, *intermuscular* or *dorsal ribs* extend out as costal processes from the transverse processes of vertebrae into the lateral skeletogenous septum which separates epaxial and hypaxial muscles. They occur metamerically at points where the myocommata, or septa between adjacent myotomes, cut across the lateral septum. Some author-ities are of the opinion that true dorsal ribs are present only in

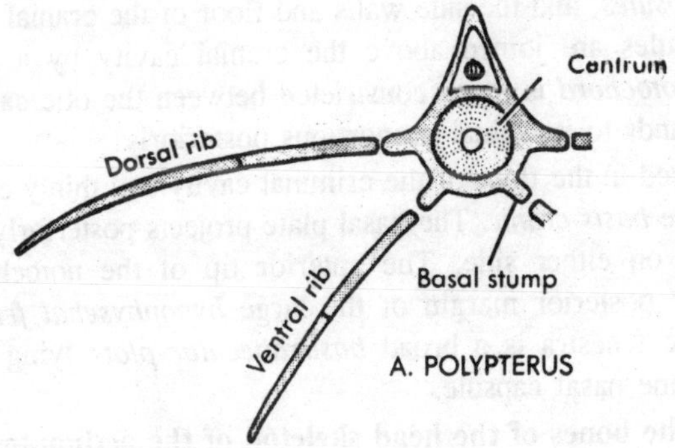

Fig. 13.7. Dorsal and ventral ribs of some fishes.

fishes. Most, however, believe that the ribs of tetrapods are homologous with the dorsal ribs of fishes.

Ventral Ribs

The ventral, or *pleural ribs* of fishes are attached to the ventrolateral angles of the centrum. They lie beneath the muscles of the body wall just outside the peritoneum and are situated along the lines where the myocommata pass toward the coelom. There is general agreement that the ventral ribs represent the two halves of haemal arches of caudal vertebrae which have split ventrally and spread apart. However, this conception may be erroneous, for the caudal vertebrae of some fishes are reported to possess both ventral ribs and haemal arches.

Additional intermediate bones having a rib-like appearance are frequently found embedded in the myocommata of fishes. In elasmobranch, small ribs, attached to the ventrolateral angles of the centrum (*basapophyses*), extend laterally into the skeletogenous septum. It would at first sight appear as though these are dorsal ribs, but it is generally agreed that they more probably represent ventral ribs which have moved secondarily into this position. The caudal vertebrae bear haemal arches with which the ventral ribs are homologous. In most of the higher fishes only ventral ribs are present. However, the chondrostean *Polypterus*, previously mentioned, as well as the salmon and a number of other teleosts, possess both dorsal and ventral ribs.

Head Skeleton

Cyclostomes

There are two types of cyclostomes, the lampreys and the hagfishes. The lamprey head skeleton consists of an *endocranium* and an extensive *visceral skeleton*. The endocranium includes a single *nasal capsul*, paired *otic capsules*, and the side walls and floor of the cranial cavity. The otic capsules are joined above the cranial cavity by a *synotic tectum*. The *notochord* is much constricted between the otic capsules; it rapidly expands to its spinal proportions posteriorly.

It is exposed in the floor of the criminal cavity and thinly covered ventrally by the *basis cranii*. The basal plate projects posteriorly along the notochord on either side. The anterior tip of the notochord is exposed at the posterior margin of the large *hypophysehal fenestra*. Anterior to the fenestra is a broad *basitrabecular plate* lying below the large midline nasal capsule.

Table 13.1. The bones of the head skeleton of the actinopterygian fishes arranged according to type and place of origin.

Cranium (brain and sense *organ capsule)*	Splanchnocranium or Visceral *Skeleton (dermal bones with)*
Endocranium (bones of chondral origin except where marked with an)	Mandibular arch–Palatoquadrate and Meckel's cartilage
basioccipital	premaxilla
exoccipital	maxilia
supraoccipital	quadratojugal
intercalare	quadrate
epiotic	metapterygoid or
prootic	epipterygoid
sphenotic	autopalatine
basisphenoid	pterygoid
orbitosphenoid (anterior	ectopterygoid
and posterior parts)	palatine
mesethmoid	
lateral ethmoid	articular
kinethmoid	retroarticular
pre-ethmoid	Meckelian ossicle mental
Eye	angular
	supraangular
sclerotic plates	dentosolenial or dentary
Dermocranium (bones of membrane origin)	coronoids
	Hyoid arch
Roof, snout, and floor	

extrascapulars
postparietal
parietal
supratemporotabular
intertemporal
nasal
internasal
vomer
parasphenoid

Circumorbital and sensory line of snout

supraorbital
postorbital (s)
jugal,
infraorbital
lacrimal (s)
lateral rostral (s)
medial rostral

Check

preopercle
suprapreopercle

opercle
subopercle
interopercle
branchiostegal rays
lateral gulars
anterior medial gular
hyomandibula
symplectic
interhyal
ceratohal (anterior and posterior parts)
hypohyal
basihyal

urohyal

—Unclassified

entoglossal

—Branchial arches

pharyngosuprabranchials I to IV
epibranchia's I to IV
ceratobranchials I to IV
hypobranchials I to IV
basibranchials I to IV.

Behind the neurocranium one or more of the "*occipital*" *arch* elements can be viewed as representing the occipital segments of the gnathostome head skeleton. The ninth and tenth cranial nerves exit behind the otic capsule, as does an occipitospinal nerve. The next segmental nerve perforates the base of the first neural arch. The side wall of the neurocranium is joined to the otic capsule at three points: above, anteriorly, and ventro anteriorly.

The dorsal opening formed by these connections is filled with connective tissue. The more ventral, larger opening is used by the trigeminal, abducens, and trochlear nerves. Anterior to this opening is a large optic foramen and just behind this a small occulomotor foramen; these may be confluent. Ventroanterior to the optic foramen is a small arterial (orbital or facial) foramen. The otic capsule has a large *interna auditory fenestra* for the common root of the seventh and eighth nerves and above this a good sized *endolymphatic foramen*. The facial nerve leaves the eighth just inside the capsule and passes forward and downward through a small canal opening between the capsule and the cranial base just behind the trigeminal foramen. The nasal capsule is thin walled and is separated from the rest of the endocranium. It encloses the posteroventral half of the *nasal sac* and is perforated by two large *olfactory foramina* separated by a rather thick strut.

Attached to the end of ocanium is a *subocular arch*. From this such a *styliform* process extends down to the side of the pharyngeal cavity. Its tip is attached by fibrous cartilage to the *cornual cartilage* extending forward below the pharyngeal cavity alongside the *piston cartilage*. Medical to the styliform processes lie small *velar cartilages*, which join below the *pharyngoesophageal tube* to support the anteriorly projecting fingers of the velum. The anterior end of the basitrabecular plate articulates with the *posterior tectal cartilage*.

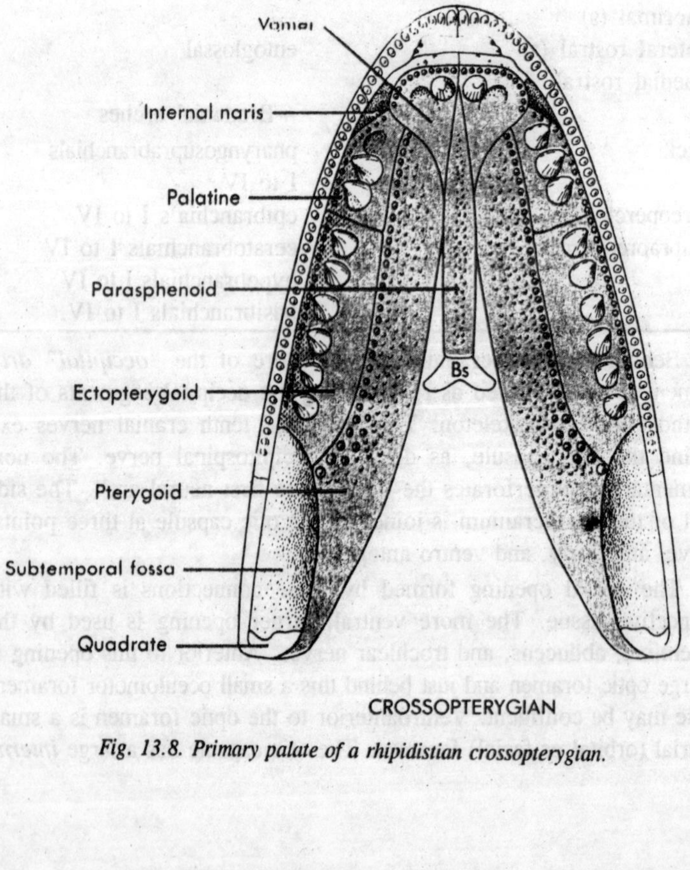

Fig. 13.8. Primary palate of a rhipidistian crossopterygian.

From the lateral margins of this *ethmoid processes* extend ventrolaterally and slightly forward as part of the subocular arches. The distal end of the ethmoid process is joined by soft cartilage to the *posterior lateral cartilage*. In front of an overlapped by the postcrior tectal cartilage in an *anterior tectal cartilage*. This has bilateral *oral processes*, which extend posteriorly and down on either side to attach to the general tissue mass above and medial to the bases of the *spinous cartilages*. Just anterior to the oral processes lie the small *anterior lateral cartilages*, which are closely bound into the general supporting mass of the tip of the tongue apparatus. The anterior tectal cartilage overlaps the annular cartilage of the oral sucker.

The *annular cartilage* supports an anterior pair of teeth and a semicircle of eight posterior teeth. Posterolaterally spinous cartilages (stylets) extend backward from the annular cartilage. These spines lie to either side of the midline *copular cartilage*. This lies below the anterior end of the piston cartilage, which bears on its tip a medial *apical cartilage* and bilateral *supraapical cartilages*. The latter support

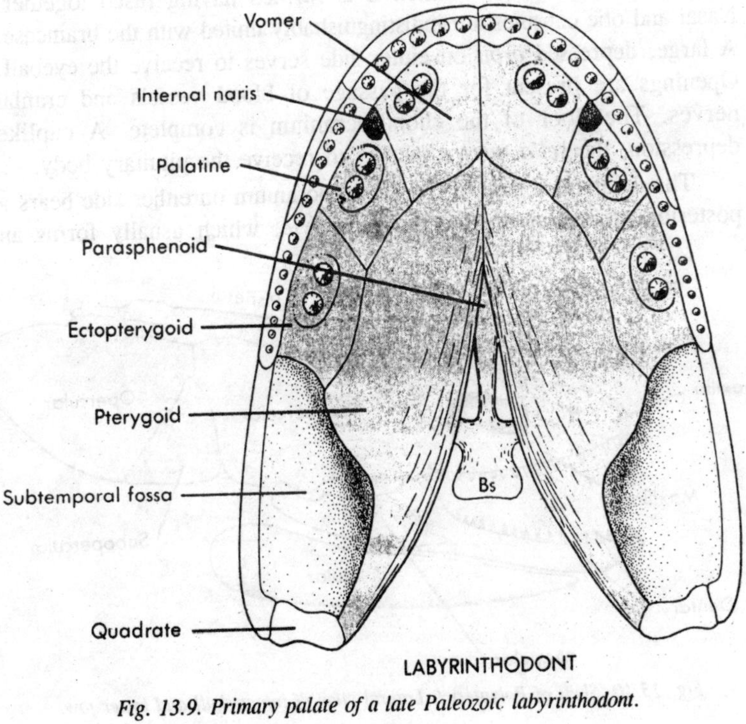

Fig. 13.9. Primary palate of a late Paleozoic labyrinthodont.

the anterior and posteriolateral *tooth plates* of the rasping apparatus. The branchial skeleton is complex in form. It is composed of soft elastic cartilage, which in preserved specimens is darker than the cartilage of the mouth area and neurocranium.

The arches are attached dorsally to a stip of soft cartilage extending along the side of the notochord and attached to the neurocranium behind the otic capsule. The basket is also attached by a band to the subocular arch just lateral to the base of he styliform process. The branchial skeleton is peculiar in that the arches are not divided into segments and in having epi and hypotrematic horizontal connectives and a pericardial basket enclosing the heart.

Gnathostome Fishes

Elasmobranchs

The skull in elasmobranch fishes is entirely cartilaginous. The cartilage is frequently calcified, however. Absence of dermal bones is presumably a secondary development and does not represent a primitive condition. The chondrocranium consists of a single piece of cartilage, the separate elements of which it is formed having fused together. Nasal and otic capsules are indistinguishably united with the braincase. A large, depressed *orbit* on either side serves to receive the eyeball. Openings are present for the passage of blood vessels and cranial nerves. The floor of the chondrocranium is complete. A cuplike depression, the *sella turcica*, serves to receive the pituitary body.

The occipital region of the chondrocranium on either side bears a posterior projection, the *occipital condyle*, which usually forms an

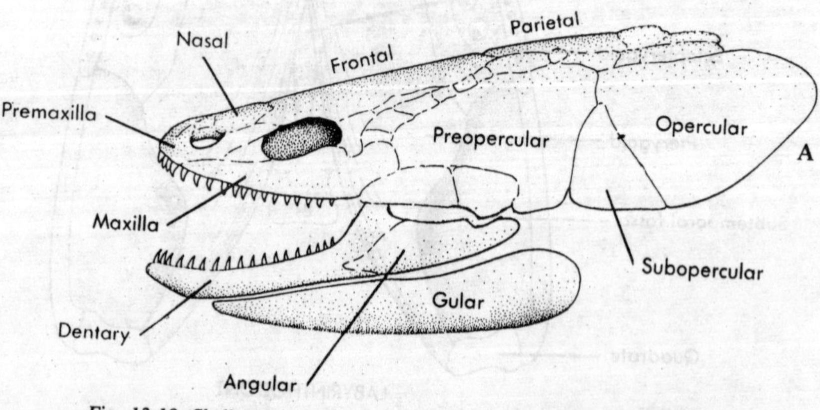

Fig. 13.10. Skull of Polypterus. Lateral view of intact skull and lower jaw.

immovable articulation with the first vertebra. Perhaps the most peculiar feature of the elasmobranch chondrocranium is the *rostrum*, an anterior, trough-like extension, formed originally from the prechordal cartilages. The presence of the rostrum causes the mouth to be subterminal and ventral in position. In the sawfish the rostrum is a very extensive projection, supplied along the sides with numerous sharp teeth and forming the formidable, sawlike weapon with which this fish is supplied. In some of the larger sawfishes the rostrum may reach 6 ft. in length and 1 ft. in width. The visceral skeleton of elasmobranchs is typically composed of seven cartilaginous visceral arches surrounding the anterior part of the digestive tract and furnishing a firm support to the gills.

The first, or mandibular arch is larger and more conspicuous than the rest. From it are derived both upper and lower jaws. On each side the mandibular arch divides into a dorsal palatopterygoquadrate bar and a ventral Meckel's cartilage. These articulate with each other posteriorly. Both cartilages are suspended, by means of a ligamentous attachment, from the hyomandibular cartilage of the hyoid arch which in turn is joined to the otic region of the chondrocranium. This is the *hyostylic* method of jaw attachment. The palatopterygoquadrate bars of the two sides make up the upper jaw. They are heavy cartilage which unite with each other anteriorly by means of a ligamentous connection.

The anterior, or palatine portion of each palatopterygoquadrate bar usually bears a *palatine*, or *orbital process* which extends upward along the inner wall of the orbit. The two Meckel's cartilages, forming the lower jaw, are united anteriorly by a ligament. Both palatopterygoquadrate bar and Meckel's cartilages bear teeth. Small *labial cartilages* of doubtful homologies are embedded in the tissue just outside the upper and lower jaws near the anterior end. The second, or *hyoid arch* is composed of three parts. An upper *hyomandibular cartilage* articulates with the chondrocranium behind the orbit in the otic-region. Below this is the *ceratohyal cartilage*. The ceratohyals of the two sides are joined together ventrally by means of a median plate, the *basihyal*. The posterior edges of both hyomandibular and ceratohyal cartilages give off numerous slender cartilaginous *gill rays* which support the tissue bearing the gill lamellae composing the first, or *hyoid hemibranch*. The number of visceral arches posterior to the hyoid varies in elasmobranchs, but usually are five. Each consists of several cartilages on each side. Named in order from dorsal to ventral borders they are (1) *pharyngobranchial*, (2) *epibranchial*, (3) *ceratobranchial*, and (4) *hypobranchial*.

The pharyngobranchials slope posteriorly and terminate just below the vertebral column. They are attached to the vertebral column by fibrous bands of tissue. The hypobranchials may meet their partners in the midventral line or are connected by unpaired *basibranchial cartilages*. The epibranchials and ceratobranchials bear the gill rays which support the remaining gill. Gills rays are wanting on the last visceral arch which, as previously mentioned, does not bear a gill. Short but strong, pointed cartilaginous projections called *gill rankers* extend into a pharynx from the inner edges of the visceral arches. They are covered with mucous membrane and are used to strain food particle, from the water.

Other Fishes

It has already been fold that the bony fishes belonging to the class Osteichthyes first appeared on earth during Devonian times, around 350 million years ago. In evolutionary terms they are much older than members of the class Chondrichthyes. The presence of a bony, skeleton, present even in ancestral placoderms, therefore, must be considered to be primitive feature which has been retained. Palaentological evidence indicates that very early in their history the bony fishes had already divided into two branches, the ray-fins, or Actinopterygii, and the lobe-fins, or Sarcopterygii.

The subclass Sarcopterygii contains members of the order Crossopterygii, primitive relatives of which were undoubtedly the ancestors of tetrapods. Since Crossopterygians were near the bottom of the ladder, so to speak of their skull structure is of particular interest in drawing parallels between the elements making up the skulls of modern fishes and those of tetrapods. The skulls of the ray-finned fishes, with the exception of *Polydon*, the spoonbill or paddlefish, are invested with numerous membrane bones, several of which are homologous with similarly placed and similarly named bones of crossopterygians and tetrapods. There has been much confusion in the past in interpreting such homologies, and some bones have been incorrectly named.

A chief difference between the skulls of the ray-fins and crossopterygians is that in the former group the front part of the skull is relatively long and the posterior, or cheek, region is short. The opposite condition was evident in the Crossopterygians, in which the anterior region was short and the posterior region tended to be elongated. Another, and significant, difference is that in the ray-fins there is normally a lack of development of a homologue of the squamosal bone present in Crossopterygians and tetrapods. Among the skulls of fishes are to be found all the transitional stages from those which are completely cartilaginous to those which are almost completely bony.

In holocephalians the skull is composed entirely of cartilage. Noteworthy is the fact that the entire palato-pterygoquadrate bar is immovably fused to the cranium. The lower jaw is suspended from the quadrate region and the hyoid arch plays not part in suspending the jaws. This is a variation of the *autostylic* method of jaw suspension. The skull articulates with the vertebral column by means of a movable joint. The hyoid arch is developed only slightly more than the remaining visceral arches. The hyomandibular portion is degenerate. The ceratohyal component, with its gill rays, supports the *operculum*, which first makes its appearance in this group of fishes.

The sturgeon, *Acipenser*, has a skull almost completely formed of cartilage. No cartilage bones are present although ossifications appear in the otic and orbitosphenoid regions. Several well-developed membrane bones have made their appearance. On the dorsal side they include frontals, postfrontals, parietals, supraorbitals, and others, which form a protective shield of dermal scales over the cartilaginous cranium beneath. On the ventral side are two other membrane bones, and anterior *vomer* and a more posterior *parasphenoid*. The rostrum, covered with scales, is prominent. The method of jaw attachment is the sturgeon is *hyostylic*. Maxillary and dentary bones are associated with the jaws, and dermal opercular bones lie over the hyomandibular cartilage. Jaws and visceral skeleton are poorly developed.

In *Polyodon*, a close relative of the sturgeon, membrane bones are less prominent. Nevertheless, the snout or paddle, which is actually the enormously elongated rostrum, is covered with dermal scales. The ancient ray-fin, *Polypterus*, exemplifies an advance over the primitive condition toward the more highly developed teleostean skull. The cartilagineous skull is retained for the most part, but a veritable armor of numerous membrane bones covers it. Among these are frontals, parietals, nasals, supratemporals, post-temporals, and *dermal* supraoccipitals. Since the palatopterygoquadrate bar is fused to the cranium, the method of jaw attachment is *austostylic*. Several important membrane bones are developed in connection with the palatopterygoquadrate bar. They include palatines and a number of pterygoids. Vomer bones are also present. The outer arch of membrane bones includes premaxillaries, maxillaries, and jugals.

In connection with Meckel's cartilage are developed dentary, angular, and splenial bones. Other membrane bones appear on the operculum. In the garpike, *Lepisosteus*, and in the bowfin, *Amia*, both cartilage bones and numerous membrane bones are present. Exoccipitals, pro-otics, and lateral ethmoids are among the cartilage bones which appear. In the gar, teeth are borne by the maxillary bones in the upper jaw. Teeth are present in *Amia* on the palatines, pterygoids, and vomers as well as on premaxillaries and maxillaries. The lower jaw of the gar consists mainly of the tooth-bearing dentary. Small angular and articular bones are also present.

In *Amia* there are several splenials, an angular, suprangular, coronoid, and articular. The dentary, bearing teeth, however, is the chief bone of the lower jaw. The *hyostylic* method of jaw attachment is found in these fishes. A rather wide variation is to be observed among the teleosts. Their skulls have more bones than those of any other vertebrates. Numerous membrane bones are present, but most characteristic is the presence of several cartilage bones which have formed from ossification centres in the primitive chondrocranium, which still persists to some degree. These bones include basioccipitals, exoccipitals, supraoccipitals, basisphenoids, pleurosphenoids, or alisphenoids (not homologous with the alisphenoids of mammals), orbitosphenoids, pro-otics, epiotics, and opisthotics. Sphenotics and pterotics, present in addition, are derived from the auditory capsules. Mesethmoid and lateral ethmoids develop anteriorly.

Among the more prominent membrane bones on the roof of the skull are parietals, frontals, and nasals. A series of *circumorbital* bones affords protection to the eyes. Several other membrane bones are present in addition to those mentioned. Forming the roof of the mouth are a large, single parasphenoid and a pair of vomers. The palatopterygoquadrate bar in these fishes ossifies to form such cartilage bones as the metapterygoid and quadrate. The posteriorly located quadrate articulates with the *symplectic*. The latter joins the hyomandibular, which in turn is fastened to the cranium. This is a variation of the hyostylic method of jaw suspension and is characteristic of teleosts. Other bones (palatine, ectoptergoid, metopterygoid) usually aid in holding the quadrate in place.

Palatines and pterygoids (ectopterygoids and endopterygoids) are membrane bones formed in relation to the palatopterygoquadrate bar. The outer arch of membrane bones includes premaxillaries and maxillaries. Both usually bear teeth, but in most teleosts the maxillaries are reduced in size, do not form part of the margin of the jaw, and do not bear teeth. Teeth are frequently present on vomers, palatines, and pterygoids. The number of bones in the lower jaw of teleosts shows a reduction. The dentary, which usually bears teeth, is the chief component. An angular bone may bc present. The articular bone is small and reportedly derived from both membrane and cartilage elements. It articulates with the quadrate. The hyoid arch is ossified in several places. A hyomandibular bone articulates with the sphenotic and pterotic bones of the auditory capsule and with the metapterygoid of the palatopterygoquadrate bar.

The symplectic bone, found only in bony fishes, and segmented off the hyomandibular cartilage during development, is interposed between hyomandibular and quadrate bones. The remainder of the hyoid arch is composed of a chain of bones. First there is a small *interhyal* which articulates at the function of hyomandibular and symplectic. This is followed ventrally and in order by an *epihyal*, a large *ceratohyal*, and a small *hypohyal*. Several stardy, curved, bony *branchiostegal rays* project posteriorly from the ceratohyal. They are probably homologous with the gular bones which appear ventral to the operculum in some primitive fishes. The branchiostegal rays support the branchiostegal membrane. In some fishes, as in the salmon, a large, forward projecting *glossohyal* (*entoglossal*), bearing teeth, serves to unite the hyoid arches of the two sides and to support the tongue. A posteriorly directed *urahyal* may connect with the basibranchial of the next visceral arch. Both these bones are probably hyoid-arch derivatives. In bony fishes the hyoid hemibranch is missing.

It may possibly be represented by opercular gill lamellae on the inner wall of the operculum. These, however, may actually be remnants of the pseudobranch located on the anterior wall of the spiracle of lower fishes. The homologies of the opercular gill are uncertain. Of importance in the bony fishes are the opercular bones. They usually consist of four large membrane bones which give support to the operculum, helping to protect the gill chamber. They lie directly posterior to the hyomandibular region. The five remaining visceral arches are often considerably reduced. They are usually composed of several separate elements name in the same manner as those of elasmobranchs. The last one consists only of a single dorsal element, the pharyngobranchial, which bears teeth but does not support a gill.

Gill rakers, or strainers, project from the inner surfaces of the visceral arches. No membrane bones develop in connection with the hyoid or the remaining visceral arches. The skull in Dipnoi is not so far advanced as the teleosts. It is cartilaginous, for the most part, but both cartilage bones (exoccipitals) and a few large membrane bones are found. The palatopterygoquadrate bar, like that of holocephalians, is indistinguishably fused to the cranium. No cartilage bones are derived from it. The method of jaw suspension is *autostylic*. Marginal teeth are lacking, but *tooth plates* are present on the vomers and pterygoids, both of which are membrane bones. In the evolution of the skull from primitive fishes to mammals, there has been a gradual reduction in number of the separate bony elements by elimination and fusion. As many as 180 skull bones are present in certain primitive fishes. The human skull consists only of 28, including the auditory ossicles. In some cases cartilage bones and membrane bones fuse together to form a single structure. The autostylic method of jaw attachment is the rule among the tetrapods.

Appendicular Skeleton

The appendicular skeleton includes the skeleton of *pectoral* and *pelvic girdles* along with the skeleton of their related fins. There are two paired fins, the *pectorals*, attached to the pectoral girdle in the anterior part of the trunk just behind the last visceral arch, and demarcating the trunk from the head; and the *pelvis*, attached to the pelvic girdle in front of the cloacal opening and demarcating the trunk from the tail. Variations, both in the paired fins and girdles, are found in various groups of fishes. In the sharks, such as *Scoliodon*, the U-shaped pectoral girdle is situated behind the fifth branchial arch and consists of the similar halves united below in the middle line to form one piece, the *scapulocoracoid cartilage*. Each half is composed of a ventral plate-like *coracoid bar* and a dorsal arched *scapular process* or *scapula*. At the junction of the two is a shallow triple facet, the glenoid cavity, for the articulation of basal cartilages of the fin. Coracoids of the two sides are united mesially.

Pterygiophores (somactidia) comprise three basalia: an anterior and lateral *propterygium*, a middle *mesopterygium*, and a posterior and medial *metapterygium*. To the basalia are attached a series of radialia. There are two radials against each of the pro- and mesopterygium, and 18-against the metapterygium. Radialia are divided into a proximal, a mestal and a distal piece each, distal pieces support horny fin-rays, the *ceratotrichia*. Small foramina over the coracoid bar are meant for the passage of blood vessels supplying the girdle. Pelvic girdle and paired pelvic fins of *Scoliodon* are situated at the end of the trunk, the former in front of the cloaca and the latter on the sides of the cloacal opening. Anterior basalia of the two sides and fused to form the pelvic girdle, which is a horizontal plate of cartilage called *ischio-public bar*. Its ends are produced slightly to the dorsal side forming the *iliac processes*.

Skeleton of pelvic fin consists of basalia, a small anterior *propterygium* and a large medial and posterior *metapterygium* (basipterygium); basalia articulate with the acetabular region of the girdle. Fifteen or more 2-segmented radials (somaetidia) are attached to the metapterygium. Distal segments of radials bear *ceratotrichia*. In the male shark, metapterygium of each pelvic fin is continued to form the skeleton of claspers for transferings sperms into the female. A groove runs on the dorsal side of the *axial cartilage* (main support) of the clasper. In the middle part of the clasper the groove serves as a closed tube, since the edges of the groove the rolled-in scroll like, carrying the skin with them. The shallow anterior opening of the groove is called *apopyle* and the posterior narrow opening, *hypopyle*. Apopyle communicates with the cloaca and a glandular sac called *siphon*. Siphon has heavy muscular walls and lies just beneath the skill in the posterior, ventral and abdominal regions. Anteriorly, it ends blindly near the posterior margin of pectoral fin and, posteriorly, it communicates with the apopyle. Siphon is used with some force in the ejection of sperms. Normally, the apopyle remains open and sea-water is drawn into siphon.

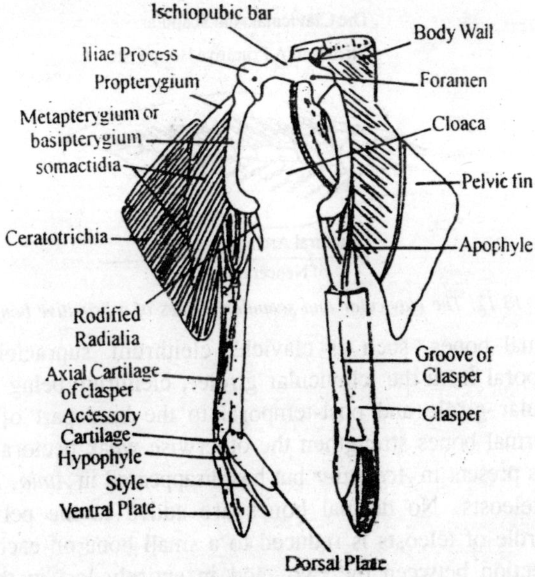

Fig. 13.11. Pelvic and fin of shark (male) clasper of one side opened out.

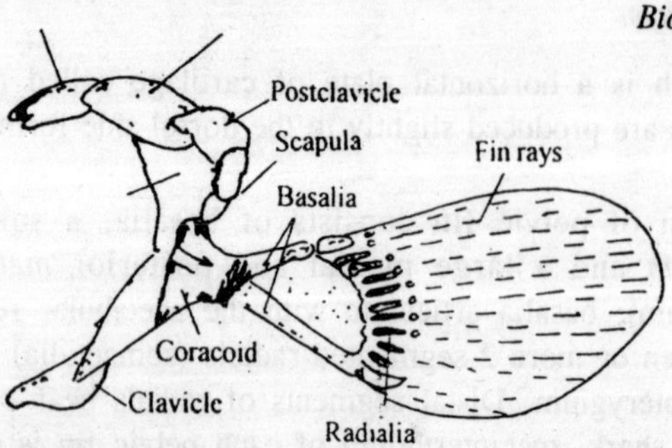

Fig. 13.12. The clavicular and scapular girdles of a primitive bony fish.

During copulation, the groove of clasper is filled with sperms and clasper is then tilted in such a way that the apopyle closes and water from the siphon passes directly into the clasper-tube ejecting the sperms already there. Distally, the groove of clasper ends in a sharp *style* bearing the hypopyle, which is covered with flattered dorsal and ventral *sheathing plates* of cartilage. A serrated (toothed) accessory sheathing plate lies immediately below the style. In teleosts and lung-fishes, pectoral girdle is usually connected to the axial skeleton by means of dermal bones. In primitive bony fishes, such as *Polypterus*, the scapulo-coracoid cartilage becomes ossified into separate scapula and coracoid bones and forms the scapular girdle.

The Clavicular And Scapular Girdles of A Primitive Bony Fish

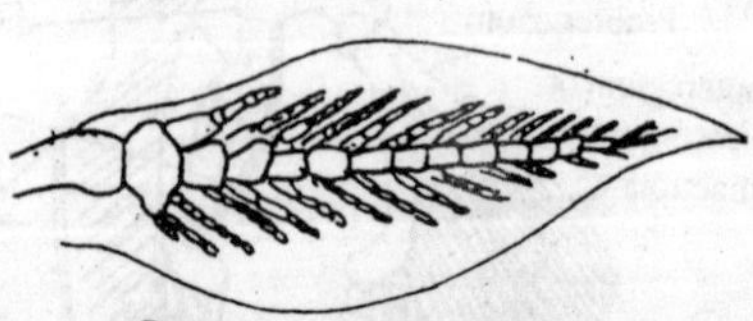

Pectoral Archipterygium of Neoceratodus

Fig. 13.12. The clavicular and scapular girdles of a primitive bony fish.

Dermal bones, such as clavicle, cleithrum, supracleithrum and post-temporal form the 'clavicular girdle', cleithrum being attached to the scapular girdle and post-temporal to the hind part of the skull. These dermal bones strengthen the otherwise weak pectoral girdle. A clavicle is present in *Acipenser* but has disappeared in *Amia*, *Lepisosteus* and all teleosts. No dermal bones are added to the pelvic girdle. Pelvic girdle of teleosts is reduced to a small bone on each side with no connection between the two, and in entirely lost in the eels. In rohu too, there is no pelvic girdle and pelvic fins articulate with a flat bone. Fin-skeleton also has gone a change from the typical form of elasmobranchs. *Polypterus* shows not much change except that various cartilages have become ossified.

In other bony-fishes, number of radials is reduced and the web of the fin is mainly supported by dermal fin-rays. In dipnoi e.g., *Neoceratodus*, the fin consists of a single segmented axial rod (the metapterygium), from the sides of which numerous secondary segmented radials arise, such a fin is called an *archipterygium*. Pelvic fin may be absent in some fishes such as eels, and in some others may be shifted far forward to be attached to the cleithra of pectoral girdle, while normally they are reduced in bony fishes. In the cod-fish pelvic fins have gone in on of pectorals, a peculiar situation indeed.

14

METACHROSIS IN FISHES

Contrary to popular opinion, it may be devoted if any animals, even insects or birds, can view with living fishes in the brilliancy and changeability of their colours. Brilliancy of coloration is most conspicuous in the teleosts. In nearly all other fishes the colours are much uniform usually sober. In many species the general colouration, and particularly the characteristic markings in the form of bars, stripes, spots and blotches, are remarkable constant in all the individuals of a particular species, but in others there is a good deal of individual variation; in the Trunk-fishes it is extremely rare to find two specimens exactly alike in the manner in which the bands on the body are arranged. Professor *Jordan* has described some of the remarkable colour variations found in a species of Sea Perch from the West Indies known as the Vaca (*Hypoplectrus*).

Generally, the ground colour is orange, with black marks and blue lines, the fins being chequered with orange, and blue, 'in a second forms, he writes, 'the body is violet, barred with black, the head with blue spots and bands.' In another forms the blue on the head is wanting. In still another the body is yellow and black, with blue on the head only. In others the fins are plain orange, without checks, and the body yellow, with or without blue stripes or spots and sometimes with spots of black or violet. In still others the body may be pink or brown, or violet-black, the fins all yellow, part black or all black.

Finally, there are forms deep indigo-blue in colour everywhere, with cross-bands of indigo-black, and these again may have bars of deeper blue on the head or may lack these altogether. For the most part, the colours of fishes, like those of any other animal, serve to conceal their owners either from their prey or their natural enemies. This is not always the case, however, for in some fishes the colours serve a totally different purpose, and attempts that have been made to explain all types of colouration in term of concealment sometimes press the matter to the point of absurdity. The fact remains, however, that in a very large number of fishes the particular hues and patterns adopted do tend to render than invisible, or, at least, very inconspi cuous in their natural surroundings. A few examples will suffice to illustrate the general principles of these concealing colours.

Colouration In Various Fishes

In an Australian Fish (*Plectropoma richardsoni*) the prevalent ground colour of the body is a brilliant carmine, with a tendency to yellow beneath, and diversified on the back and sides with ultramarine spots of almost sapphire-like intensity. Certain Australian species of *Beryx* (*B. affinis*, and *B. mulleri*) have a similar ground-colour when freshly caught, but with various opalescent tints, chief blue and lilac reflections. In *Polynemus vereker* the ground colour is chrome yellow, with darker markings, the pectoral and caudal fins are bright orange, the remaining fins being a lighter shade of the same tint, and by contrast the long free filaments of the pectoral fins are a bright

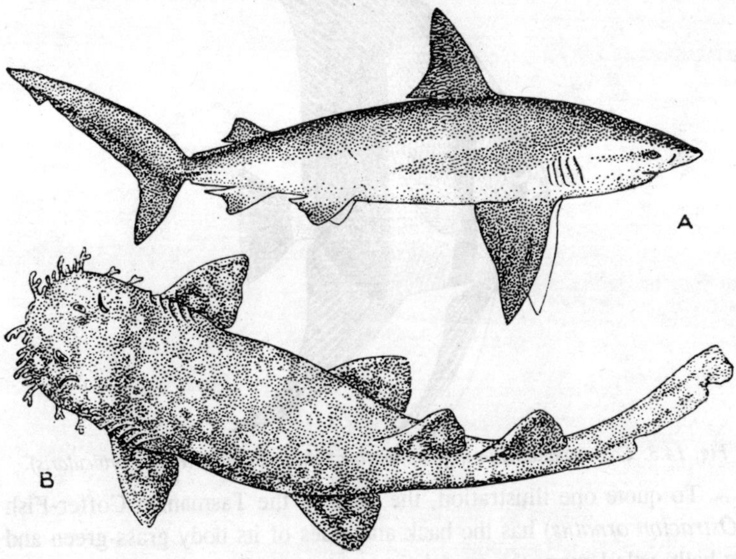

Fig. 14.1. Colouration in pelagic and bottom-living sharks. A—Sandbar shark (Carcharinus milberti); B—Carpet shark (Orectolobus barbatus).

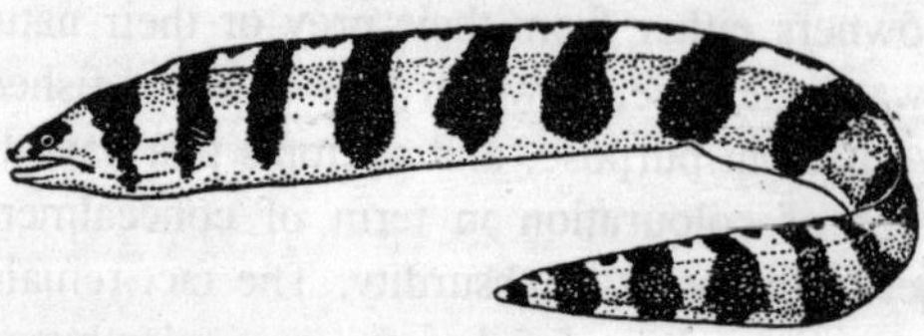

Fig. 14.2. Colour patterns in tropical marine fishes—Muraena or Moray (Gymnothorax petelli).

vermilion red. The Velvet-Fish (*Holoxenus cutaneous*), also a denizen of Australian seas, has a dominant colour of brilliant scarlet vermilion, or a mixture of vermilion and orange. The skin has no scales and presents a singular pilose of velvety appearance. It is, however, in some of the Pacific Trigger-Fishes (e.g. *Monacanthus*) and Coffer-Fishes (species of *Ostracion*) that the eccentricities of colouration are perhaps most strikingly manifest, for not only are the prevailing colours of the most brilliant description, but the presence of differently coloured bands or stripes, often arranged in complex patterns, adds greatly to the gorgeous and singularly bizarre appearance of these fishes.

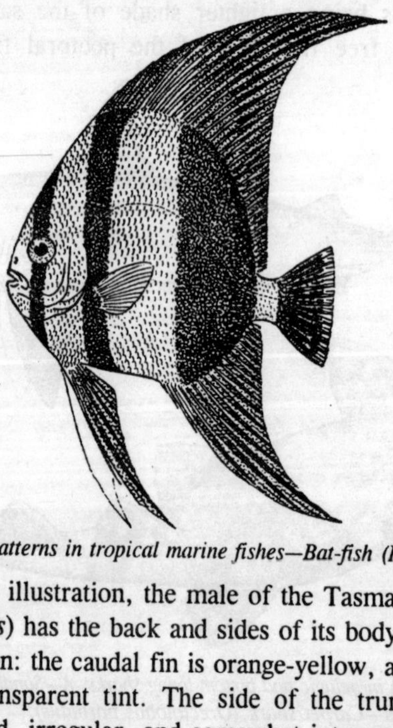

Fig. 14.3. Colour patterns in tropical marine fishes—Bat-fish (Platax orbicularis).

To quote one illustration, the male of the Tasmanian Coffer-Fish (*Ostracion ornatus*) has the back and sides of its body grass-green and its belly pale lemon: the caudal fin is orange-yellow, and the remaining fins a neutral transparent tint. The side of the trunk and head are traversed by broad, irregular, and somewhat interrupted bands of the most brilliant ultramarine blue, the edges of which are sharply defined by dark chocolate-brown lines. Two or three of the blue body-bands are continued on to the caudal fin, where they curl into characteristic loop-like patterns. The lemon-yellow of the belly is further variegated by a reticulated pattern in pale blue. In the female, formerly regarded as a distinct species, the ground colour is not green but a pale pinkish-grey, or dove-colour, with local flushes of a more decided pink, and the belly is a pure yellow. The blue stripes of the male are represented in the female of comparatively unbroken bands of a rich reddish-brown which, at the bases of the pectoral and dorsal fins, form an irregular spiral pattern. In both sexes the pattern of the longitudinal bands is never precisely the same in any two individuals. Scarcely less brilliant is the colouration of those Teleosts, notably species of Pomacentridae and Chaetodontidae, which frequent the coral reefs of the East Indian Archipelago and the Pacific and feed on the coral polypes, and of many of the Wrasses (Labridae).

Fig. 14.4. Colour patterns in tropical marine fishes—Butterfly-fish (Pomacanthodes semicirculatus).

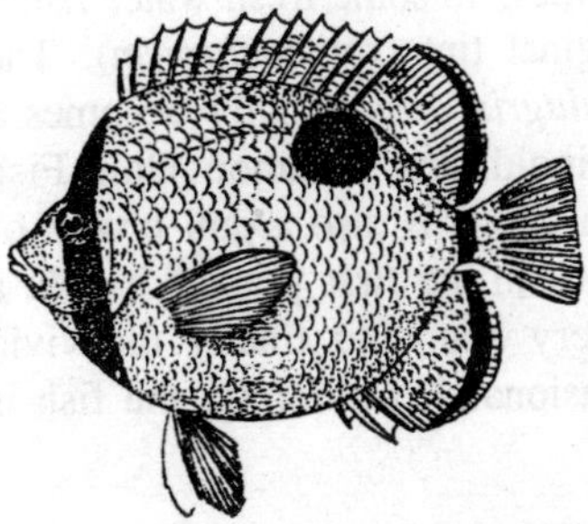

Fig. 14.5. Colour patterns in tropical marine fishes—Butterfly-fish (Chaetodon unimaculatus).

Many other groups, such, for example, as the Percidae, Cirrhitinae, and the Pipe-Fishes (Syngnathidae), include species in which the colouration is vivid and often beautiful, although less striking than is the case with the fishes mentioned above. As illustrating the opposite

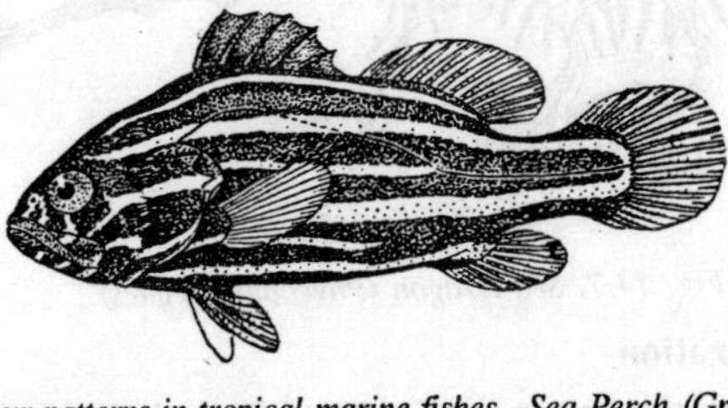

Fig. 14.6. Colour patterns in tropical marine fishes—Sea Perch (Grammistes sexlineatus).

extreme in the scale of colouration, between which and the brilliant tints just described every conceivable gradation exists, mention may be made of the colourless appearance of those fishes which, like the Kentuckian Blind Fish (*Amblyopsis spelaea*), are denizens of subterranean rivers; and, omitting a few species in which the colouration is almost brilliant, the prevalent sombre tints, dark brown or black, rarely relieved by spots, bands, or other distinctive markings, of the fishes inhabiting the abyssal waters of the deep sea.

Changes In Colouration

Changes of colouration sometimes take place, which either have no discernible relation to age, condition, or surroundings, or are brought about by domestication; and in individuals of the same species there is often a wide range of colour-variation, which is sometimes, but not always, associated with particular localities. In some fresh-water fishes a yellow colour may replace the original tint (*xanthochroism*). The usually dull greenish Tench (*Tinca vulagris*) occasionally becomes a bright orange-yellow. Another Cyprinoid, the common Gold-Fish (*Cyprinus auratus*), in its wild state in China is also a dull brown or green, but, when domesticated, assumes in the first year of its life a black colour (melanism), then a silvery hue, and finally the vivid ruddy golden colour of the adult; occasionally, but rarely, the fish is an albino and value of colouration.

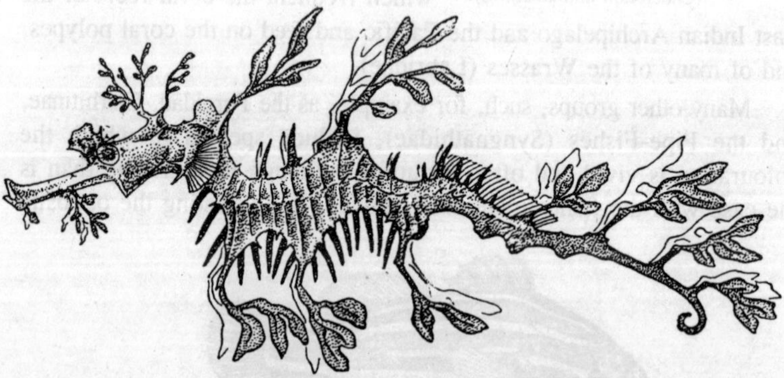

Fig. 14.7. Sea Dragon (Phycodurus eques).

Protective Colouration

Individuals of the same species vary in colour according to the capacity of the water they live in, becoming darker in muddy or peaty water, and brighter and lighter or shallower or clearer water. Trout caught in a stream with a gravelly or sandy bottom are lighter in colour than those obtained from a muddy stream, and it is well-known that the same fish changes colour as it passes from the one background to the other. In a lake in County Monaghan, Ireland, the Trout are darker on that side which is bounded by a bog, but are of the beautiful and sprightly variety generally inhabiting rapid and sandy streams on the opposite side where the bottom is gravelly; and narrow as the lake is, the two kinds of Trout appear to confine themselves to their respective areas. Trout obtained from a stream near Ivy Bridge have become much lighter since the pollution of the water by white china clay. As an illustration of the necessity of vision to such colour-changes, it may be mentioned that blind fishes cannot vary their tint in this protective fashion.

A blind Trout living upon a light sandy bottom differed from its fellows in being much darker and more conspicuous. Dark Trout have been observed among their light-coloured brethren in a chalk stream in Hampshire, but the former were invariably blind, probably, as their larger size indicated, through age. Of other forms of protective resemblance, reference may be made to the bottom-feeding Fiat-Fishes (Pleuronectidae) many of which have the upper surface of the body coloured with various shades of brown, speckled with black or light specks or blotches, in harmony with the prevailing tints of the sandy banks which usually from their feeding-ground. When disturbed these fishes court concealment, and render themselves still less conspicuous by partially burying themselves in the sand.

Many of the Skates and Rays, which have a white ventral surface, have the back mottled and coloured in accordance with the colour of the sea-bottom, but in this case it is possible that the advantage lies in enabling the fish to secure passing prey by concealing its own

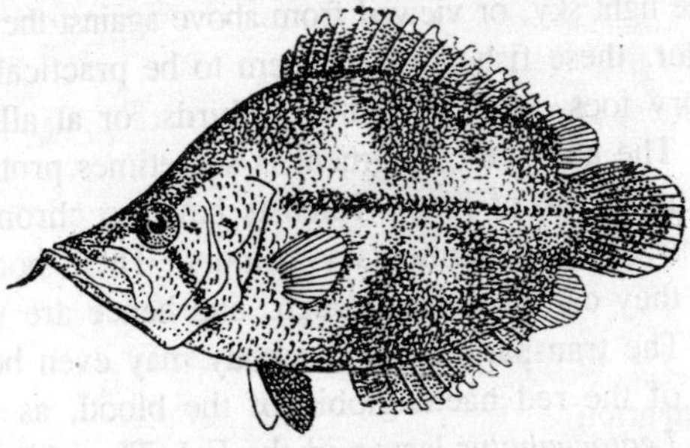

Fig. 14.8. Monocirrhus polyacanthus.

whereabouts. Striking examples of protective colouration occur among the Pipe-Fishes and Sea-Horses (*Syngnathidae*), which usually frequent groves of *Zostera*, Fucoids, and other sea-weeds. A British species of Pipe-Fish (*Siphonostoma typhle*), while lives among the blades of the sea-grass, *Zostera*, is olive-green in colour, and is a typical example of protective resemblance both in colour and in the slender elongated shape of the body.

Similar protective resemblances are noticeable among the Sea-Horses, the colouration varying with the general hue of their environment of sea-weed; but the climax is certainly reached by the singular Australian species, *Phyllopteryx eques*. In this fish the skin is produced into numerous long, flattened, branched filaments, which are prolonged from the extremities of spine-like outgrowths of the dermal skeleton, and marked by alternate bands of brown and orange, thus resembling both in shape and colour the fronds of the surrounding fucoids and other marine algae amongst which the fish lives. Many of the fishes frequenting the coral reefs of the East Indian and Pacific areas, especially those belonging to the Teleostean families Chaetodontidae and Pomacentridae, have a most brilliant and vivid colouration, frequently marked by bands or stripes of different tint. So far from rendering these fishes unduly conspicuous, there can be little doubt that, by harmonising with the striking and varied colours of the anemone like coral polypes, their colouration is distinctly protective; and it is interesting to note that similar colour-patterns have been independently reproduced in both families.

Even the reef-frequenting Flat-Fishes (Pleuronectidae) have the usually sombre upper surface ornamented by vivid colours and striking patterns. Pelagic fishes, like the Herring, Mackerel, Flying-Fish (*Exocoetus*), and many others, often have the belly and sides silvery or white, and the back dark green, black, or steely blue. Seen from below against the light sky, or viewed from above against the background of the dark water, these fishes would seem to be practically invisible to their predatory foes, whether fishes or birds, or at all events not easily detected. The absence of pigment is sometimes protective. The surface-swimming larvae of many Teleosts have no chromatophores, and therefore no obvious pigmentary colours. Their bodies are so translucent that they can be seen through, and hence are visible only with difficulty. The transparency of the body may even be increased by the absence of the red haemoglobin of the blood, as is the case with the pelagic *Leptocephalus*-larvae of the Eel. The iridocytes of the

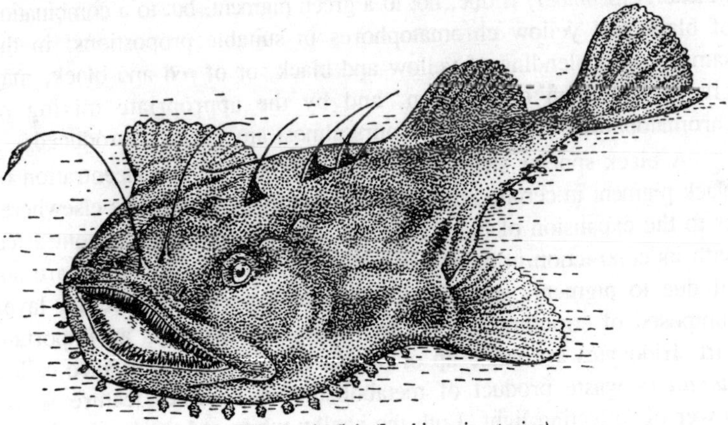

Fig. 14.9. Anglerfish (Lophius piscatorius).

reflecting tissue may also disappear under the influence of changed surroundings.

The larvae of various species of *Onus* (Gadidae) are silvery on hue during their pelagic career, owing to the presence of iridocytes in the skin, but on becoming mature they change to a dull dark colour, and live under stones or in holes and crevices in the rocks. During the change of habit the reflecting tissue (argentium) is lost, and the needful chromatophore are acquired. Instances of the loss of pigmentary colours, owing to the cessation of the controlling influence of natural selection, are to be found in the absence of chromatophores on the white under surface of the flat-fishes, where such colours are useless but not necessarily harmful, and in the colourless, cave-inhabiting Fishes of which the Blind-Fish (*Amblyopsis*) of North America may be taken as an example.

Causes of Colouration

The colours of a fish are mainly due to the presence in the dermal layer of the skin of numerous pigment-containing cells known as *Chromatophores*. Each of these is a much branched, small sac with thin walls. By expansion or contraction the contained pigment granules may be concentrated into a minute spot, or flattened out to form a relatively large and irregular sheet, the cell itself retaining its shape since only the pigments move. The granules of pigment deposited in each of the chromatophores may be either red, orange, yellow or black. Other shades are produced by the blending of two or more of these primary colours. The exquisite green colour on the back of the

Mackerel (*Seomber*) is due, not to a green pigment, but to a combination of black and yellow chromatophores in suitable proportions: in the same way, a blending of yellow and black, or of red and black, may give a brownish colouration, and by the appropriate mixing of chromatophores of different colours almost any shade is produced.

A black spot or stripe may be due either to the concentration of black pigment in certain regions and its comparative absence elsewhere, or to the expansion of black pigment in particular areas as contrasted with its contraction in other parts of the body. But the colours are not all due to pigment, and the presence of a peculiar reflecting layer composed of structures known as iridocytes also plays an important part. Iridocytes are made up of opaque crystals of a substance called *guanin* (a waste product of metabolism) whose chief feature is the power of reflecting light. Both the chalky white and the bright silvery appearance of fishes is due to the manner in which light is reflected from iridocytes. By interference these same colour elements are also responsible for the prismatic hues and brilliant iridescence characteristic of so many fishes.

The relative abundance of either type of colour agent varies greatly, not only in different species, but in different parts of the same individual, chromatophores being most abundant in the dark back

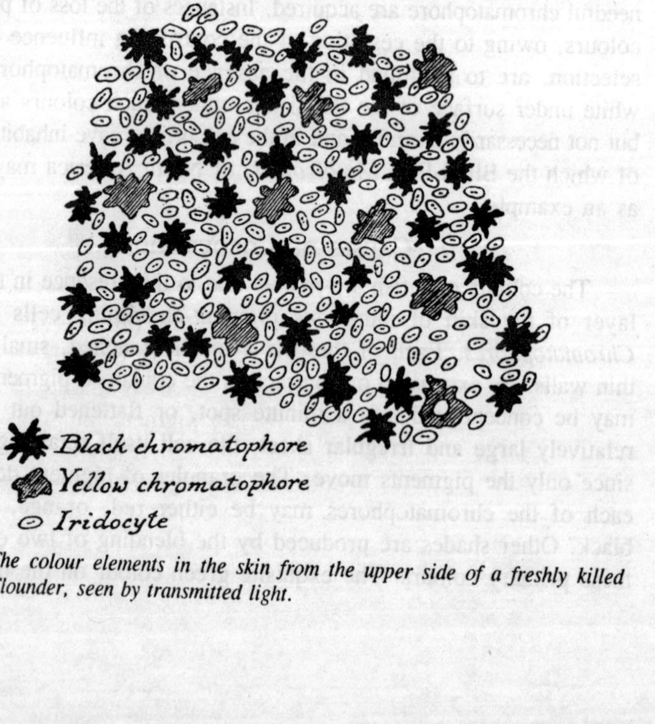

Fig. 14.10. The colour elements in the skin from the upper side of a freshly killed Flounder, seen by transmitted light.

region, whereas, in the pale belly the iridocytes play the chief part. Thus, the relative abundance of chromatophores and the kind of pigments they contain, the manner in which they are distributed in the skin, as well as the iridescence and the reflecting powers of the iridocytes, all play their part in determining the characteristic colour of a fish. The dark bluish-grey colour of the back of a Whiting (*Micromesistius merlangus*) is due largely of the abundance of black and yellow chromatophores in this region, but these are much less numerous on the sides and absent altogether on the pale belly.

The iridescence and silvery appearance of the sides are due to the iridocytes lying above the scales, combined with the reflecting but non-iridescent layer of similar structures below the scales. The dead white of the abdominal region owes its appearance to the different reflecting power of another deep layer of iridocytes in this region known as the *argenteum*, and to the absence of chromatophores. The colour changes described in the foregoing pages, whether slow, rapid, or instantaneous, are all due, therefore, to the action of the chromatophores. If their pigment be expanded, the particular shade of colour is intensified: when the pigment is contracted it may shrink to a mere dot and thus diminish the vividness of the colour involved. This latter process may even change the colour, for yellow chromatophores become orange when the pigment contracts, and orange or red appear brown or black.

The prevailing hues of the body are altered, not only by the expansion or contraction of the pigment, but also by an increase or decrease in the number of chromatophores, or by an alteration in the manner of their distribution in the skin. The initial stimuli leading to colour change are undoubtedly received through the eyes, but full details of the physiological mechanisms involved in the movement of pigments within the chromatophores have still to be worked out. It seems that both nervous and hormonal control is involved.

Rapid colour changes are probably under nervous control but slower, more stable changes may be effected through the action of hormones. There is certainly some evidence pointing to the direct action of light on the production of pigment in areas normally unpigmented. Conversely, lack of light can result in a reduction of pigmentation. For example, if the Catfish (*Ameiurus*) or the Flounder (*Poralichthys*) are kept in an aquarium with blackened sides and top but with an illuminated bottom, numerous dark chromatophores appear in areas of the body normally lacking pigments cells; if a blinded

Ameiurus is kept in darkness, reductions in pigmentation take place due to the lack of illumination.

An almost complete absence of pigment is found in blind cave-fishes but even here light may have an effect because blind African Carp (*Caecobarbus*) living in caves with some light have more pigment than those living in darker grottoes. Incidentally, the pale pink colour of many cave-fishes is due to blood in the numerous superficial blood vessels showing through the transparent skin. Some fishes exhibit a condition known as *xonthochromism*, sometimes occurring in the wild state, but generally brought about by the artificial conditions associated with domestication. In such individuals the black or brown pigment is entirely wanting and the whole body has a golden colouration, or, if the orange and red are also undeveloped, a uniformly silvery hue.

The familiar Gold-fish (*Garassius auratus*) is a native of Eastern Asia, an in its natural habitat has the greenish and brownish colours of the other Cyprinids, only the domesticated varieties exhibiting the golden or silver liveries. Individuals that have escaped from artificial ponds and regained the rivers often revert of the original colouration. Golden Tench (*Tinca*), Golden Orfe (*Idus*), and Golden Trout (*Saimo*) are other well-known varieties produced by fish culturists, but Trout of this type are sometimes found in the wild state, and golden-coloured Eels (*Anguilla*) are by no means rare. The relation between the incidence of light and the distribution of pigment in the skin has already been mentioned in connection with obliterative shading, and it is interesting to find that in certain species of the African Cat-fish *Synodontis*, which have the remarkable habit of swimming with the belly upwards the normal colouration is reversed, the lower parts being dark and the back place.

Chromatophores Terminology

According to the colour of pigments they contain, the chromatophores of fishes have been commonly classified as *melanophores* (brown or black), *erythrophores* (red), *xanthophores* (yellow), *leucophores* (white), and *iridophores* (reflecting). Both the leucophore and iridophore contain colourless pigments (primarily *guanine*), but the former contains small crystals which can move back and forth in the cytoplasm, whereas in the latter three are larger crystals incapable of movement and usually stacked in layers. Sometimes more than one type of pigment can be found in the same chromatophore. For instance, *Goodrich* et al. (1941) described the *xantho-erythrophore* of the swordtail, showing that the center of the cell contains yellow carotenoid pigments, whereas in the cellular processes there is an accumulation of the red pteridine. Frequently, the pigment-containing bodies are not unicellular chromatophores but combinations of such cells. These have been referred to as "*chromatosomes*" by *Parker* (1948).

The most commonly found chromatosome is probably the *melaniridosome*, a complex of melanophore and iridophore. In view of the fact that the suffix "some" has more commonly been used to denote cytoplasmic inclusions such as ribosomes, lysosomes, or melanosomes, this kind of term may be replace with more proper expression, e.g., "*compound chromatophore*". As far as the melanin-containing cells of vertebrates are concerned, a recommendable terminology has recently been proposed (*Fitzpatrick* et al., 1996). In this terminology, the *melanocyte* is defined as a cell which synthesizes a special melanin-containing organelle, the *melanosom*. The new definition of the melanophore, with which we are much concerned here, is essentially the same as that familiar to most zoologists, *i.e.*, a type of melanocyte which participates with other chromatophore in the rapid colour changes of animals by intracellular displacement of melanosomes.

In addition, the melanoblast is a cell which serves at all stages of the life cycle as the precursor of the melanocyte and melanophore. The terms "*melanosome*" and "*premelanosome*" are more or less unfamiliar to students of zoology. In most cases, the commonly used term "melanin granule" may be replaced with "*melanosome*", which is defined as a discrete melanin-containing organelle in which melanization is complete, more or less uniformly "electron dense" by electron microscopy; with *tyrosinase* activity usually not demonstrable. "*Premelanosome*" means all distinctive particulate stages in the maturation of melanosomes. Their electron density is variable, and they possess an active tyrosinase system after the onset of melanin synthesis. For describing other kinds chromatophores than the melanophore, and number of terms have been employed (*Parker* 1948; *Bagnara*, 1966). However, most of the less commonly used words can be grouped into a few generally accepted terms without causing much inconvenience. As for the cellular inclusions carrying pigment, no definite usage of terms has been adapted. This may be largely because of the lack of definitive information about the chemical nature of pigments inside these cells.

In recent years, attempts to identify these pigments chemically in various chromatophores have been in progress (*D.L Fox*, 1957; *Hama*,

1963). These studies will surely result in a finalization of the nomenclature of these intracellular granules in near future. For instance, *Matsumoto* (1965a) has recently designated the pigmented particle in the erythrophore of the swordtail as the "*pterinosome*", since it primarily contains pteridines.

Chromatophore Pigments

Following types of chromatophore pigments have been reported in fishes.

Melanins

The brown or black pigments widely found in fish species are *melanins*. They are highly polymerized compound derived from tyrosine. The synthesis of melanins involves first the oxidation of tyrosine to 3,4-dihydroxyphenylanine (Dopa) and then to Dopa quinone. The enzyme concerned is *tyrosinase*. Dopa quinone is then polymerized to form melanins, which are usually found attached to a protein. Melanin synthesis takes place in melanocytes, young melanophores, and possible even in mature melanophores. The subcellular site of synthesis is known as a *premelanosome*, the early stage of melanosome development. Usually melanosomes are only to be expected in adult melanophores wherein melanization has been completed. However, *melanogenesis* can take place in mature melanophores because morphological colour changes may involve an increase in melanin content without a corresponding increase in number of melanin-containing cells.

Indeed, *Bikle* et al. (1966) found that among melanosomes presumably immature melanin granules, premelanosomes, are present which contain a number of vesicular structures and dense particles inside the limiting membrane. Recently, there have been some quantitative studies on fish melanogenesis. Thus, *Kosto* et al. (195) incubated isolated *Fundulus* fin samples with Dopa and obtained a significant correlation between O_2 Consumption and the degree of darkening, Recently, as sensitive radiometric tyrosinase assay has been developed by *Chen* and *Chavin* (1965). Working with skin of the black moom goldfish, *Chen* and *Chavin* (1966) made a contribution by modifying the classic metabolic pathway of melanogenesis. The incorporation of tyrosine and Dopa into melanin without decarboxylation was demonstrated by using carboxyl-labelled compounds.

Chen and *Chavin* (1967) have further shown that in general the dorsal; integument of fishes contains higher tyrosinase levels than the ventral. In four varieties of goldfish (white, xanthic, gray, and black) the total tyrosinase activity increases with increasing pigmentation, and the activity increase occurs mostly in the particulate fraction. This observation is rather inconsistent with the finding by *Tomita* and *Hishida* (1961a) that tyrosinase and Dopa-oxidase activities in adult *Oryzias* skin are higher in light-coloured varieties than in the brown variety. *Hishida* et al. (1961) showed that the light-coloured varieties of *Oryzias* have amelanotic melanophore and the application of sulfhydryl poisons or some other agents can bring about their melanization. Thus, they concluded that tyrosinase exists in an inhibited state in the amelanotic melanophore. In recent years, it was found in goldfish skin (*Kim* et al., 1962), *Fundulus* embryos (*Spitz* and *Burnett*, 1968), and in a variety of other fishes including lamprey, shark and eel (*Chen* and *Chavin*, 1967) that the molecules of the enzyme tyrosinase do not have the same physical and chemical characteristics. Thus, there are multiple forms of the active enzyme. In a few species (the hagfish *Eptatretus*, the bichir *Polypterus*, the garpike *Lepisosteus*, and the African lungfish *Protopterus*) tyrosinase was found only in the particulate fraction (*Chen* and *Chavin* 1967).

Carotenoids

Carotenoids which occur in fishes were especially well documented in the article by *D. L. Fox* (1957). Although some exceptions were known, it had long been thought that all yellow and red pigment of *xanthophores* and *erythrophores* were *carotenoids*. These are highly unsaturated hydrocarbon compounds and consist of a chain of carbon atoms with a ring structure at one or both ends. In addition to carbon and hydrogen, xanthophylls or their esters involve oxygen. In many species large amounts of xanthophylls including taraxanthin, lutein, and astaxanthin have been detected, while carotenes, mostly carotene, were less frequently found in the integument (*D L. Fox*, 1957). All carotenoids are insoluble in water but soluble in organic solvents. Hence, the term "*lipophore*" has been rather widely used to denote both xanthophores and erythrophores. It is known that animals do not have the ability to synthesize carotenoids. Therefore, those found in fishes must have come either directly from plant sources or from their prey. They incorporated pigments are transferred to xanthophores or erythrophore, where a great amount can be stored.

Some species contain a variety of carotenoids, whereas others contain only specific ones (*D L. Fox*, 1957). The latter fact can be explained by their ability to convert ingested carotenoids into the type characteristic of the species. For instance, *Summer* and *Fox* (1935) demonstrated the *Fundulusparvipinnis* is able ot convert dietary carotenoids into as integumentary xanthophyll. More information was lately added about the transfer of carotenoids by *Takeuchi* (1960), who suggested that carotenoids in larval *Oryzias* xanthophores may have come from the pigments found in yolk. It is also worth noting that during early stage of differentiation, xanthophores do not contain carotenoids, while in later life these pigments begin to accumulate in the cell.

Hama and *Hasegawa* (1967) found that xanthophores of blue and white-colour varieties of *Oryzias* do not contain carotenoids and suggested the use of "acarotenoid xanthophores" to denote such cells. The carotenoid-containing cells of the wild or orange-red varieties may therefore be called "*carotenoid xanthophores*". The subcellular site of carotenoid deposition in either xanthophores or erythrophore is not clear. Since larger granules about 0.5 m in diameter are claimed to be the storage parcels solely for pteridines, the small vesicles described by *Loud* and *Mishima* (1963) might be the possible organelles. Very recently, *Bellamy* (1966) reported that yellow and orange pigments of the chromatophores of the painted bomber, *Serranus scriba*, are xanthophylls, and they are stably associated with the particulate fraction of tissue homogenates. Although their presence has not been observed in electron micrographs as yet, the fat yellow droplets described by *Hama* and *Hasegwa* (1967) in xanthophores might possibly be storage compartments.

Pteridines

Much attention has recently been directed to the presence of *pteridines* in the chromatophores of fishes as well as of amphibians. These compounds are soluble in water closely related to purines and flavins; they have both a pyrimidine and an associated pyrazine rings. Both coloured and colourless pteridines are present in the chromatophore, Drosoptering including drosopterin, isodrosopterin, and neodrosopterin, are red, white sepiapterins, comprising sepiapterins and isosepiapterin, are yellowish.

The colourless leucopterins are generally divided into two groups, *i.e.*, blue or violet fluorescent. They include biopterin, rana-chrome 3, xanthopterin, and iso-xanthopterin. *Hama* (1963) and *Matsumoto* (19651a, b) reported an intimate association of pteridines with yellow or red pigmentation, emphasizing that coloured pteridines actually serve as functional pigments of both xanthophores and erytrophores. The coloured pteridines are usually accompanied by an array of colourless pteridines.

Although the role of these colourless pteridines in pigment cells is still unclear, *Hama* (1963) recently suggested their possible participation in melanin or carotenoid formation in chromatophores. As mentioned previously pteridines are primarily contained in the pterinosomes. *Matsumoto* (1965b) reported that in early larval stages of cyprinid fish the pigments responsible for integumentary colour are exclusively pteridines. He found that both late larval and adult xanthophores contain pteridines as well as carotenoids, and that at least in larval stages sepiapterins act as the actual colouring agent in xanthophores. Similar changes of pigment substances during development have also been observed in *Oryzias*. Their results indicate that the larval coloured pterinosomes, so-called "*sepiapterinosomes*," may be transformed gradually into non-colored adult-type pterinosomes.

Chemically, the amount of the coloured pteridines also decreased during this process becoming almost absent in the adult. That is, in adult life, fat and/or carotenoid pigments are mainly responsible for the colour of xanthophores. Melanophores also seem to contain a considerable amount of colourless pteridines comprising xanthopterin, isoxanthopterin, and biopterin. On the other hand, *Matsumoto* (1965b) reported that the appearance of leucophores during the development of cyprinid fishes is not accompanied by a modification in pteridine fractions.

Purines

Among the purines responsible for the white or silvery tone of the integument of fishes, guanine definitely predominates. Its abundant accumulation is found in special chromatophores the *Leucophores* or *iridophores*. In the cytoplasm of these cells, guanine exists in minute crystalline or granular form, or sometimes, as thin platelets. Thus, these cells have often been referred to as "guanophores." Actually, *Hitchings* and *Falco* (1944), and *Neckel* (1954) formerly demonstrated chemically the abundant presence of guanine in the fish integument. At least in iridophores the molecules of purines seem to be orderly arranged within individual platelets. Recently, however, other kinds of purines than guanine have also been reported. For example, *Ziegler-Gunder* (1956) showed the occurrence of hypoxanthine in the integument of cyprinid fishes. The presence of both guanine and hypoxanthine has been confirmed in the belly skin of presmolts, smolts, and grilse of the coho salmon, *Oncorhynchus kisutch*, and in the integument of parrs and smolts of the Atlantic salmon, *S. salar* (1954) detected uric acid in the skin of the hairtail, *Trichiurus manumela*.

15

Organs of Sound Production

Contrary to popular belief sound-producing or vocal organs are by no means uncommon in Fishes, especially in certain families of teleosts. It is not always easy, however, to discriminate between involuntary, abnormal, or accidental sounds, and those due to the action of special vocal organs. There are, moreover, some Fishes which observations have shown to utter highly characteristic sounds, although the precise nature of the sound-producing mechanism is at present unknown; while other Fishes appear to possess organs which, on anatomical grounds, are perhaps vocal in function, although nothing is known of the nature of the sounds they emit.

Here those organs only will be considered which, either with certainty or with some degree of probability, may be regarded as vocal structures. Two international symposia on marine bioacoustics covered many aspects of the field, including sound-production in fishes, crustacean, and other marine organisms. A serious problem encountered in obtaining recordings of aquatic animals under natural conditions in that the listener is often working blindly and thus the identification of the sound producer often becomes impossible. A major approach to the solution of this problem has been to accompany sonic observations with visual observations by means of underwater television. A unique installation of this sort is now located off Bimini, Bahamas (*Steinberg* et al., 1962; *Steinberg* and *Koczy*, 1964; *Kronengold* et al., 1964).

Under Water Accounts

Any study of sound production or sound detection in fishes necessitates an understanding of the acoustic properties of water as a medium. Since water is much denser than air, the velocity of sound in

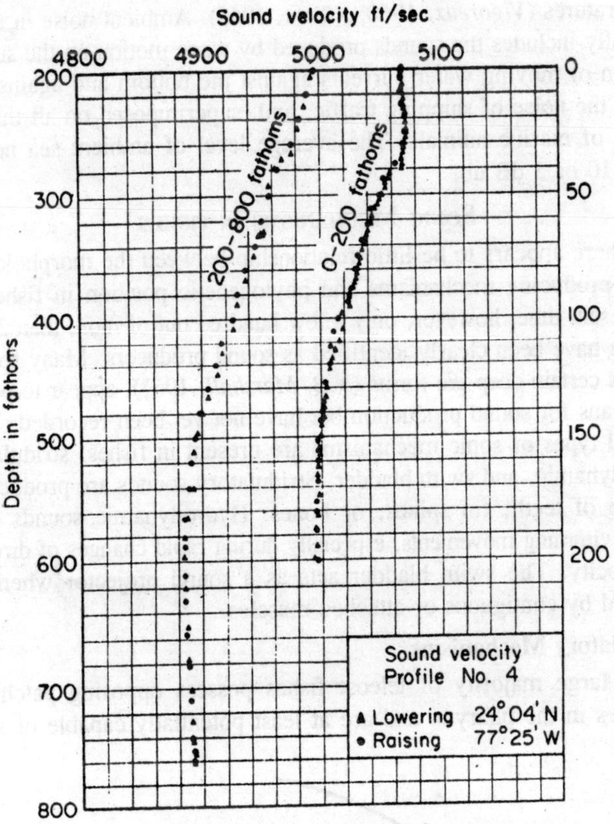

Fig. 15.1. Velocity of sound in the sea.

water is almost 1500 metres/sec while in air, it is about 330 metres/sec. In air, sound velocity is affected slighy by humidity, temperature, and barometric pressure. In water, temperature and pressure are independent variables in shallow areas, while at greater depths pressure affects the temperature.

The curves that relate sound velocity to depth, therefore, become quite complex. Salinity increases sound velocity, and in the oceans sound velocity may attain 1540 metres/sec. Since water is about a thousand times denser than air, more input energy is required to initiate the propagation of sound in water. However, once the sound is propagated, the acoustic energy will be transmitted faster in water. This transmission is further enhanced by the reflection of sound from the water surface (up to 99.9% is reflected back), from the sea bottom, and from interfaces that are formed by layers of water at different temperatures (*Vigoreux*, 1960; *Albers*, 1965). Ambient noise in the sea normally includes the sounds produced by wave motion on the surface, friction of moving water currents against the bottom and against each other, the noise of shipping traffic, and, superimposed on all that, the noises of marine animals. The average level of ambient sea noise is about 10 or 5 dB ub.

Sonic Mechanisms in fishes

There appears to be little relationship between the morphology of sound-producing mechanisms and phylogenetic position in fishes. At the present time, however, only a few hundred out of more than 20,000 species have been clearly identified as sound producers. Many species, such as certain deep-sea forms (*N.B. Marshall*, 1967), appear to possess the means for sound production but have not yet been recorded. Three general types of sonic mechanisms are present in fishes: stridulatory, hydrodynamic, and swim bladder. Stridulatory sounds are produced by friction of teeth, fin spines, or bones. Hydrodynamic sounds result from swimming movements, especially during rapid changes of direction or velocity. The swim bladder acts as a sound projector when it is vibrated by contiguous or attached muscles.

Stridulatory Mechanisms

A large majority of teleost fishes possess opposing patches of denticles in the pharynx and are at least potentially capable of sound

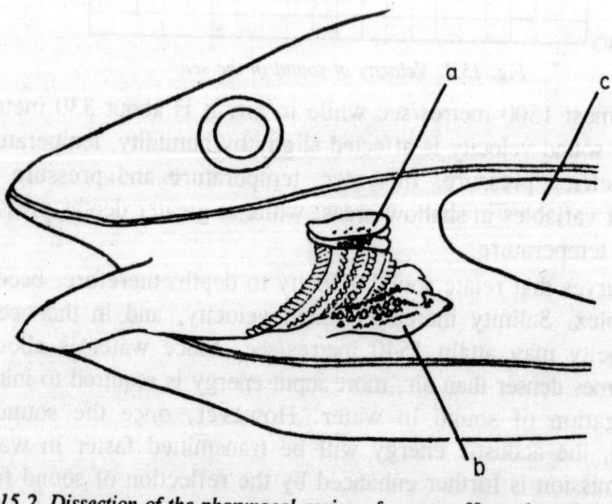

Fig. 15.2. Dissection of the pharyngeal region of a grunt, Haemulon, shows the location of pharyngeal denticles (a and b) and the swim bladder (c).

production during feeding. Some species stridulate pharyngeal denticles in connection with other activities such as alarm and territoriality. The best known sound producers of this type are members of the family Pomadasyidae—the grunts. *Burkenroad* (1930) described the sounds and the mechanism in some detail for the white grunt, *Haemulon plumieri*. He also noted that if the swim bladder were deflated, the character of the sound became "dry" and lost its "gruntlike" quality. He concluded that the swim bladder acts as a "resonator."

A similar arrangement has been described for other members of the family, and the margate fish, *Haemulon album*, has been observed to emit sounds, probably produced by this mechanism, in connection with feeding and schooling behaviour (*Cummings* et al., 1966). The function of the swim bladder as a possible resonator was also demonstrated in triggerfishes (*Salmon* et al., 1968). The occurrence of pharyngeal tooth stridulation has been reported from a wide variety of species and families of teleostean fishes. The courtship and territorial sounds of the croaking gourami, *Trichopsis vittatus*, are evidently produced by pharyngeal denticles and this small species offers many possibilities as an experimental animal in this field. It is also probable that the sounds of priacanthids are produced in this manner.

Almost any predatory species of fish is likely to produce sounds when feeding, and even some herbivorous forms that browse on sessile plants and animals can emit sounds when crushing rocks and corals. Incisor types of teeth are capable of biting through the exoskeletons of crustacea and thus produce strong metallic sounds. Given any food of

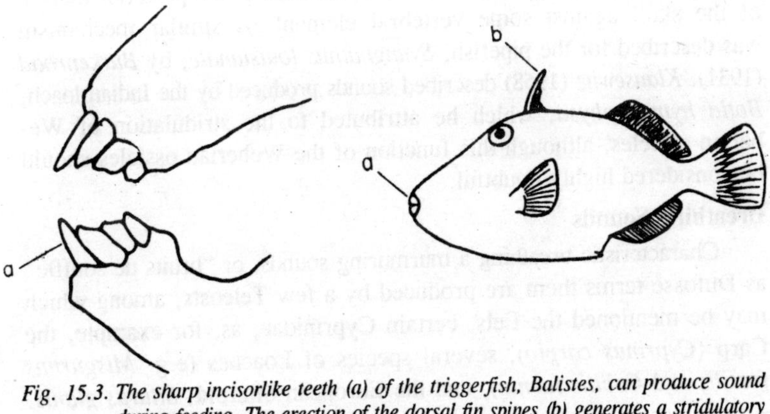

Fig. 15.3. The sharp incisorlike teeth (a) of the triggerfish, Balistes, can produce sound during feeding. The erection of the dorsal fin spines (b) generates a stridulatory sound.

moderate hardness, the action of teeth will produce sounds. Sometimes fish will gnash their teeth without the direct presence of food. *Fish* (1954) even listed some sharks and rays that produce sounds when feeding, despite the fact that elasmobranches are not known to be sound producers in any specialized sense (*Backus*, 1963). Burkenroad (19311), Fish (1954), and Moulton (1958) listed a number of species that produce sounds by means of modified molariform teeth.

Sound production of this type is particularly common among the members of the order Tetraodontiformes (puffers, filefish, etc.) and the family Scaridat (parrot fishes). Molariform teeth are also present in the family Sparidae (porgies), and Fish (1954) included the scup. *Stenotomus chrysops*, as a sound producer. Another common sparid is the pinfish, *Lagodon rhomboides*, whose sounds have been described by *Burkenroad* (1931) and *Caldwell* and *Caldwell* (1967), *Knudsen* et al. (1948) listed over a dozen additional species that were found to produce sounds when feeding. Stridulation by the movement of special fin rays and spines has been reported in a few species.

The sea catfish, *Galeichthys felis*, and the gafftopsail catfish, *Bagre marinus*, produce high-pitched squeaks when the enlarged pectoral fin spines are moved and several species of freshwater catfishes produce sounds in a similar fashion. The erection of the specialized dorsal fin spines in triggerfish produces a stridulation (*Schneider*, 1961) and it is possible that low intensity sounds may be produced by sticklebacks, *Gasterosteus aculeatus* and *Apeltes quadracus*, when dorsal fin spines are moved. Sounds produced by the friction of adjacent bones were reported from able mechanism is the stridulation of the posterior margin of the skull against some vertebral element. A similar mechanism was described for the pipefish, *Syungnathus louisianale*, by *Burkenroad* (1931). *Klausewitz* (1958) described sounds produced by the Indian loach, *Botia hymenophysa*, which he attributed to the stridulation of Weberian ossicles, although this function of the Weberian ossicles should be considered highly doubtful.

Breathing Sounds

Characteristic breathing a murmuring sounds, or "bruits de souffle" as Dufosse terms them are produced by a few Teleosts, among which may be mentioned the Eels, certain Cyprinidae, as, for example, the Carp (*Cyprinus carpio*), several species of Loaches (e.g. *Misgurnus fossilis* and *Cobitis taenia*), and the European Siluroid, *Silurus glanis*. According to Dufosse these sounds originate in some cases from the expulsion of gas from the air-bladder through the ductus pneumaticus and mouth, and in others, as in *Misgurnus fossilis*, they are produced by the rapid ejection through the anus of bubbles of air previously taken in at the mouth.

Sounds Produced through the Agency of Muscles Connected with the Air-Bladder

In addition to its usual function as a hydrostatic organ or "float" the air-bladder is often modified in various ways in different Teleosts, and adapted for use as a sound producing organs. In the South American Siluroid, *Auchenipterus nodosus*, the transverse processes of the fourth vertebra ate bent downwards and backwards, and at the same time become converted into flexible and highly elastic springs. Their distal extremities expand into oval bony plates which are imbedded in the anterior wall of the air-bladder, and often cause the latter to bulge inwards. From the occipital region of the skull arise two powerful muscles which pass backwards to their insertion into the anterior faces of the two springs.

By the contraction of these muscles the springs, and consequently also the front wall of the bladder, are drawn forwards; but directly the muscles relax, the elasticity of the springs causes them to move backwards to their former position, carrying with them the wall of the air-bladder. Hence it follows that the rapid alternating contraction and relaxation of the muscles will import a vibratory movement to the anterior wall of the bladder and to the gaseous contents of that organ, with the result that a sound is produced. As a rule, those Fishes in which an elastic-spring-mechanism is present have the air-bladder subdivided by internal septa into a series of chambers freely communicating with one another; and no doubt the intensity of the sound is greatly increased by the vibratory movements of the gases across the free edges of the septa, and from one chamber to another.

The elastic-spring type of vocal organ is apparently restricted to the Siluridae, and besides occurring in *Auchenipterus* is found also in the South American genera *Doras*, *Oxydoras*, *Rhinodoras*, and *Euanemus*; in the African genera *Syndontis* and *Malopterurus*; and in at least four species of the Indian genus *Pangasius*. There are also a few Teleosts in which the air-bladder is provided with special muscles, but, instead of being connected with elastic springs, the muslces extend from the skull, and are inserted directly into the wall of the bladder or, without being in any way attached to the skeleton, the muscles simply invest some portion of the surface of the air-bladder. In other fishes the air-bladder, without possessing special muscles of its own, may, nevertheless, be partially invested by tendinous, or partly muscular and partly tendinous, extensions from the muslces of the body-wall or may be intimately related to certain muscles connected with the pectoral girdle. Whatever the precise relation of the air-bladder to its muscles it is probable that the physiological effect is in most cases the same.

By the rapid alternating contraction and relaxation of the muscles, some part of the wall of the bladder becomes alternately compressed and relaxed in such a way as to initiate a series of vibratory movements in the gases of that organ, and so produce definite sounds. In not a few of the fishes the cavity of the bladder is subdivided by external constrictions or by internal septa, or is complicated by the development of lateral, tubular, caecal branches; and hence the vibratory movements of the gases will be greatly strengthened by their passage across the edges of the septa, or the apertures of the caeca, and the intensity of the resultant sounds also increased. It will be readily understood that the nature and quality of the sounds omitted by different fishes will necessarily vary with the shape of the air-bladder, the number and arrangement of the internal septa and the caeca, and the strength and disposition of the contracting muscles. In a few Teleosts (Triglidae and Zeidae) sounds are said to be produced by the rapid vibration of an annular, or centrally-perforated, muscular diaphragm, which stretches across the cavity of the air-bladder.

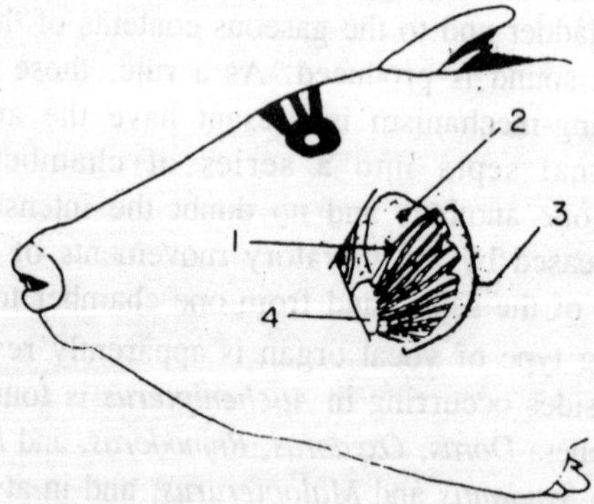

Fig. 15.4. Sound production in a triggerfish, Rhinecanthus rectagulus, takes place by rubbing the pectoral fins against the swim bladder: (1) pectoral fin spine, (2) drumming muscle, (3) pectoral fin rays, and (4) fleshy muscular lobe of pectoral fin.

Nevertheless, it must be strongly emphasised that, while in some fishes the air-bladder and its muscles undoubtedly constitute a vocal organ, there are many others in which the bladder can only be inferred

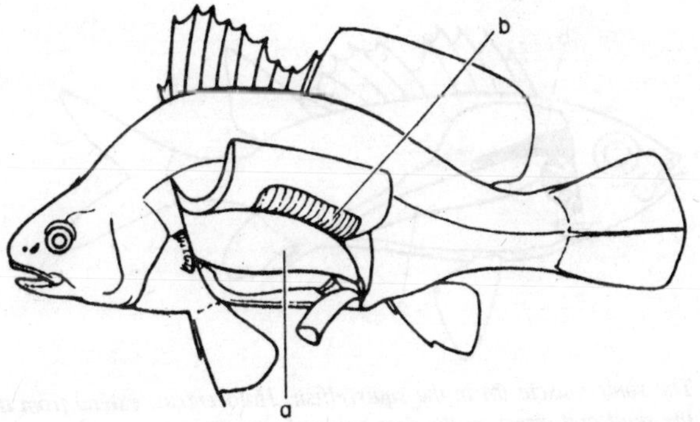

Fig. 15.5. The swim bladder (a) in the drumfish, Aplodinotus, is vibrated by a pair of sonic muscle (b).

to be sound producing from its general agreement in anatomical structure with the same organ in fishes where its vocal function has been clearly proved. By one or other of these various methods the air-bladder is either known to be sound-producing, or is believed with good reason to be such, in the following Teleosts, and many others: Certain species of the South-American genera of Siluridae, *Pimelodus*, *Sorubim*, *Platystoma*, *Piratinga*, *Centromochlus*, and *Trachelyopterus*; species of the South-American family Characinidae; *Amblyopsis spelaea*, the blind fish from the Mammoth Cave of Kentucky (Amblyopsidae); among the Syngnathidae, the short-snouted Sea-Horse (*Hippocampus brevirostris*); of the British Coasts; certain Sclerodermi, such as the Trigger-Fishes, *Balistes vetula*, *Triacanthus brevirostris*, *T. biaculeatus*, and *Monacanthus pardalis*, and also some "Coffer Fishes" (e.g. species of *Ostracion*); some Gymnodontes (species of *Diodon* and *Tetrodon*); a few Serranidae (e.g. species of *Therapon* and *Pristipoma*); species of *Holacanthus* (Chaetodontidae) and in *Holocentrum sogho* (Berycidae); such Sciaenidae as the "Drum" (*Pogonias chromis*), the "Maigre" (*Sciaena aquila*), which has sometimes been taken in British waters, *Umbrina cirrhosa*, *Otolikhus regalies* and *Micropogon undulatus*, and, with more or less probability, many other species of the same family; one species of Zeidae, the John Dory (*Zeus faber*); *Batrachus tau* among the Batrachidae; several species of Gurnarda (Triglidae) belonging to the genera *Prionotus* and *Trigla*; the so-called Flying Gurnard.

Dactylopterus volitans (Dactylopteridae); the Indian species *Ophiocephalus marulius* and *G. gachua* (Ophiocephalidae); amongst the Gadidae, the Cod (*Gadus morrhua*) and the Haddock (*G. aeglefinus*); in such Zoarcidae as the blind fish (*Lucifuga subterranea*) from the

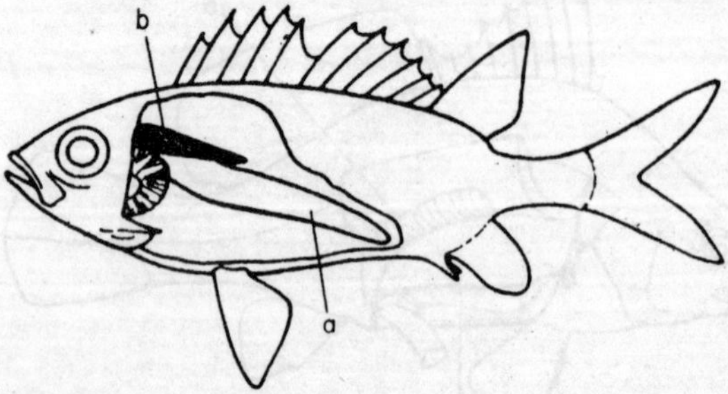

Fig. 15.6. The sonic muscle (b) in the squirrelfish, Holocentrus, extend from the rear of the skull and insert on the first pairs of ribs. The ribs are tightly laced to the swim bladder (a).

subterranean waters of the caves of Cuba, and in some Ophidiidae (e.g. species of *Ophidium*). In fishes other than Teleosts, instances of normal sound-production by special vocal structured are rare. No recorded instances are known in the Cyclostomes or the Elasmobranchs, but there is evidence that sounds are emitted by *Polypterus* among the Crossopterygii, and by the Dipnoids *Neoreratodus*, *Protopterus*, and *Lepidosiren*, although it is not certainly known how they are produced, or that they may not be the accidental concomitants of the inspiratory or expiratory action of the lungs in breathing.

As to the nature of the sounds produced by the air-bladder and its muscles in different Teleosts, a few examples may be given. The sound produced by the elastic-spring-apparatus of a recently caught. *Doras maculatus*, has been described as a "deep growling tone", which may be distinctly heard at a distance of 100 feet when the fish in out of the water. Under like conditions the air-bladder and its muscles, in a species of *Platystoma*, emit a similar sound. On the other hand, the sound produced by the elastic springs of the Electric Siluroid (*Malopterurus*, *electricus*) has been compared to the hissing of a cat. The Sea-Horse (*Hippocampus brevirostris*) utters a monotonous sound analogous to that of a tambour, which is characteristic of both sexes, but is more intense and frequent in the breeding season.

The "Coffer Fish" (*Ostracion Trigonus*) emits a growling sound, as also does the "Globe Fish" (*Tetrodon honckenii*) when taken out of the water. The air-bladder and its muscles in the "Drum" (*Pogonias chromis*), constitute the most powerful sound-producing organ yet found in any fish. The sounds emitted by the "Drum" are better expressed by the word drumming than by any other, and have frequently been heard by persons in vessels lying at anchor on the coasts of the United

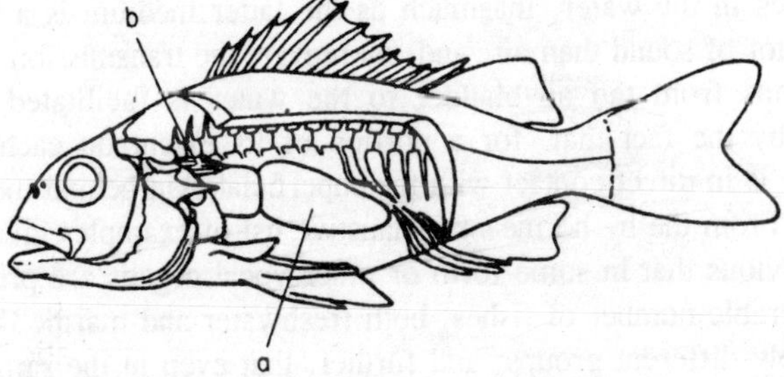

Fig. 15.7. The swim bladder (a) in the tigerfish, Therapon, is vibrated by a pair of sonic muscles (b).

States, where these fishes abound. The "Drum" begins its drumming noise in the spawning season in April, but is rarely heard afterwards.

The "Maigre" (*Sciaena aquila*), whose musical performances are perhaps responsible for the Homeric fable of the song of the Sirens, is remarkable among fishes for the variety of its sounds, which have been compared to bellowing, purring, buzzing, and whistling. The sound is often so intense that it may be heard when the fish is at a depth of 18 metres, and the ear of the observer two metres above the water; and it has been recorded that by listening for these sounds, shoals of Maigres have been successfully netted. They rarely emit sounds when isolated; but in shoals, during the breeding season, they do not cease to make sounds with a vigour and a persitency which apparently must soon wear out their strength. One of the Indian Horse-Mackerels (*Caranx hippos*) grunts like a young pig when captured, and the sound is repeated whenever it is moved, as long as vitality remains.

A West Indian species of the same family (*Argyriosus vomer*) has been observed to produce a like sound, while an Egyptian *Caranx* (*C. rhonchus*) is known to the Arabs as the "Chakoura" or "Snorter." The sounds produced by the different British Gurnards, such as the Grey Gurnard (*Triglagurnardus*), the Piper (*T. lyra*), the Elleck or Cuckoo Gurnard (*T. cuculus*), and the Tub-Fish (*T. hirundo*), have been compared to snoring, a sonorous and prolonged grunting, crooning (whence, perhaps, the term "crooner," by which the Grey Gurnard is known in Ireland), and croaking. The John Dory (*Zeus faber*) also utters sounds analogous to those of the Gurnards.

Among the Dipnoi *Lepidosiren* is said to make a growling sound, and *Neoceratodus* a grunting noice which may be heard at night for some distance. Whatever the nature of the vocal mechanism, it is highly probable that the sounds produced by fishes travel to considerable distances in the water, inasmuch as the latter medium is a far better conductor of sound than air, and, moreover, the transmission of sound-vibrations from the air-bladder to the water is facilitated in many fishes by the fact that, for a portion of its extent on each side the bladder is in direct contact with the superficial skin behind the pectoral girdle. From the by no means exhaustive list of examples given above, it is obvious that in some form or other vocal organs are present in a considerable number of fishes, both freshwater and marine, belonging to widely different groups; and further, that even in the same species (e.g. *Doras maculatus* and other Siluridae), both stridulation and the action of extrinsic muscles on the air-bladder may be utilised as a means of sound-production.

Certain Teleostean families like the Siluridae, the Sciaenidae, and the Triglidae, seem to be distinguished above all others by the prevalence of some form of vocal organ. According to Sorensen, the first mentioned of the three families includes no less than 68 species, which utilise the air-bladder alone as a sound-producing organ. Nevertheless, there still remain many Teleostean families, rich in genera and specise, and with an almost world-wide geographical distribution, in which such organs have not yet been found.

Mechanics of Swim Bladder Sound Production

If we consider the swim bladder and its associated skeletal and muscular systems as an underwater loudspeaker, the energy source for the emitted sound is the contraction of specialized muscles. In no case is there a muscular antagonist to the action of the sonic muscles. In the catfishes the "elastic spring" maintains tension on the some muscles and insert on the first pairs of ribs. These muscles contract against the tension of hypaxial muscles, vertebral ligaments, and connective tissue lacings to the roof of the swim bladder (Hazlett and Winn, 1962; Winn and Marshall, 1963). In the sea robins and toadfishes, where the sonic muscles are intrinsic to the swim bladder, the elasticity of the bladder walls and the internal pressure of the distended bladder will tend to returns the muscle fibers to normal length after each contraction. In the sciaenids, the return of the sonic muscles to normal shape after contraction is a function of the swim bladder tension, in part, and probably a function of the elasticity of the lateral body wall musculature. In catfishes. Tavolga (1962) showed that the fundamental frequency of the sound is a direct translation of the frequency of contraction of the muscle. Packard (1960) recorded sounds and muscle action potentials simultaneously in the pigfish. *Congiopodus*, from New

Zealand, and he found that the contraction frequency coincided with the sound frequency.

A similar observation was made by Barber and Mowbray (1956) in the Sculpin and Schncider (1964b, 1967) in *Therapon* Winn and Marshall (1963) found that each sound from a squirrelfish, *Holocentrus rufus* consisted of upto 5 pulses, about 10 msec apart. This corresponds to a fundamental frequency, of about 100 Hz. Muscle action potentials coincided with the pulses. Artificial stimulations of nerves leading to the sonic muscles can show that the muscles are capable of unusually rapid contraction and recovery cycles. In the gaff-topsail catfish. *Bagre marinus*, the sonic muscles took upto 12 sec to tetanize at 150 pulses/sec. and up to 3 sec. at 200 pulses/sec. (Tavolga, 1962). Using the same stimulating equipment as described earlier (Tavolga, 1962), the nerves leading to the sonic muscles were stimulated in the toadfish. *Opsaus tau*, the slender sea robin.

Prionotus scitulus, the squirrelfish, *Halocentrus ascensions*, and the red hind *Epincphelus guttatus*. *Opsanus* and *Prionotus* muscles were found to be most resistant to tetanization, but all could be stimulated at pulse frequencies over 100/sec without tetanization (Tavolga. 1964b). Among vertebrates, the occurrence of such fast-acting muscles is unusual. Extrinsic eyeball muscles are known to reach 350 contractions/sec, but the majority become tetanized when stimulated at 50 pulses/sec. The sonic muscle of *Opsanus* has a contraction—relaxation cycle of 10 msec. (Skoglund, 1959), which would theoretically limit its response to about 100 contractions/sec, but Skoglunds observations were limited to single, twitches.

Sonic muscles in the squirrelfish show fusion at frequencies above 200/sec. while normal white muscle begins to show summation at 50 and fusion at 100 pulse/sec. Pearson (1936, traced out the course of the VIIIth nerve and showed that the peripheral sensory supply to the lateral line system was derived from the acoustic nerve VIII. In spite of this report and the earlier descriptions of (Herrick 1898), some recent textbooks and manuals of comparative anatomy still give the innervation of the lateral line and head canals as the facial (VII), glossopharyngeal (IX), and vagus (X) nerves, based upon gross anatomical observations. This error has been pointed out most recently by van Bergeijk (1976a).

Swim Bladder and Hearing

According to Griffin (1950, 1955) and Pumphrey (1950), a fish is essentially transparent to water-borne sound and its only acoustic discontinuity is the swim bladder (or other air chamber). Sound reception under water requries the presence of a transducer constructed of material very different in acoustic properties and density from the surrounding medium. Air bubbles are known to be excellent reflectors and resonators (Horton, 1959, Meyer, 1957) and certainly the swim bladder can serve efficiently as transducer. Jones and Pearce (1958), N.B. Marshall (1951), and Midttun and Midttun and Hoff (1962) have shown that fish swim bladders are effective sonic reflectors and that 50% or more of impinging sound energy is returned by the bladder, while a smaller percentage is reflected by the rest of the fishes body. Klekoper and Roggenkamp (1959) demonstrated that damage to the swim bladder in the catfish. *Ictalurus*, raised threshold by 20 dB or more.

It is quite probable however, that some portions of the fish, such as the skull, may also serves as acoustic discontinuities and thus permit sound reception by bone conduction, although the swim bladder still appears to be the most obvious and efficient sonic transducer that the fish possesses. If the above is correct, then fishers with swim bladders should have better hearing than those without. Furthermore, those species in which the swim bladder is acoustically coupled to the inner ear should have the highest auditory sensitivity and broadest range. It appears that the Ostariophysi possess the lowest auditory threshold and highest upper frequency limits. This is undoubtedly a function of the Weberian apparatus which couples the auditory signal received by the swim bladder to the inner ear in a manner analogous to the operation of the middle ear ossicles in mammals.

Other air chambers can serve in similar fashion, as, for example, the branchial cavity in the labyrinthine fishes (Schncider, 1941). Among nonostariophvisines there are a number of forms in which the swim bladder has anterior extensions which are either coupled directly to the perilymphatic fluid (as in many clupeids) or attched to the occipital region of the neurocranium (Froese, 1938; Crasse, 1958; Tracy, 1920); Wohlfahrt (1936, 1938) described long, thin anterior extensions of the swim bladder in clupeids. These terminate in gas-filled capsules enclosed in bone and coupled to the perilymph through in elastic fenestra. Although auditory thresholds-using psychophysical methdos are not yet available for any clupcid fishes. Enger (1967a) obtained action potentials generated in the medulla in response to acoustic stimuli in the herring, *Clupea harengus*.

The tentative audiogram so generated showed a rather flat frequency response over a range of 30-1200 Hz with a threshold of —20 to —25 dB ub, and a sharp increase to +35 dB ub at 4000 Hz. In contrast to the neurophysiological technique used by Enger (1967a), a number of workers have resorted to conditioning techniques in which the fish are trained to make some behavioural response in the presence of the test sound. By means of avoidance conditioning. Tavolga and Wodinsky (1963) showed that the squirrelfish. *Holocentrus ascensionis*, has a low threshold and broad frequency response spectrum. It is probable that this is related to the contiguity of the anterior end of the swim bladder to the skull, as described by E.M. Nelson (1955).

Species with reduced or absent swim bladder should have poor hearing but the evidence for this is sparse, Bull (1928) was unable to condition a blenny. *Blennius*, to respond to sound. In *Gobius*, Dijkgraaf (1949-1952b) showed an upper frequency limit of only 500 Hz. and he postulated that most sound reception in this species took place through lateral line or cutaneous tactile senses. On this basis, sharks and other clasmobranes should be virtually deaf, yet the studies of Kritzler and Wood (1961) and D.R. Nelson (1967a) and the electrophysiological work of Lowenstein and Roberts (1951) show this is not true. The validity of the statement that the fish is acoustically transparent needs to be reexamined, and the possibilities that the skull, vertebral column, and other portions of the body can act as acoustic discontinuities should be investigated. Furthermore the contribution of the lateral line to hearing needs clarification.

Lateral Line and Hearing

Fish possess two sensory modalities for the detection of underwater vibrations. In addition to the inner ear, they possess a series of integumentary sense organs collectively known as the lateral line system. This system will be covered in detail by Flock in Chapter of this volume. Most of the motonical studies on the lateral line fishes have been summarized by Disler (1960), and the microscopic anatomy of the individual receptors was described by Dijkgraaf (1952a), Cahn and Shaw (1962), and many other authors (cf. Cahn, 1967). As reviewed by Dijkgraaf (1963a), some of the early literature on the function of the lateral line was contradictory, but strong indications were that it responds to minute water currents and to frequencies below 500 Hz (Kleere, Koper and Roggenkamp, 1959). Electrophysiological studies by Suckling and Suckling (1950). Suckling (1962), Harris and van-Bergeijk (1962), Jielof *et al.* (1952), and Kuiper (1956) have established the fact that the lateral line functions as a tactile receptor specialized for the detection of water displacements.

Dijkgraaf (1917b, 1967) developed the idea that the lateral line could function as a sort of low frequency sonar system to detect the presence of nearby obstacles. Acoustic energy is present in two forms: a pressure wave and a displacement. The pressure wave (far-field) is usually measured by a hydrophone, and its intensity drops off as the square of the distance from the source. Displacement (near-field) is detectable only close to the sound source since its intensity drops off as the cube of the distance. Harris and van-Bergeijk (1962) demonstrated the importance of the near field in acoustic detection by fishes. They showed that the lateral line is strictly a near-field receptor, and operates only at low frequencies. At effective near field would exist at distances of about one-sixth of a wavelength, as calculated by van-Bergeijk (1964). It is clear from the neurophysiological studies of Kuiper (1956), Jielof *et al.*, (1952).

Suckling and Suckling (1950, 1964), and others that the lateral line organ is a displacement detector, with a threshold in the order of magnitude of about 20 A (Harris and van-Bergeijk, 1962). Based on additional behavioural studies, Dijkgraaf (1963a, 1964, 1967) concluded that these structures primarily respond to "current-like water disturbances." He also made the point that the lateral line organ are not acoustic detectors, i.e., they do not respond to a propagated pressure wave. Many behavioural studies, however, have referred to the lateral line as responding to "sound". As pointed out by van-Bergeijk (1967c), Dijkgraaf's original contention (1934) that the lateral line organ can only detect water motion is perfectly correct. However, the near-field effect is an inseparable component of acoustic energy, and the entire, lateral line system acts as a near-field acoustic detector.

The functional similarity of the inner ear and the lateral line can be illustrated by the following: (1) Both inner ear hair cells and lateal line organs have a fundamenal structural similarity, even to the polarity of response and arrangement of kino and stereocilia; (2) both inner ear and lateral line are supplied by branches of the acoustic nerve; (3) both are essentially displacement detectors (van-Bergeijk, 1967a). The lateral line system remained a near-field detector, while the inner ear, by virtue of the proximity of the swim bladder, became a far-field detector (van-Berjeijk, 1967a). Both can be considered as acoustic sense organs.

Advantages

The advantages which Fishes drive from the possession of sound-producing organs are sufficiently obvious. A characteristic feature in the reproduction of most fishes is the general absence of any process of conjugation between the sexes, the eggs being fertilied in the water after their extrusion from the body of the female, and, consequently, any device which will facilitate the formation of shoals during the breeding season must be of great advantage to the species by largely increasing the chances that the ova will be fertilised, and thus secure the more successful propagation of the race. Hence it may be concluded that the vocal organs of fishes are a means to this end, and that the sounds they produce are in fact recognition-sounds which enables fishes of the same species to congregate together at periods when reproductive activity is greatest. This view is in harmony with much that is known of the habits of these fishes, especically with the fact that particular sounds are often characteristics of particular species, and that the sounds are produced most frequently and with greater intensity during the breeding season than at any other time.

While useful to all fishes that possess them, vocal organs are, no doubt, specially serviceable to those fishes which, from the nature of their habitat, can make but little use of their eyes; and this fact may perhaps explain the prevalence of such organs in the Siluridae, which are frequently bottom or ground-feeding fishes, and often live in muddy waters. The sounds emitted by fishes may also, in some instances at least, be warning sounds. Many of the sound-producing fishes are provided with exceptionally strong spines either in connexion with the median and paired fins, as in many Siluridae, or on the general surface of the body, as in *Diodon hystrix*. Such spines are very effective weapons for offensive or defensive purposes, and are capable of inflicting very serve wounds. The natural enemies of these fishes learn by experience or instinct to associate particular sounds with the possession of dangerous spines, and warned by the sounds, they refrain from attacking the onwers of the spines, to the mutual advantages of both.

16

MOVEMENT IN WATER

Fishes are variously adapted for swimming. The body is spindle shaped, thicker infront than behind and is perfectly streamlined for movement through water. The body is covered with a layer of muscous which reduces drag on it. Typically, the shape of the body is fusiform but variations occur when the fish becomes adapted to specialized mode of life. Some species have laterally compressed body as in the clupeidal, others possess dorsoventrally depressed body as the skates, rays and many cat fishes, while some are elongated and eel like. Fishes swim with such apparent ease that it is hard for us to realize (until we dive ourselves) just how difficult it is to move rapidly through water, and how well adapted fishes are for locomotion in a dense medium. It is not surprising that our own underwater designs are often very similar in shape to fishes, for the density and viscosity of the medium dictate the same streamlined form: the solid of least resistance, which the great pioneer aerodynamicist *Cayley* took from the body of a trout, looks just like a modern submarine hull.

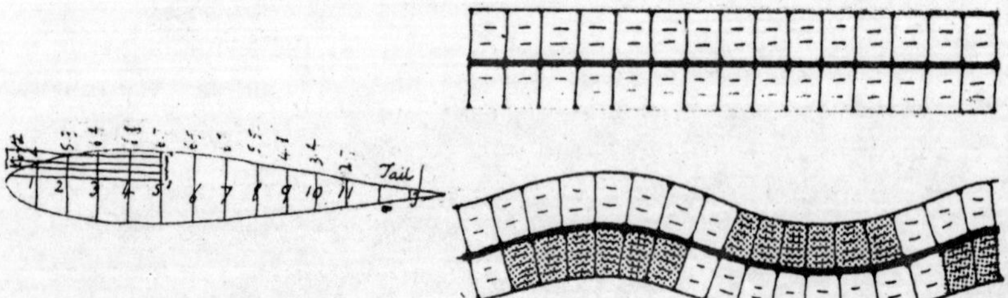

Fig. 16.1. Left: Cayley's drawing of a trout; right schematic diagram illustrating bending of a body with flexible backbone by segmented muscle blocks.

Yet there is an essential difference between the fish design and our own which makes the analysis of fish swimming a difficult and intriguing problem. Submarines are rigid vehicles which use rotating propellers to provide the thrust required to overcome the drag on their hulls. It is possible, therefore, to make rigid models and to test them in water tunnels to collect data which can be used to estimate the drag they incur and the power needed to drive them along. Data from such tests and from experience with actual submarines and air craft form the basis for theoretical analyses of the hydrodynamics of rigid bodies.

Unfortunately, for biologists, fishes are usually far from rigid, and they generate thrust not with propellers, but in quite a different way, usually involving change in shape of the body as they pass through the water. Various fishes have adopted special methods of locomotion (not all for use in water) in which they use paired and unpaired fins to row, flap, or undulate themselves along, but the great majority of fishes propel themselves by a combination of two processes—the backward passage of transverse waves along the body, and lateral movements of the caudal fin. In such fish as scombroids, thrust is generated only by the caudal fin, and the body flexes rather little, whereas in eels, or in dogfish, there are large-amplitude body waves passing towards the tail. Most fishes fall somewhere between these two extremes, sending smaller amplitude waves down the body which terminates in a substantial caudal fin, as in gadoids or salmonids.

In general, the fastest swimming fishes rely on the oscillation of high aspect ratio caudal fin, attached to the body by a narrow caudal peduncle; slowly-swimming fish undulate through the water like eels, and as we shall see, this is a less efficient process. This flexible method of thrust generation makes it impossible to apply directly hydrodynamic data from the performance of rigid bodies to the analysis of fish swimming. We could, for example, make a model of an eel, or stiffen an actual eel in some way, and measure its drag in a water tunnel, but we should not expect that the drag we measure would be the same as that which must be overcome by the swimming fish, nor that the surface flow patterns around the model would be even remotely like those around the living fish.

Although the hydrodynamics of fish swimming are thus more complex, and less well understood than those of rigid bodies, good progress has been made in the past decade, both in obtaining kinematic and respirometric data from swimming fishes set up in water tunnels, and in working out suitable mathematical models of fish swimming.

We can now make reasonable estimates (at least for small fishes) of the power needed for swimming, and of the efficiency of the process. In this chapter, we shall look first at the machinery which fishes use to move their tails and bodies; next see how these movements generate propulsive thrust; and finally, look into the efficiency of swimming, and at the adaptations which some fishes have devised to reduce the cost of swimming—some of these remarkably resemble devices found on modern aircraft.

Swimming Styles

We have seen how the flexing movement impels the fish: now let us look at the way the waves of flexure spread out in the water, producing the forward impetus necessary for swimming. By analysing the movements of various species in slow motion, we can see their different ways of swimming. It varies from the style of the tunny in which only the caudal fin moves, to that of the mackerel and the shark which displays the waves of flexion far more, and to the style of the eel, in which the waves convulse the whole animal. Both the eel and the tunny move in their surroundings without any difficulty, but the eel has virtually no caudal fin while the tunny has a very well-developed one.

To understand how different types of fish move, we must analyse the forces that create the movement. The undulations of the eel-shaped body produce a series of movements in the water which are obliquely directed to both sides of the fish's rear. This oblique movement can be divided into two components, one perpendicular to the fish and the other directed to the rear; the perpendicular forces cancel each other out, because those on one side compensate those on the other, but the rearward force makes the fish go forward. The swimming of spindle-shaped fishes is not very different: they reduce the oscillations of the body, but the impetus is supplied equally by the rear part and the caudal fin. If we watch the oscillating movement of the caudal fin, we shall see that the force it exerts on the water is directly oblique to the rear and, as in the case of the eel, can be divided into two parts. The tail and the body contribute to the movement to a different extent: in the eel it is only the body, in the tunny only the tail. We know that the tunny is one of the fastest fish, while the eel is rather slow in its movements.

Myomeres and Swimming

When the muscle fibers of a myomere contract they shorten. If the myomeres on one side of a fish are shortened, the head and tail bend towards that side and the myomeres on the opposite side of the body stretch. The action of a muscle segment is affected by two forces. One of these is muscular, curving the body; the other is resistance which straightens the body. Thus alternating, serial contractions of the myomeres on the two sides, flex the fish, or successive parts of the fish, into bows. The number of contractual waves passing down the body range from about 5 per minute in sharks (Squaliformes) to 170 in mackerels (Scombridae). Slow contractual waves are of high amplitude, whereas rapid waves are of low amplitude. The wave resulting from the serial contractions passes smoothly along the body, from head towards tail, because of the overlapping of the muscle segments as in the gnathostomes. When they are strong enough to overcome equilibrium, the muscular forces acting against the forces of resistance, result in forward swimming.

In the myosepta, the expression of forces such as these is seen in the development of strengthening intermuscular bones in many teleosts. If the body wave moves from tail to head, the fish swims backward. This is a notable capacity of eels and other elongated fishes which have anguilliform locomotion. The lampreys are an exception since they cannot produce reverse waves; their compensation is in great flexibility, enabling with case the tightest hairpin turn. Other fishes are not able to do this so well because the body waves of the trunk are not large enough to produce sufficient thrust. As a result, non-anguilliform fishes most often use the fins for reversing. The principle of the action of the forces that are applied by a contraction of a myomere may be diagrammatically represented. In this figure, AB represents the midline of the body, with A being towards the head.

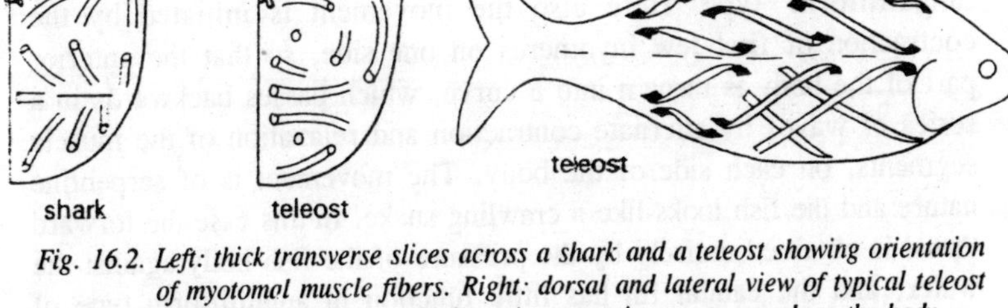

Fig. 16.2. Left: thick transverse slices across a shark and a teleost showing orientation of myotomol muscle fibers. Right: dorsal and lateral view of typical teleost showing course of muscle fibers in successive myotomes along the body.

Horizontal sections through the forwards and backward flexures of a myomere are represented respectively by the rhomboids *abcd* and *efgh*. A myomere exerts its pressures to bend the body at these flexures.

At the flexures, the myomeres are attached to the midline of the body at the angles *Odc* and *Oef* whereas the muscle fibers in the myomere are more or less paralled to *AB*. When the muscle fibers contract, the pull on the anterior septum of the forward flexure and on the posterior septum of the backward flexure. *P* represents the pull of the muscle fibers on the midline. Muscle contraction draws the flexures towards one another as the midline bends about the central point, O. Upon flexion, the myomeres and the midline AB bend to the arc shown by the broken line. Serial contractions of the myomeres form a smooth wave along the body as a result of the overlapping and nesting of the myomeres, as already described.

Types of Swimming

In the fish, three common types of swimming have been studied:

Carangiform Type

When a fish wants to move forwards, the first action is the contraction of first few myomeres of the anterior end of the body on one side only, so that the head is bent sharply to one side. This is followed by alternate contraction and relaxation of the successive myomeres, first on one side of the body and then of the other, from head towards the tail and the curve of the body passes backwards. The fish thus shows undulating movements and its body during swimming is thrown into a 'S' shaped curve. The actual forward thrust is effected by the pressure of the tail of the fish against the surrounding water. This is the most common method of locomotion in fishes and is called the 'carangiform' type.

Anguilliform Type

In an elongated type of fish such as the eel, the locomotion is of 'anguilliform' type. Here also the movement is initiated by the contraction of first few myomeres on one side, so that the anterior part of the body is thrown into a curve, which passes backwards in a series of waves by alternate contraction and relaxation of the muscle segments, on each side of the body. The movement is of serpentine nature and the fish looks-like a crawling snake. In this case the forward thrust is obtained entirely by the pressure of the fish body against the water, and the caudal fin has little function in anguilliform type of movements. Hence, the caudal fin is very much reduced or absent in eel-like fishes. Instead, the hinder part of the body is laterally compressed and the blade-like structure thus formed, serves to provide a largc amount of thrust to the fish. Fishes that have long, laterally compressed body as the Ribbon-fish (*Trichiurus*) also swim very actively by undulating movements of their body.

Ostraciform Type

The third type of locomotion is known as 'ostraciform' as in the "trunk fish." In this case, the head and body is enclosed in a hard and rigid bony case and the tail with a fan-like caudal fin projects behind. In such fishes, undulating movements of the body are not possible and the dorsal and the anal fins form the chief propelling agents for slow movements. Rapid swimming takes place by lashing movements of the tail. The anguilliform, carangiform and the ostraciform are the three main types of locomotion in fishes, but many intermediate types are also found to suit different species.

Generation of Thrust

Caudal Fin Oscillations

The earliest method of generating thrust was certainly that still used by lampreys and eels, that is, the propagation of waves back down the body, but it is simplest to begin with the much more advanced caudal fin oscillations used by tuna and other fast fishes, which have virtually separated the propulsive machinery from the thrust-generating apparatus: they are much closer to our own rigid machines. Tuna are essentially well-streamlined masses of myotomal muscle, attached to a rigid high aspect ratio lunate foil by a narrow neck. The myotomal muscles are attached to tendons, which pass across a flexible joint in this narrow caudal peduncle, so that as they contract the foil is oscillated from side to side across this joint, acting much like a

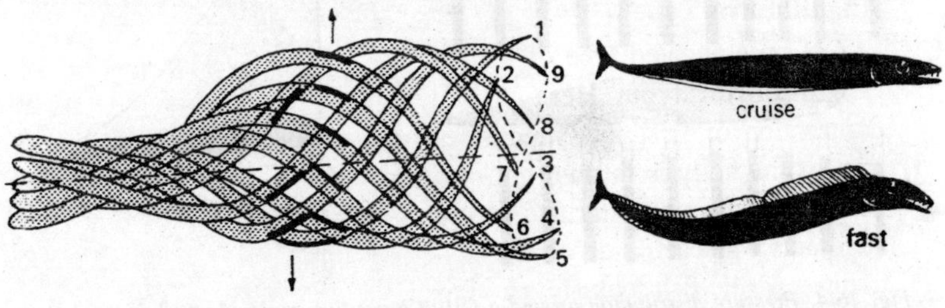

Fig. 16.3. Left: drawings from frames (1-9) of film of young Anguilla, superimposed to show short sections of the body as planes inclined to axis of progression when the eel moves forwards. The oblique dashed line is the axis of progression. Right the two swimming pattern of the trichiurid.

propeller (through an oscillating rather than a rotating one). Because it oscillates rather than rotates, the caudal foil operates alternatively with the same surface forming the upper and lower surface of the foil, and since it comes to a halt, as it were, at the end of each stroke, the thrust produced is oscillatory, though smoothed by operation at high frequency, up to the remarkable frequency of 10 tail beats per s. How does this oscillatory foil generate thrust, and how is it specially adapted to do so? Here we are on firm ground, for the process is the same as that by which aeroplane wings generate lift, or propellers thrust.

Circulation, Lift and Thrust

The forces which act on a foil as it passes through a fluid arise from the displacement of fluid by the foil, they are of two kinds. Frictional forces (resulting from viscous effects) can for the moment be neglected, since they are only indirectly concerned. Lift or thrust forces can only result from a net difference in the pressure of the fluid on either side of the foil. As an aeroplane wing passes through the air, it is both sucked upwards from above, and pushed up from below, for there are higher pressures below the wing and lower pressures above. How do these pressure differences arise? The relationship between pressure and velocity in the flow of an incompressible fluid is given by

$$P + \frac{1}{2}\rho V^2 = \text{constant}$$

where P=static pressure; ρ=density, and V=velocity. The quantity $\frac{1}{2}\rho V^2$ (often abbreviated to *q*) is the *dynamic pressure*, the force experienced if one puts one's hand of the window of a moving car, and it is this which is the fundamental source of aerodynamic or

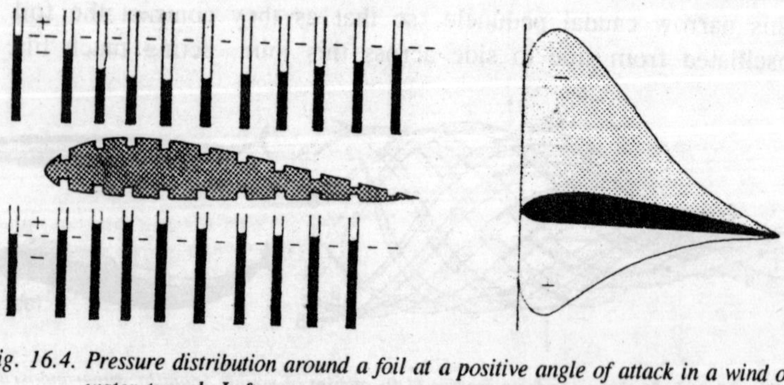

Fig. 16.4. Pressure distribution around a foil at a positive angle of attack in a wind or water tunnel. Left: manometers connected to openings in upper and lower surfaces show pressure above and below ambient. Right: typical pressure distribution around such a lifting foil.

hydrodynamic forces. As we shall see, the dynamic pressure appears in formulae for the calculation of thrust, life and drag. Since, the density of the fluid does not change at the speeds at which fish swim, clearly the pressure distribution which one can measure over a foil in a fluid must result from differences in the velocity of flow over the two surfaces, and the velocity is higher above a lifting foil than below it. If small particles like polystyrene beads are distributed in water, and a foil is moved through the water, the displacement of the water caused by the passage of the foil can be observed by taking high-speed cine films.

Let us follow the displacement to two particles A and B from their initial positions at rest before the foil arrives, until their final positions after it has passed. We find that they follow curved paths of opposite sense, that is, they are forced to *circulate* by the passage of the foil. Since the displacement of particle A above the foil is greater in the same time than is that of B below it, the algebraic sum of the two results in a net clockwise rotation of *circulation*, and because it is in this sense, the relative velocity of fluid above the foil will be greater than that below, it, and lift will be generated. We can see that the stronger the circulation the greater the lift.

Reduction of Drag Associated with Lift and Thrust

The circulation over the main part of the foil continues downstream in a series of tip vortices at the tips of the foil as it moves through the fluid, giving rise to an upwash just behind the back of the tip of the foil, and a downwash across the span, decreasing to the root of the foil. This process requires energy, and the energy that has to be expeneded to maintain the tip vortices is a component of the flag the lifting foil incurs as it passes along, called vortex drag (D). It is not hard to guess that as the tip vortices depend on the circulation, the greater the circulation, the greater the amount of vortex drag which is incurred. Like aeroplanes, fish take advantage of two possibilities to reduce this drag.

First, since a given lift or thrust can be produced by a weak circulation over a long span, or by a strong circulation over a short span, doubling the span can halve the vortex drag incurred, and for this reason, the caudal fins of fast swimming fishes are long and thin, as are the lifting pectoral fins of fast fishes like tuna, which are denser than water. Their aspect ratio, AR ($span^2$/fin area) is as high as 6; high as can be arranged without sacrificing structural strength. Secondly, it can be shown that for a given aspect ratio, vortex drag is least if the spanwise lift distribution is elliptical; this can be conveniently achieved by making the plan form of the foil elliptical, and this is the shape both of the caudal foils of scombroids, and of the lifting pectoral fins of fast sharks and many large fishes like marlin or swordfish. High AR tail fins of this kind, and the similar lifting pectoral fins, are designed to minimize vortex drag, and are suitable for high-speed cruising. But they are not very suitable for rapid manoeuvering, or for rapid accelerations, and this is why fishes which need these capabilities have caudal fins that are much broader and of lower aspect ratio. To provide thrust or (in the case of an aeroplane or dense fish) lift, the foil has to be operated at an angle of attack to the fluid, or the circulation of fluid above and below the foil will be equal, and no lift or thrust will result.

At small angles of attack, fluid will flow around the foil in an ordered way, but if the angle of attack is increased above a certain limit, the flow around the upper part of the foil will separate from it, leading to loss of lift and greatly increased drag: the foil will stall. A long thin foil of high AR is much more prone to stall as the angle of attack is increased than is a short wide foil of low AR, and it is when accelerating or manoeuvering that the angle of attack of the caudal fin is greatest. Thus the scombroid or albatross elongate foil is not suitable for salmonids or for pheasants, which require rapid acceleration from rest. In many respects, then, the thrust-generating foils of tuna show similar design features to aeroplane or bird wings, or propellers. Most fishes are less highly modified for rapid swimming than are tuna, and gain part or all of the thrust they need by a rather different process.

Efficiency

Efficiency is essentially measured by the ratio of input to output, and there are various measures of effciency depending on what input/output properties we want to consider. We could compare the swimming fish to our own machines, such as motor cars, which gain power (as does the swimming musculature) by converting chemical to mechanical energy, and then have to transmit this power to the road to drive themselves along. Engineers define two principal kinds of efficiency for such machines.

$$\text{Mechanical efficiency } (\eta_m): \frac{\text{brake horsepower}}{\text{indicated horsepower}} \times 100$$

This is the ratio of the power developed at the pistons *vs.* the power available at the output shaft after frictional losses in bearings, gear trains, etc. In practice, the mechanical efficiency of a motor car is usually around 80-85%.

$$\text{Thermal efficiency } (\eta_t): \frac{\text{work done}}{\text{heat supplied}} \times 100$$

This is the ratio of one kind energy, heat energy, supplied during the combustion of fuel in the cylinders, to the useful energy obtained in the form of work done by the engine driving the wheels. Diesel engines have thermal efficiencies around 40%, petrol engines rather less. Overall efficiency, H, will be the product of these two, so that the overall efficiency of a well-designed diesel lorry will be something like 30%. How do fish compare with this? In the case of the fish, N may be considered as the product of the efficiency of the caudal propeller (usually called η_p), and that of the conversion of chemical to mechanical energy in the muscles (η_m). Values for H during sustained aerobic swimming by salmonids work out between 5% at low speeds to around 20.22% at the maximum speeds sustainable.

Calculations for a different biological machine, man, during sustained activity, give similar values. Obviously, if we can measure H from oxygen consumption during swimming, and if we knew either η_m or η_p, we could obtain both. But unfortunately η_m has not been measured for any fish owing to difficulties with the way in which myotomal muscles insert into compliant myocommata. Still, fish muscles are fairly similar to those of other animals in which the measurements enabling η_m to be determined have been made, and by substituting these values, *Webb* (1971) calculated that propeller efficiency in trout rose to a maximum around 75% at a swimming speed of 2 body lengths per s. It seems very probable that scombroids and other high-speed fish, swimming by oscillating lunate tail foils, will have a higher η_p than this, and thus that overall efficiency H will be higher than for trout. So fish compare quite favourably with our transport designs, and indeed, with ourselves!

SPEED

Fish vary widely in overall size (from larvae only a fer millimetres long to the 12 m whale-shark), and the speed of only a very few within this range has been studied, but it seems generally true that fish smaller than half a metre long are capable of maximum speeds of about 10 times their body length per second. It does not seem likely that larger fish are capable of such a performance (which would give the whale-shark a maximum speed of 422 kph!), although they can travel at relatively high speeds. Accurate measurements of the speed of large fish are hard to come by, but measurements have been made up to 42 kph of line run-out on hooked tuna about 1 m long. A novel approach to the estimate of maximum swimming speeds has recently been made by *Wardle* (1975) who reasoned that the maximum contraction velocity of the white muscle could be related to swimming speed, since swimming speed is closely related to tail beat frequency. His observations have only so far been made on the muscles of fish under a metre long, but they give maximum swimming speeds in good agreement with those observed. In the case of tuna (*Wardle* and *Videler*, 1979) where maximum contraction speed of the white muscle is similar to that of other fish, but whose maximum swimming speeds seem well established as much higher, it is possible that this is achieved by reducing body flexure and maintaining a constant angle of attack of the caudal tail as it sweeps across. Wardle's concept of stride length is an interesting one, but the case of tuna requires further investigation.

Swimming Adaptations In Different Fishes

Swimming adaptation in different fishes have been studied in a variety of fishes as stated under the following heads:

Sharks and Rays

Three examples selected from amongst the Sharks and Rays will serve to illustrate this point. The Blur Sharks and their allies (*Carcharinus*) possess slender, perfectly streamlined bodies, conical heads, pointed snouts, and powerful muscular tails, the Carpet Sharks (*Orectolobus*) have stout thick-set bodies, considerably flattened from above downwards, massive heads, broadly rounded snouts, wide mouths, much reduced tails and comparatively small dorsal fins; the Rays (Raiidae) have very broad, flat bodies, the head, trunk and enormously expanded pectoral fins beings united to form a circular or quadrangular disc, from which the feeble tail with its tiny dorsal fins projects as a slender appendage.

The Blue Shark is an inhabitant of the open sea, feeding almost exclusively on other fishes; it is essentially a strong, speedy fish, every line of its body adapted for rapid progress in pursuit of prey. The Carpet Shark, on the other hand, relies on cunning rather than speed to obtain a meal, lying in wait on the sea floor until the prey comes within reach of its jaws. The loss of swimming power is here compensated for by the remarkable manner in which the Shark resembles its surroundings, its appearance when at rest being that of a weed covered rock. The uniform steely blue coloration of the Blue Shark is replaced by a beautiful variegated pattern which harmonises closely with the sea bottom.

The Ray, another sluggish, ground-living fish, also depends to a large extent on its general resemblance to the surroundings to escape observation by enemies. Its flattened form is admirably adapted for this particular mode of life, but, as will be sea later, an unusual method of locomotion enables this fish to move much greater rapidity than would seem possible from its appearance. Some still more specialised members of this order have acquired other protective devices in addition to their coloration. For example, the Torpedo or Cramp-fish (*Torpedo*) has powerful electric organs, and the Sting-ray (*Dasyatis*) has one or more strong, saw-edged and poisonous spines on its tail.

Bony Fishes

Among Bony Fishes the former type is rare, but the well-known Angler-fish or Fishing-frog (*Lophius*), in which mimetic resemblance and cunning in obtaining a meal have been brought to a pitch of perfection, and the little Bat-fish (*Ogcocephalus*), with the upper surface of its body protected by a covering of hard bony warts, provide excellent examples. The laterally compressed body, on the other hand, unknown in Selachians, is common in Bony Fishes. Often the body is shortened as well, and flexibility being no longer an absolute necessity, many of these forms have developed heavy protective armour of some kind.

The brilliant little Butterfly-fishes (Chaetodontidae) of tropical coral reefs are excessively quick in their movements, in spite of their short, deep, flattened bodies, and they rely largely on their agility to escape being eaten, coupled with the fact that their deep bodies and strong, spiny fins make them awkward mouthfuls to swallow. The beautiful Angel-fish (*Pterophyllum*) of the rivers of South America, a familiar object in aquaria, has a very much compressed and almost circular body, and the large fins have some of the rays drawn out into lengthy filaments. It is a very slow swimmer, spending most of its time suspended almost motionless in mid-water, and relies on its disruptive coloration to escape detection.

The Flat-fishes (Pleuronectiformes), a group which includes such well-known edible forms as the Halibut, Turbot, Plaice, and Sole, all have very flattened bodies, and, like the Rays, spend much of their time on the sea floor, where their mottled coloration harmonises with the ground on which they lie. The Plaice and the Skate are often lumped together as 'Flat-fishes,' but it is obvious that the resemblance is purely superficial. Both fishes have taken to a life on the bottom, where a flattened body is a decided advantage, but the Skate has become flattened from above downwards, whereas the Plaice is compressed from side to side. In other words, the colourless lower surface of the skate which rests on the bottom is the true lower or *ventral* side, whereas in the Flat-fishes this surface represents the right or left side.

The Globe-fishes or Puffers (Tetraodontidae) and their relatives the Porcupine-fishes (Diodontidae) provide examples of fishes with shortened, rounded bodies, in which the consequent loss of swimming power is compensated for by the development of armour in the form of spines or small prickles. In addition to their spiny covering, these fishes possess the power of swallowing water or air thereby inflating the body like a balloon. When thus inflated they float passively with the currents, more often than not upside down. If taken from the water a Puffer will generally inflate at once, but if slow to begin it can be persuaded to swell up by gentle tickling. When inflated, and with the spines erected, the fish is adequately protected against most predatory enemies, who would find it difficult to bite, much less to swallow. Observation made by *Dr. Beebe*, however, indicate that such protection is not always complete. He watched a number of little Porcupine-fishes, and saw that when they were threatened by a large Gar-fish, four feet in length, they bunched together for protection, giving the appearance of one large, round and prickly fish; occasionally, however, a single individual would become detached from the mass, whereupon it was promptly seized and devoured.

The allied Trunk-fishes (Ostraciontidae) are also slow-swimming creatures, living at or near the bottom of the sea, and rely for protection on their armour, which here takes the form of a rigid bony case. Also related to the Porcupine-fishes and Puffers are the large and grotesque Sun-fishes (Molidae), of which there are three genera all widely distributed in warm seas. The Round-tailed Sun-fish (*Mola*) has a deep, circular and somewhat compressed body looking as though the tail end had been amputated just behind the high dorsal and anal fins, a feature to which the popular name of 'Head-fish, refers. Such as body is probably well-adapted for more or less passive drifting in ocean currents.

The Round-tailed Sun-fish reaches a length of eight feet or more and a weight-estimated at more than a ton. It is a sluggish fish, often observed basking or swimming lazily at the surface of the sea. Underlying the skin, which is very tough and leathery, is a layer of a hard, gristly material some two or three inches thick—ample compensation for any loss of locomotive power! At the other extreme are fishes with long bodies, which may be rounded as in the Eels (Anguilliformes) or very much compressed, as in the Ribbon-fishes (Trachypteridae) and Cutlass-fishes (Trichyiuridae). From their shape one would hardly expect such fishes to be other than slow swimmers, but, as will be shown below, the adoption of a particular method of locomotion gives them a greater speed than the short-bodied forms mentioned above.

The peculiar shape of the Eel's body is almost certainly associated with its habit of living in soft river bottoms, wriggling in and out of the mud, creeping through reeds, or insinuating itself into holes and crevices as do its relatives in the coral reefs. Some of the Eels carry the elongation of the body to such an extreme that they have the appearance of a piece of slender whipcord, and the fins are often much reduced. Such a filiform type of body is characteristic of the curious Snipe-eels (Nemichtyidae), oceanic forms which sometimes descend to considerable depths. When seen swimming at or near the surface, these Eels are not infrequently mistaken for snakes. It may be noted here that similarity in eel-like form is not necessarily indicative of close relationship, but the result of convergent evolution. The so-called Synbranhoid Eels, for example, are not closely related to the true eels, and the same type of body in the two groups has evolve in response to the demands of similar environments.

In the Sea Horse (*Hippocampus*) the form of the body is unique, the head being at right angles to the trunk in a manner suggestive of a horse, and the trunk itself is definitely curved. The tail is also unique in that it is prehensile and can be used by the fish to anchor it to moving or fixed objects. The body is protected by a series of bony, rink-like plates, and in some species the spines or membranous processes with which these are ornamented serve to break up the outline and so render the fish inconspicuous when swaying to and fro among aquatic vegetation.

The Sea Horses are defenceless creatures, and depend largely on this mimetic resemblance to escape from predatory fishes. The locomotion of fishes provides the biologist and physicist with a number of interesting problems, and has also attracted the attention of the marine engineer, some of whose mechanical inventions own their inception, at least in part, to observations made upon living animals. Although actual swimming forms the main subject for consideration in this chapter, it must be remembered that this is by no means the only method in use, and 'walking' or creeping over the sea floor, skipping about on sand or mud, burrowing, wriggling on dry land, leaping and flying, are also indulged in by some fishes. These are however, rather in the nature of specialised developments, and, since, they are often accompanied by a modification of certain organs may be conveniently considered in later pages. Flying, for example, which involves the modification of the pectoral fins, is discussed in the chapter devoted to fins.

Evolution of Swimming

The various vertebrates here grouped together as fishes include a very deverse assemblage of forms, but there is, nevertheless, a basic similarity in their swimming movements, however different these may apper to be at first sight. The earliest vertebrates probably swam by means of simple rhythmic contractions of the trunk and tail muscles producing certain definite contortions of the body; by the pressure of different parts of the body in succession against the surrounding water the animal was driven forward. Most fishes have retained the primitive arrangement of the great body muscles, the myomeres, as they are called which forms a series of blocks or segments, arranged in pairs one behind the other and separated by partitions. In this respect, fishes differ from all land vertebrates, in which the main muscle masses are more or less concentrated on the fore and hind limbs, these being the normal organs of locomotion whereas the corresponding pectoral and pelvic fins of fishes more often than not perform quite different functions, such as balancing and steering.

We have already noticed the essential similarity in the shape of the body in the Cetaceans and fishes, but, owing to its different ancestry, the arrangement of the body muscles in a whale is quire unlike that found in a fish, and the swimming movements themselves are in a different plane, being up and down instead of from side to side. Two primary methods are employed by fishes to produce forward movements while suspended in a fluid medium: (1) body movements due to alternate expansion and contraction of the myomeres, and (2) movements of the appendages (fins); a third method is by the action of jets of water expelled from the gill-openings during the process of respiration. The first method is the most common and of the greatest importance, the others being, for the most part, auxiliary to it. It must be borne in mind, however, that in the majority of fishes the three are inter-related, and may all be used at different times, or when at the same time.

Locomotion solely by means of fin movements, for example, may be employed when slow progress only is wanted, but should danger threaten or prey appear in sight, body movements quickly come into play, and at the same time the increased rate of breathing assists in the general strengthening of the forward thrust. The Trout (*Salmo trutta*) uses body movements for forward progression, and will serve as an excellent example of the first method of swimming. When the Trout wishes to move forward the first action which takes place is the contraction of the first few myomeres at the front end of the body on one side, only resulting in the throwing of the head sharply to one side. The successive segments then alternately contract and relax from the head towards the tail, and the curve or flexure of the body is passed backwards, so that at any one movement the body is thrown into a gentle 'S'-shaped curve. The accompanying illustrations of a swimming fish give a good idea of the manner in which the body is undulated, and show how the flexure may be traced backwards from the head towards the tail.

The actual forward thrust is effected by the pressure of the fish's tail against the surrounding water. In order to use its tail in this way the front end of the fish must be braced against the water. It will be seen that during any one cycle three points of the body are braced against the resistant medium. The reaction from the most posterior point is the propulsive one, whilst those from the two anterior points (S_1 & S_2) are at right angles to the fishes body and do not exert any force along the path of movement. These are the chief bracing points and are those which prevent undue oscillation from the forward path. To understand the action of these body movements on the surrounding medium it will be convenient to study an elongated type of fish such as the Eel (*Anguilla*), and compare its locomotor methods with those of the Tout. The movement is again initiated by the contraction of the first few myomeres on one side. The anterior part of the body is thus thrown into a curve, and this curve is passed backwards in series of waves by the alternated contraction and relaxation of the serial muscle segments.

The movement is mechanically the same as that of a long rope held at one end and given a smart jerk with the hand at right angles to its axis. This results in a wave passing down the rope, the curves gradually decreasing in size and eventually dying out because the initial action of the hand was the sole agent of propulsion. But, in the living fish each successive muscle segment gives an added impetus to the wave, and so soon as the first wave has started backwards, a second follows, but on the opposite side, and so on. Here the forward thrust is attained almost entirely by the pressure of the fish's body against the water contained in the spaces between the curues. With this elongate form the fish gains much greater pressure areas from its sides than does the Trout, but, at the same time, it naturally loses the important terminal effect of the tail-fin. Indeed, we find that in all eel-like fishes the caudal fin is either very much reduced or wanting altogether. In most Eels the anterior part of the body is cylindrical in cross-section, whereas the hinder part is distinctly compressed the feature has a mechanical advantage, since a blade-like structure which presents its surface more effectively to the water naturally provides a greater amount of thrust than a rounded one. Mechanically, this part of the Eel's body functions like the expanded caudal fin of other fishes.

The elongate Ribbon-fishes (*Regalecus*), and other fishes with long bodies greatly flattened from side to side, undulate them into curves which are even more ample than those of the Eel, the extreme ribbon shape making this excessive bending comparatively easy to perform. It is of interest to note here that fishes with rounded, elongate bodies can move over rough surface out of water by applying the same locomotor forces normally used in swimming. Indeed, even a Trout can move across a board which is liberally provided with pegs against which the fish can exert a backward force.

Thus it will be seen that movement in a fish is based on the same physical principles as all other types of locomotion. Namely, the exertion of a force against the surroundings which in turn apply an equal but directionally opposite force against the body which then moves in that direction. If the locomotion of the Eel be regarded as one extrene type of body movement, that of the Trunk-fish (*Ostracion*) undoubtedly represents the opposite extreme, the Mackerel, Trout and other fishes being intermediate between the two types. In the Trunk fish with its head and body enclosed in a hard and inflexible bony case, from which the fleshy tail with a large fan like caudal fin at the extremity projects freely backwards, undulations of the body are clearly impossible. Normally, the dorsal and anal fins form the chief propelling agents, but where greater speed is required the fish swings the tail vigorously form side to side, the movements being brought about by the alternate contraction of the muscles on either side of the fleshy part of the tail.

A Trunk-fish swimming in this way may be likened to a small boat propelled by means of a single oar sculled from the stern, but the Hydrodynamic principles involved are somewhat more complex because of the tail fin's greater flexibility. The three types of body movements here described, and exemplified by the Eel, Trout, and Trunk-fish respectively, are arbitrarily chosen examples. Among fishes we fined such a complete gradation from one extreme to the other that it is not easy to say where one beings and the other ends. The extremes are methods employed by comparatively slow-swimming forms, mostly living close to the shore, whereas those of the Mackerel and Trout are of the highest efficiency and pre-eminently suited for high speed.

With the sole exception of fishes such as the Sea Horses, in which the locomotor emphasis is placed entirely on the fins, all existing forms fall somewhere within the series described above. Turning to the second of the primary methods of locomotion (by the action of fins) it may be noted that movements essentially the same as those of the body may be localised in one or more of the fins, and the same kind of series occurs, ranging from a serpentine, undulating motion like that of the Eel to a fan-like waggle which recalls the tail movements of the Trunk-fish. It has been remarked that the caudal fin is operated primarily by the action of the muscles of the body and tail, but army fishes are capable of moving slowly forward by means of wave like movements of the fin itself, the waves travelling at right angles to the longitudinal axis of the body.

In most fishes the shape of the fins, and more especially that of the caudal, provides a very good index of speed and agility, the same type of fin occurring in quite unrelated groups of fishes whose swimming habits are similar. It is impossible to enter into the mechanical possibilities of the different shapes of caudal fin, but it may be said that as a general rule fishes with large tails, the hinder margins of which are square-cut (truncate) or rounded, are comparatively slow swimmers, and, although able to accomplish sudden short bursts of speed, they are incapable of swimming for long periods at a high speed, as are those species provided with deeply forked or lunate tails. Such fishes have the upper and lower lobes of the fin long and pointed, and the fleshy part of the tail, known as the caudal peduncle, is nearly always very narrow and not infrequently strengthened by one or two fleshy keels on either side as in the '*Sword-fishes*'.

The Bonito (*Katsuwonus*) reckoned to be among the swiftest of all fishes, provide an excellent example of this type of caudal fin—crescent-shaped, without flesh, almost without scales, composed of bundles of rays, flexible, yet as hard as ivory. To other median fins, the dorsal and anal may also be used by certain fishes as propulsive organs, especially in those forms whose bodies have radically departed from the streamline shape. The fins may act in conjunction with the caudal fin or as a substitute for it. By appropriate use of the muscles controlling the fin-rays a series of wavelike movements can be produced in the fin, similar to those seen in the body of the Eel. In the Bow-fin (*Amia*) of North America, for example, undulating movements of the long dorsal fin are often used to propel the body slowly forward and the Electric Eel (*Electrophorus*), in which the dorsal fin is wanting, employs the long anal fin in a similar way.

The File-fish or Leather-jacket (*Monaanthus*) has both dorsal and anal fins placed a little obliquely, and makes use of both simultaneously for forward progression. Other fishes, such as the Globe-fishes or Puffers (Tetraodontidae) and Porcupine fishes (Diodontidae), move by flapping the short dorsal and anal fins in a fan-like manner. The Sea Horse (*Hippocampus*), which characteristically swims in an upright position, glides slowly through the water by means of rapid wave-like movements passing along the dorsal fin, which has the appearance of a tiny propeller revolving in the middle of the fish's back. The related Pipe-fishes (Syngnathidae) swim in a similar way, but their bodies being elongate and more flexible they are able to make more rapid progress at times by throwing the body into S-shaped curves.

Flat-fishes, when moving about on the sea floor, often make use of the long dorsal and anal fins which fringe the upper and lower edges of the body to obtain a grip of the ground, and by undulating these fins are able to progress at a fair speed. We must now consider the part played by the paired fins in locomotion. The pelvics may be dismissed at once, as these merely assist in maintaining, stability, and rarely, if ever, serve as organs of propulsion. The pectorals, on the other hand, are often used for locomotor purposes, particularly in those fishes of slow or moderate speed. In slow-moving fishes these fins are generally spatulate in shape, and may produce forward movements of the body by a simple synchronised flapping, as in some of the Wrasses (Labridae).

In others, of which the File-fish (*Monacanthus*) and the Porcupine-fish (*Diodon*) are good examples, wave-like motion similar to that described in connection with the caudal fin, is employed. This type of motion is particularly well-marked in the Rays (*Raia*) and their allies, in which the pectoral fins are very much enlarged and constitute practically the sole organs of locomotion. It will be noticed, however, that the waves travel in a vertical plane instead of a horizontal one—up and down instead of from side to side. In a few species, notably among the Danisel-fishes (Pomacentridae), the pectoral fins seem to be operated after the manner of oars, being brought forward almost edgewise and pulled back broadside on. In fishes of high speed the shape of the fins is generally long and falcate (*i.e.* sickle-shaped), and these are probably used mainly for changing course, and for braking, but scarcely ever for propulsion. Turning when in motion would appear to be effected largely by appropriated movements of the fins, and particularly of the pectorals and pelvics, but body movements also play their part. Stops are nearly always made by using the pectorals in the manner of brakes, but some forms pull up by reversing their primary locomotor apparatus. Finally, there remains the third factor in locomotion, namely, the jets of water from the gill-openings during respiration. These exhalations may play some part in driving the body forward, but the effect varies with different fishes, being of greater importance to some and of little or none to others.

A particularly powerful jet is usually expelled when a fish commences any swimming movement, thus giving an added impetus to the initial muscular efforts of the body in getting under way. Unless a fish is actually resting on the bottom, it is by no means as easy as it would appear to remain stationary in the water. Breathing cannot be suspended for a moment, and although this respiration may be comparatively slow as compared with that taking place when the fish is swimming, the exhalant jets of water are of sufficient strength to move the body forward, and some sort of action is necessary to counteract their force. Observation of a fish resting in mid-water in an aquarium shows that the pectoral fins are in more or less constant motion backing water, to counteract the forward thrust produced by the respiratory jets.

The speed attained by fishes has always been a subject of much speculation. Despite modern of research there is still little information on the speeds of different species. Gray is of the opinion that the maximum speed larger fishes (such as Salmon) could maintain for about twenty seconds would be 10 m.p.h. To keep up a speed of 5 m.p.h. a nine-inch trout would have to exert a backward thrust of about a half to a third of its own weight. It has been estimated that salmon can maintain a speed of 8 m.p.h. That the '*swords*' of swordfishes have been found deeply embedded in the timbers of ships is sometimes cited as evidence for the great speed attained by these species. However, this is susceptible to another interpretation. To quote Professor *Gray*... 'A little reflection will show, however, that the energy set free when the swordfish strikes the side of the dinghy, does not come from the movements of the fish's tail but from what is called the kinetic energy stored in the animal's whole body. If a fish weighing 600 lb and travelling at 10 m.p.h. runs into the side of a boat and is thereby brought to rest in a distance of 3 ft, the average force applied to the boat is about one-third of a ton, and the whole of this force is applied over an area—the tip of the sword—of about one square inch. The blow would be the same as the blow of a sledge hammer weighing 10 lb and meeting the boat at a speed of about 80 m.p.h. And, if a swordfish of 600 lb travelling at 10 m.p.h. meets, end-on, a wooden ship travelling at 10 m.p.h. in the opposite direction, the average force applied at the point of the sword is about 4½ tons.

Speed and Resistance

We have seen that the fastest fish are those with tapering bodies and a considerable amount of muscle, but to function the muscles need fuel and combustion like any motor. In vertebrates the 'fuel' is provided by the food which is taken in and absorbed by the blood, mainly in the form of fats and sugar. 'Combustion' is made possible by the oxygen taken in by respiration. There are, however, physiological limits to this combustion. Numerous observations have been carried out which show that a fish can sustain the maximum speed of which it is capable for a short time. The muscular fatigue that we notice after great physical exertion shows that our 'motor' is not in an ideal 'machine' to continue carrying out the work required. Fish feel tired in the same way, but underwater also some are in better training to combat fatigue than others.

Observing a trout (*Salmo trutta*) and a herring (*Olupea harengus*) of the same size, we find respective maximum speeds of 5.8 km an hour for the trout and 2.6 km an hour for the herring. But the trout rarely swims at this speed and lives rather a quiet life, while the herring swims tirelessly in search of food. There thus exists a fairly loose relationship between the maximum speed, used by the fish only in times of necessity, and its cruising speed which can be maintained for a period of time. To flee from the jaws of a predator, quick acceleration is more necessary than speed. Some fishes have amazing performances: carp, minnows, pike, cover the first five centimetres, starting from scratch, with an acceleration of 40 m/sec^2. Over such a short time, their acceleration equals that of a racing car.

Buoyancy

Dynamic Lift

As stated already that some fishes are denser than the water which they swim, and have to generate dynamic lift mainly by using their outspread pectorals as lifting foils. This process inevitably generates frag, which makes a significant contribution to the total drag and energy requirement of such fishes. For example, a mackerel weighing around 25 g in sea water has to generate some 1.2×10^4 N of dynamic lift during level swimming—mackerel effectively are climbing a 1 in 15 hill all their lives! But increased energy expenditure is not the only drawback to keeping aloft in the water by generating dynamic lift. Dense fishes must maintain a minimum forward cruising speed to generate enough lift to prevent themselves sinking, so they cannot hover or swim backwards.

The swimming rhythms of spinal sharks (35—40 tail bears min^{-1} in spinal *Scyliorhinus*) are apparently related to this minimum swimming speed. For benthic fishes that rest on the bottom and only swim occasionally, dynamic lift generation is appropriate, but for other fishes, a better solution would seem to be to store light materials to provide *static* lift (as do airships and submarines) and thus avoid the drawbacks of generating dynamic lift.

Static Lift

Most of the materials making up a fish are denser than water. For example, much of the body consists of locomotor muscle (density 1050—1060 kg m^{-3} in common marine fishes). Skeletal tissues loaded with heavy mineral salts are correspondingly denser (2040 kg m^{-3} for typical teleost bones). Nevertheless, water is so much denser than air that even without special stores of low density materials, fishes are not so much denser than the water in which they swim. They have the option (denied to birds) of storing sufficient low-density material that they can make themselves the same density as the water; they can achieve neutral buoyancy and need expend no muscular energy to keep station in the water. Fishes use two quite different materials to provide static lift. Gas is efficient in giving lift, since its density is very low, and most teleosts possess gas-filled swimbladders. Usually the swimbladder is around 5% of the volume of marine fishes, and around 7% in fresh-water fishes, providing enough lift for neutral buoyancy.

Fish swimbladders cannot significantly resist changing in volume as the fish swims up and down in the water and the ambient pressure changes, indeed, the swimbladders of almost all teleosts obey Boyle's

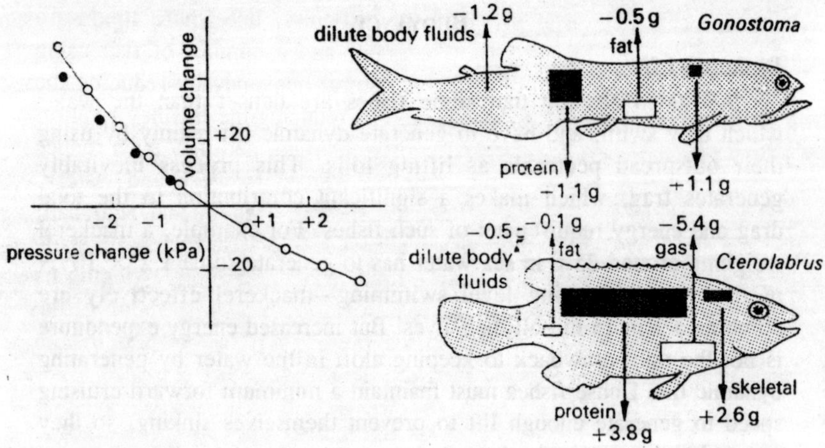

Fig. 16.5. Left: volume changes of airbubble and fish swimbladder as ambient pressure changes. Right: buoyancy balance sheets for a mesopelagic fish (Gonostoma) and a shallow-water acanthopterygian (Ctenolabrus).

law perfectly. So if a fish with a gas-filled swimbladder is to remain neutrally buoyant at different depths, it must secrete of absorb gas to keep the swimbladder at constant volume as the ambient pressure changes. To regulate the *mass* of gas within the swim-bladder in this way requires complex mechanisms of great physiological interest.

Fats and oils are also used by fishes as sources of static lift. These are much less efficient and much bulkier for a given amount of lift. However, they have the great advantage that the lift provided varies little with depth, because changes in ambient pressure have relatively little effect on the volume of the fat or oil: if a fish using lipid as a source of static lift is neutrally buoyant at the surface of the sea, it will be close to neutral buoyancy at the sea bed, even at considerable depths. In the short-term then, lipid provides fewer problems of buoyancy regulation than does gas, but in the longer-term, difficulties arise, because the lipids stored to provide lift may have other functions as well. For example, sharks which store oil in the liver and muscles may have to draw on this store as a fuel for continuous swimming, or as a food reserve for developing embryos or for the adult.

Where lipid is the sole source of static lift, there will certainly be great complexities in the regulation of lipid metabolism, and the fish will find it difficult to adjust its density rapidly to cope with the short term density changes resulting from feeding and parturition. Fish using gas as a source of static lift can afford to support the heavy components of their bodies without difficulty, but where lipids are used, the safety margin is not so great; and a number of fish using lipid for static lift can only achieve neutral buoyancy by reducing the dense components of their bodies and so reducing the lift they require. Their skeletons are reduced and poorly calcified, and they even reduce the amount of protein in the muscles, which are weak and watery. Figure shows the buoyancy 'balance sheets' (Denton and Marshall, 1958) of a typical shallow-water teleost using as a source of static lift, and of a deep-sea form using lipid, and in addition, reducing its dense components. The extreme cases of this 'evasive' solution of the buoyancy problem are the deep-sea predatory fishes which live in midwater, such as *Saccopharynx* and the deep-sea angler fishes.

As Denton and Marshall (1958) pointed out, they are little more than floating traps, with weak skeletons and watery muscles; they attract their pray with luminous lures and need not pursue it actively like the fish of the upper ocean. To some extent reduction in density of body components has taken place during teleost evolution, from the thick-scaled 'ganoid' fishes to the thin-scaled clupeids of today. Many fishes also have planktonic larvae whose dense components are reduced, and which have spaces for low density dilute body fluids.

Lipid as a Source of Static Lift

In four unrelated fish groups (and perhaps in others) enough lift is provided by stored lipids to achieve neutral buoyancy. Interestingly enough, the lipids stored are not biochemically similar, but they are all of particularly low density and so most efficient in providing lift. Fish lipids vary in density from around 930 Kg m^{-3} (Cod liver oil, and the oils of dogfish livers) to the wax esters of myctophids, gempylids, and *Latimeria* (densities around 860 Kg m^{-3}), and the hydrocarbons of some sharks (860 Kg m^{-3}). The difference in specific gravity between cod liver oil, and the wax esters or shark hydrocarbons may not seem very striking, but 1 g of the less dense oil will provide 0.1675 g of lift in sea water whose density is 1027.5 Kg m^{-3}, whereas a gram of the denser oil will only provide about half as much lift (0.0975 g). It is thus well worthwhile for the fish to store these lighter lipids rather than the more common triglyceride metabolic reserve, as all the fish that achieve neutral buoyancy using lipids actually do.

Wax Esters

A little squalene is also found in the living coelacanth *Latimeria*, but most of the massive amounts of lipid stored are wax esters (Nevenzel *et al.*, 1969). These make up 30% of the wet weight of the ventral musculature and over 60% of the wet weight of the swimbladder (which contains only lipid, no gas), and in the pericardial and pericranial tissues, the percentage is even higher. We know very little about the habits of *Latimeria*, but its fins certainly seem unsuited to provide dynamic lift, and the only specimen examined alive floated at the surface, so that it seems safe to assume that it is neutrally buoyant in life. Wax esters are probably stored to provide lift in many families of mesopelagic teleosts, but the only ones examined so far have been the gempylids and a few of the numerous kinds of myctophids.

The gempylid *Ruvettus* is loaded with low-density oil (density 870 Kg m^{-3}) which has purgative properties, hence the common name of castor oil fish. The cranial bones have been modified as oil tanks, and are the least dense tissues of the body, This large (1 m) predatory fish ranges from depths of 15 m to over 500 m and is very close to neutral buoyancy, feeding on smaller fishes which undertake diurnal vertical migrations. The most remarkable of these are the myctophids, some undertaking daily a double journey of 500 m up and down in the water column. Although all myctophids have gas-filled swimbladders as larvae, in many species the swimbladder shrinks and becomes invested with more and more lipid as the fish gets older, until no gas remains. As in *Latimeria*, lift from gas is replaced by lift from lipid, which may eventually make up 15° of the wet weight of the fish. As we should expect, the species with a high lipid content are neutrally buoyant, and store low-density wax esters. Why should many myctophids have abandoned gas as a source of static lift? Although the evidence is only circumstantial, it seems a reasonable guess that it is because there are difficulties in regulating buoyancy over a wide depth range when gas is the source of lift.

Those species which as adults have much low density lipid and are neutrally buoyant, have a greater depth range than the juveniles, which still have gas in the swimbladders and similarly, species which have gas-filled swimbladders as adults undergo less extensive vertical migrations than those which rely on lipid only. On the whole, the evidence is that lipid storage for static lift is a secondary phenomenon in myctophids, and that the most 'advanced' species in the family are those which have abandoned gas altogether as adults; it is interesting that amongst the family there are fishes showing all stages in this changeover adapted to different life-styles.

Insufficient Static Lift for Neutral Buoyancy

So far, we have considered fishes of different groups which store low density lipids to attain neutral buoyancy. What of fish which use lipid for static lift, but are not neutrally buoyant? Many teleosts are certainly in this category. Mackerel vary in density at different times of ycar because they have more or less lipid stored in the muscles; such a system is evidently a primitive one, for lipid is not set aside, as it were, to maintain a constant density. Rather, the lipid store is at the mercy of metabolic demand, and reduction in density, valuable as it must be, is simply a side effect of the storage of lipid for metabolic purposes. We know more about elasmobranchs. Baldridge (1969) and Bone and Roberts (1969) showed that a wide range of elasmobranchs, including large pelagic sharks, are each of a characteristic density, some being fairly close to neutral buoyancy, others being very dense. Liver lipid is an important factor in determining the density of some species, but in most, it is the density of other tissues (amount of fat in the white muscle, mineralization of the skeleton) which determines the density of the species.

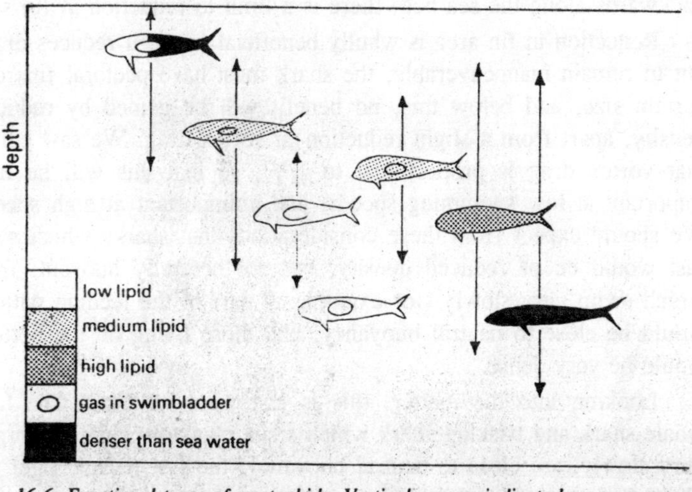

Fig. 16.6. Functional types of myctophids. Vertical arrows indicate known or assumed extent of vertical migration.

Not only of different elasmobranchs store different amounts of lipid in their livers, but the stored oil is least dense in the least dense species. It seems that elasmobranchs have managed to set apart the lipid used for density regulation, whether in the muscles or liver, from metabolic stores, so that the fish can regulate its density to a characteristic figure, whatever metabolic demands are made. How this is done remains to be discovered. At the beginning of this section, the advantages of neutral buoyancy were extolled, and it seems odd to find fishes using static lift to maintain a fixed density below that of neutral buoyancy. The regulatory mechanisms are there, and at first sight, it seems a harder problem to maintain a constant density (say around 1.03) than to regulate to neutral buoyancy, and what is more, dynamic lift must still be generated, with the penalties that this incurs. However, it is not hard to see why many pelagic sharks use lipid to reduce their density, but not so far as to become neutrally buoyant. Shark fins very a good deal in flexibility, and in their stiffening by skeletal elements, but the basic design, unlike that of teleosts, makes them impossible to retract: they are not of varying geometry. If we bear this in mind, we can see how considerations of dynamic lift generation lead to different densities in different species, depending on their mode of life. Reduction in density by lipid storage means that less lift need be generated in level swimming, so that the shark can either reduce the size of its sometimes, as does *Heterodontus*, to creep backwards along the sea bed, there is a limit to reduction in fin size.

Reduction in fin area is wholly beneficial (since it reduces drag), but to remain manoeuverable, the shark must have pectoral fins of a certain size, and below this, no benefit will be gained by reducing density, apart from a slight reduction in vortex drag. We saw earlier that vortex drag is proportional to $1/V^2$, so that this will be most important at low swimming speeds, and unimportant at high speeds. We should expect from these considerations that sharks which swim fast would be of reduced density, but not neutrally buoyant; those which swim very slowly (for example as part of the feeding pattern) would be close to neutral buoyancy; and those living on the bottom, would be very dense.

Looking into the matter, this is just what we find. The huge whale-shark and basking shark which sieve plankton while swimming very slowly, are close to neutral buoyancy; the fast pelagic tiger and blue sharks are of reduced density, but are not so close to neutral buoyancy: and the bottom-living dogfish and rays of shallow water have no special arrangements for static lift. Since teleosts have different designed fins, which *can* vary their geometry, the arguments above do not apply, and we should not expect to find any teleosts using lipid to reduce their density to a constant figure not close to neutral buoyancy: so far, none have been found.

Gas as a Source of Static Lift

Because fish swimbladders obey Boyle's law to obtain constant lifts as they swim up and down in the water, fish are obliged not only to store gas, but also to secrete and absorb it. In the sea, the ambient pressure increase by 1 atmosphere (101.3 kPa) for every 10 m increase in depth, so that a fish living at the surface (with the swimbladder gas at atmospheric pressure) will have the volume (and lift) of the swimbladder halved if it descends to 10 m, and vice versa. It is hardly surprising that the regulatory mechanisms cannot cope effectively with such large changes, and that many surface-dwelling pelagic fish have either lost the swimbladder altogether (tunas and *Scomber*) or have a reduced swimbladder (*Scomberomorus*). Gas is not a suitable source of static lift for this mode of life, unless the fish remains strictly at the surface, as do hermirhamphids and flying fishes. At greater depths, however, changes in ambient pressure have relatively alter the volume of its swimbladder by 1/40th, and the same applies with greater force, the greater the depth at which the fish lives.

The distribution of fish with gas-filled swimbladders in the sea has been examined by Marshall (1960) who has found that they occur

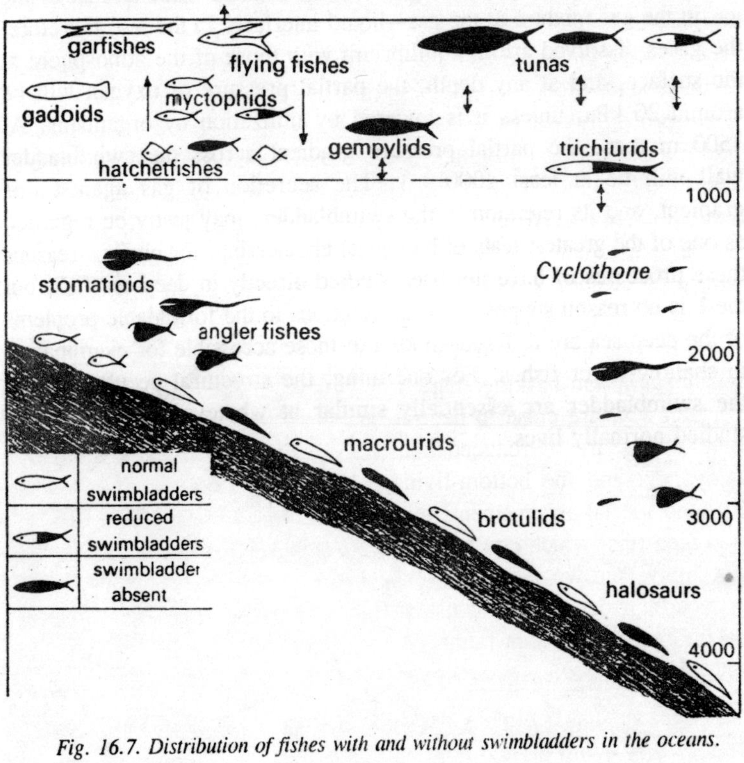

Fig. 16.7. Distribution of fishes with and without swimbladders in the oceans.

down to the greatest depths, but are absent from the upper layers unless they live just at the surface. The deeper the fish lives, the less significant are changes in volume with depth, but the more formidable are the problems of gas secretion and retention. Yet fish with gas-filled swimbladders have been caught at great depths. The macrourid *Nematonurus armatus* has several times been caught below 4000 m, on one occasion at 4500m. At this depth the ambient pressure is 450 atmospheres (45.6×10^3 kPa). The swimbladder gas in deep-sea fish is mainly oxygen so that the partial pressure of oxygen (PO_2) in the surrounding seawater, from which oxygen has to be collected across the fills and then transferred to the swimbladder via the circulation, the partial pressure of oxygen is several orders of magnitude less than this.

The partial pressure of a gas in a gas mixture is the pressure it would exert if the other gases were removed. Thus PO_2 in a 50:50 oxygen—nitrogen mixture at 2.0×10^3 kPa will be 1×10^3 kPa. The partial pressure of a gas throughout a liquid is the same as that of the gas in the gas mixture at the gas—liquid interface. In the sea, therefore, the gases dissolved are in equilibrium with those of the atmosphere at the surface, and at any depth, the partial pressure of oxygen will be around 20 kPa, unless it is lowered by utilization by organisms. At 4500 m then, the partial pressure gradient across the swimbladder wall will be at least 2000 : 1. The secretion of gas against this gradient, and its retention in the swimbladder, may justly be regarded as one of the greatest feats of biological engineering; for obvious reasons these processes of have not been studied directly in deep-sea fish, but there is no reason suppose that the solutions to the formidable problems of the deep sea are different in *kind* to those accessible for examination in shallow-water fishes. For one thing, the structural peculiarities of the swimbladder are essentially similar at whatever depth the fish studied normally lives.

17

ADAPTATIONS IN FISHES

Adaptation means the ability of animals to modify themselves according to the environment. Adaptation of all organisms to the environment may be of great importance for the survival in the struggle of existence. When conditions become unfavourable, the animal must do one of the three things: (i) Adopt themselves to the new conditions, or (ii) Migrate to better environment, or (iii) May become extinct. The question how adaptation arises that enables organisms to fit so perfectly into their envircnment, is of great interest.

In general, weak organisms are soon betted out in the struggle for existence by the process of natural selection. The fishes live in an astonishing variety of habitats: from the perpetual night of the abyss to the lighted surface spaces of the oceans. They are found along wave-battered rocky shores and coral reefs, in sandy bays and muddy estuaries, and in the mid-depths of the oceans where there is no bottom, no surface, no shore. They are to be found in great rivers and vast lakes, in rushing torrents and still canals, in stagnant ponds and farm ditches, and in caves deep underground and as deep as the abyss. They live happily beneath the pack ice of polar seas and in tropical swamps. They are to be found in the hot springs of Arabia where the water temperature may reach as high as 40°C (104°F) and, even, more astonishing, in the thermal soda springs of Lake Magadi in Kenya, where the temperature of the water, or rather of the soda liquor, ranges from 27°C to 49°C (80.6°F to 120.2°F).

With very few exceptions—the Dead Sea, which is too salty to support any form of life, is a notable one—whatever there is water, there will be fish. As is only to be expected, there is within this huge group, conditioned by this fantastic variety of habitats, a vast diversity of fishes. We are accustomed to think of the cigar-shaped fish as the typical fish. But, adapted to particular surroundings or to specialized ways of life, there are flat fishes, round fishes, ribbon fishes, tubular fishes, and so on. There is, in fact, no such thing as a 'typical fish', at least, so far as shape is concerned.

Adaptation in Shape and Symmetry

The most striking instance of adaptation in shape and symmetry is that of the flat-fishes or Pleuronectidae. These fishes habitually lie on the ground upon one side of the body, some species on the left side some on the right. In adaptation to this abnormal position the symmetry of the two sides is in many respects altered. If a flat-fish is held in a vertical position and compared with any ordinary fish many of the paired organs will be seen to be present in their usual position: on each side there is a lateral line, a pectoral and pelvic fin, and a gill-cover or operculum. The position of the gills shows which is the venral edge of the body, and it is seen that the marginal fins are continuous dorsal and ventral fins, which in many species, e.g., the plaice or turbot, are distinct from the tail-fin, but in some species they are continuous with the latter. The direction of the mouth also is, as in other fishes transverse to the median plane of symmetry, so that half of the jaws is on one side, half on the other. In plaice and sole, however, the two sides of the mouth are unequally developed, the jaws and teeth being much more strongly developed on the left or lower side where they are most used for seizing food.

The most remarkable departure from symmetry is in the eyes both of which are on the side which it naturally uppermost, that is to say the right side in a plaice or sole, the left side in a turbot or brill. The eye which is nearest the dorsal edge is that which belongs to the lower side, although it is separated from the latter by the dorsal fin: the other eye, the ventral, which is nearer to the mouth, belongs naturally to the upper side. The asymmetrical feature next in importance is the difference between the two sides in colour: the lower side is white, the upper side coloured, the colours are dark brown ranging to black, with sports or markings ranging from yellow through orange to red; the plaice, for instance, is characterized by conspicuous red spots. There are then three chief peculiarities in which a flat-fish resting on the ground with its original plane of symmetry in a horizontal position, differs from an ordinary fish which swims with its plane of symmetry vertical: (1) the position of both eyes on the upper side; (2) the absence of pigment from the lower side; (3) the extension of the dorsal fin along the line outside the dorsal eye, instead of between the eyes as in other fishes.

In addition to these there are minor differences between the upper and lower sides: the scales or tubercles are less developed on the lower side, the pectoral fin on the lower side is usually smaller, and as already mentioned the jaws and teeth are usually larger on the lower side, reduced on the upper. Obviously these peculiaries of structure are all adaptations to the peculiarity of position and mode of life, and they serve as excellent examples of what we mean by adaptation. The flat-fishes are evidently descended from ordinary symmetrical ancestors which swam in a vertical position, and in the course of evolution their structure has undergone the modifications which are more suitable for their peculiar position and habits. To have a flattened shape and to lie upon the sand or gravel at the sea-bottom, is a mode of life which offers distinct advantages to a fish in the way of food or concealment from enemies, and it has been adopted by other fishes as well as flat-fishes. Skates, for instance, and the anger have adopted this mode of life, but they have always rested on the ground in the vertical position, and either were originally or have become flattened dorso-ventrally without alteration of their original symmetry. The ancestral flat-fishes must have been originally flattened from side to side, and therefore to obtain the advantage of contact with the ground were compelled to lie on one side. In this position the lower eye would be useless and therefore the head has been modified so as to bring that eye round to the upper side.

Anatomy shows that the change is produced by a twisting or torsion of the skull in the region of the eyes through an angle of 90°, and this twisting has taken place without affecting the symmetry of the posterior region of the skull, or the jaws and anterior region. The interorbital septum of the skull has been bent round from the middle line to the right side, in the sole for example, and the eyes lie on either side of the septum as usual. The bones which were originally below and external to the left eye have become larger a new connection on this side has been formed between the bones of the anterior and posterior region of the skull. Along this bony ridge which is now in line with the original dorsal edge of the body, the dorsal fin has extended, thus dividing the originally left eye from the left side of the face. The jaw-bones, instead of being twisted in the same manner, have actually become much larger on the lower side, so that the asymmetry of the mouth is in the opposite direction to that of the eyes. An excellent example of the rounded shape is the huge *Sun-fish*, which may grow to a length of 10 ft and attain a weight of more than 1000 kg. The body is slightly compressed (that is slightly flattened from side to side) and is a very deep oval in shape.

The Sun-fish, which is quite well known in British waters in summer, is sometimes called the 'Headfish'; this is an appropriate name, for it does have the appearance of being composed of an enormous head the two fins, and no more. There is a large anal high dorsal fin and, immediately opposite to it, a large and deep and fin. The body ends abruptly behind these fins, giving the impression that the tail has been amputated. Actually, of course, there is a rounded tail-fin, but this is included in the deep oval and is not immediately apparent. Obviously, a body of this shape is not designed for swift swimming. The *Sun-fish*—the name springs from its habit of basking on the surface—is indeed, a singularly sluggish creature and probably does little more than drift, more or less passively, with the ocean currents. But there must be a purpose in the space, and it has been suggested that it is designed for a deep diving. It is not known why the rounded shape should be an aid to deep diving, but it is a fact that the Sun-fish, normally a surface-swimming species, has been taken with the remains of deep-sea fishes, living at depths of 100 fathoms (200 mts), in its stomach.

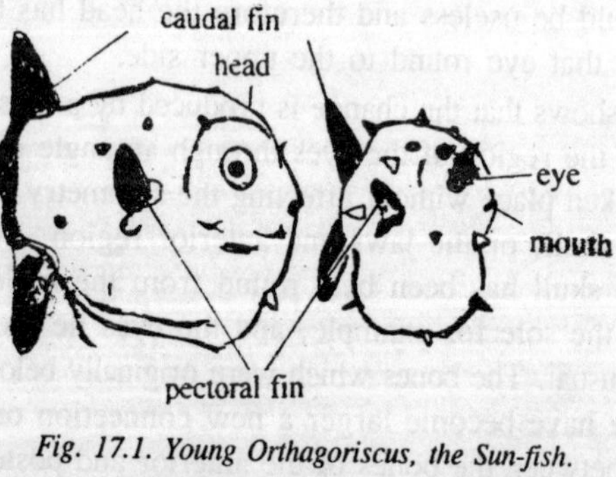

Fig. 17.1. Young Orthagoriscus, the Sun-fish.

Other fishes with shortened rounded bodies are the *Puffer-fish* and the *Porcupine-fish*. These, too, are slow swimmers and would be easy prey for swift predators because of no bodily armour of spines. The Sun-fish, lacking spines, has an armour of tough gristle some 7.5 cm thick. The allied *Trunk-fish* also has a shortened, but rectangular rather than rounded, body and is another very slow swimmer which relies on armour; for protection in this case, a rigid bony case or box. But the beautiful *Angle-fish*, so popular in aquaria, has no armour at all on its very compressed body and spends most of its time suspended almost motionless in mid-water. It evidently relies for protection upon its close resemblance to the water plants among which it lives. At the opposite extreme to the shortened rounded fishes are those with long narrow tubular bodies. The best known of these is, undoubtedly, the common Eel.

The Ribbon fishes are also basically tubular fishes, but their bodies have been so laterally compressed that the original tube has been superceded by the ribbon from which they derive their name. This body shape has been evolved to suit particular living conditions and has resulted in a loss of speed, but nothing like the same extent as in the shortened rounded fishes. The typical cigar-shaped fishes occupy a position midway between these two extremes. But between them and the extremes at either end, there is every conceivable body shape.

Perhaps the most extraordinary is that of the Sea Horse, another very popular aquarium exhibit. In the Sea Horse, the head is bent at right angles to the body, giving the impression (an erroneous one) that there is a definite neck. The body is curved and bears a number of membranous projections, giving it a broken outline. The tail is prehensile—a feature unique among fishes—and is used by the fish to attach itself to plants. The Sea-Horse swims in an upright position, propelling itself by means of its fins, and is, therefore, a very slow swimmer. It possesses nothing that can be described as armour and is completely defenceless. But the upright position and the broken outline do help to render it inconspicuous among the upright swaying water plants amongs which it lives and to which it attaches itself by its tail.

Adaptations in Locomotion

The use of the tail as chief organ of locomotion is not suitable for a fish gliding along the ground, and we find both in skates and flat-fishes that the fish moves by undulations of lateral fins at the margins of the flattened body. The tail or caudal fin in both cases, has become reduced, and the lateral fins much increased in extent. In the skates it is the pectoral fins which are lateral and are thus increased, in the flat-fish lying on its side the marginal fins are the dorsal and ventral, which have become increased so much that they extend from the base of the tail nearly to the anterior extremity. The dorsal fin in the sole extends to the anterior extremity of the head, in the turbot nearly as far, and even in the plaice and its allies as far as the middle of the dorsal eye. The ventral fin cannot extend so far forward, as it cannot pass in front of the anus, but the latter opening itself is shifted forward as far as possible so as to be near the gill-openings, and the pelvic fins occupy more or less of the small extent of margin left between the anus and the gill-apparatus. The dorsal and ventral regions of the flat-fish are thus much more similar than in other fishes, and in fact a new though imperfect symmetry is produced about the middle line of the upper side, which has to some extent replaced the original symmetry between the two sides; the bilateral symmetry characteristic of a vertical fish has given place to a dorso-ventral symmetry in the horizontal fish. The most remarkable adaptation for locomotion among fishes it that of the pectoral, and to some extent the pelvic fins also, in the flying-fishes.

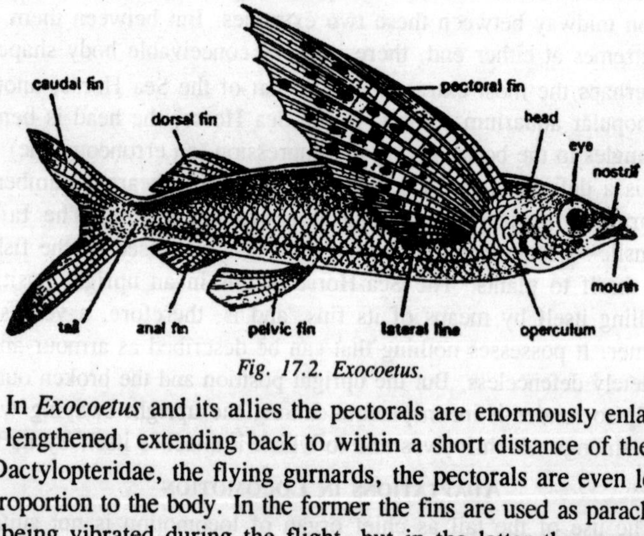

Fig. 17.2. Exocoetus.

In *Exocoetus* and its allies the pectorals are enormously enlarged, and lengthened, extending back to within a short distance of the tail. In Dactylopteridae, the flying gurnards, the pectorals are even longer in proportion to the body. In the former the fins are used as parachutes, not being vibrated during the flight, but in the latter they are stated by observers to be moved rapidly up and down like the wings of a butterfly. Recently a fresh-water flying-fish belonging to a tocally distinct family has been discovered in the Congo region, namely *Pantodon buchholtzii*, belonging to the soft-finned sub-order Malacopteygii. The pectorals in this case are not so much elongated, but they are united to the body by membrane, and the pelvic fin-rays are separate and elongated as filaments; this little fish is only three and a half inches long, and was taken in its flight over the water by a butterfly net.

Among the Gobies, the mud-skippers of tropical shores are adapted for hopping about the surface of the mud, using their pectorals, of which the bases are very muscular, as short legs. The head is raised on the pectorals in an attitude resembling that of the frog, and the eyes project upwards from the head, and are capable of free movement, whence the names *Periophthalmus* and *Boleophthalmus* given to the genera. They not only hop actively on the mud but climb on to mangrove roots, and move so quickly that it is difficult to catch them. They are unable to live under water more than a limited time, and in their habitual mode of life and tail-fin, which is richly furnished with blood-vessels, serves as an accessory organ of respriation. *Periophthalmus australis* is common on the shores of Queensland and grows to a length of twelve inches.

In the various families of anglers, forming the sub-order Pediculati, the pectorals are modified for use as limbs rather than as fins, none of these fishes swimming freely in the water. The bones at the base of the fin are reduced to two or three in number and greatly lengthened, so as to resemble the bones of the forelimb radius and ulna) of a terrestrial vertebrate, and they form a kind of wrist joint with the broad part of the fin. In the angler and other forms which live on the bottom, they are used for shuffling along the round. The pelvic fins are similarly modified, and indeed these are really more to be compared to fore-limbs, for they are situated in front of the pectorals and nearer to the middle of the ventral surface. The resemblance in attitude and locomotion to a frog or toad is greatest in the Malthidae, e.g., *Malthe vespertilio*, in the West Indies.

Adoptations due to Light

The close relation between the distribution of pigment on the stain of a fish and the incidence of light is remarkably shown by the peculiar pigmentation and the peculiar habits of the various species. The colour on the upper side is doubtless some advantages to the fish in masking it from enemies or from prey, and the benefit is much increased by the faculty which the fish possesses of changing the intensity of the colour to bring it into harmony with that of the ground on which it rests. The absence of pigment on the lower side, on the other hand, has never been proved to be of any benefit of the fish. The suggestion that this colour harmonizes with the light coming from above when the animal is seen from a lower level, and is therefore useful in concealing the fish, is open to two objections. Firstly, the fish is not attacked from below, and always seeks protection by covering itself with sand, and secondly, when the fish is observed from below it appears as an opaque object shutting off the light, and is therefore conspicuous. There is very strong evidence that the reason why the lower side has lost its pigment is that it is shaded from the light. We do not know exactly how the action of light produces pigment, but in fishes at last we have good reason to believe that this action is necessary for the production of pigment in the skin.

Fishes, like the cave-fishes *Amblyopsis* and *Lucifuga*, which live in the dark, are colourless, and fishes generally are colourless on the ventral side. The present writer, moreover, has shown that if flounders are kept alive in an apparatus which exposes their lower sides to light, pigment is gradually developed on the skin of the lower sides. The experiments were carried out at the Laboratory of the Marine Biological Association at Plymouth during the years 1890-1893. The apparatus consisted of a shallow tank the bottom of which was made of plate glass, on to which light was directed upwards by means of an inclined mirror. Young flounders soon after their metamorphosis were placed in this tank, are after a period varying considerably in different specimens, pigment precisely, similar to that of the upper side began to show itself on the lower side.

In some specimens kept under these conditions from one to three years the lower sides became almost as completely pigmented as the upper. In this case then the utility principle, in other words the idea of the theory of natural selection, is not applicable, and the question arises whether we are to call such a feature an adaptation or not. If we define an adaptation as a peculiarity of structure which is advantageous to the animal the word is not applicable to the absence of colour from the lower side of flat-fishes. But as the latter feature is due to the conditions of life, the others may have been produced in a similar way. The distortion of the eyes may have been caused in evolution by the constant efforts of the fish to direct the lower eye towards the light, that is to the upper side, together with the greater use of the muscles and bones on the lower side in connection with the jaws, but the latter condition does not seem to be necessary, for in many species, e.g., turbot and halibut, the jaws appear to be equally developed on the two sides.

The extension of the dorsal and ventral fins may have been due to the greater use made of them. If this were true these two features would be due to the conditions of life acting not directly, as light acts on the skin, but indirectly by stimulating the functional exercise of the parts affected. It seems certain that all three peculiarities are in some way due to the conditions of life, and if we confine the term adaptation to these structural adjustments which have a definite function, a purpose, as it were, we require another term to indicate changes due to the conditions of life whether advantageous, indifferent, or even harmful. The best term for such changes is modifications, and we may define modifications as changes of structure produced in evolution from a more primitive condition, by changes in the conditions of life. Many species of *Synodontis* which live in the Nile and other African rivers, but especially *S. Membranaceus* and *S. Batensoda*, are in the habit of swimming at the surface of the water in an inverted position, with the belly upwards. These same species have the pigmentation also reversed, the ventral surface is dark brown the dorsal a light silvery grey. The peculiar habit attracted the notice of the ancient Egyptians who have represented the fish frequently in this unusual position.

By the expression "distribution of pigment" used above is meant the presence of pigment on one part and its absence on another; the cause of the particular colours and markings such as spots and stripes in various patterns which are so conspicuous in some fishes in another question which is much more difficult to answer. The reason for the inverted position assumed by *Synodontis* seems to be unknown, but it may at least be said that the diminution of pigment on the dorsal side, which becomes the lower in the inverted position, is not known to be of any benefit to the fish.

Adoptations in Respiratory Organs

The simplest and probably the most primitive condition of the branchial organs, at least in the true Pisces, is that which is presented by the Elasmobranchs. Here, there is a series of wide tranverse gill-slits at the sides of the pharynx, opening on the surface of the body and separated from one another by complete septa. The internal border of the septum is supported by a cartilaginous bar or "arch" from which radial rods of cartilage extend towards the outer edge of the septum. The actual respiratory organs are folds of the mucous membrane forming narrow leaf-like "lamellae" on the anterior and posterior surfaces of the septa; these lamellae are situated almost parallel to one another, radiating from the internal border of the septum towards the outer border, and projecting into short processes at their outer ends. The reduction of the septum is associated with the development of the gill-cover, which consists of a fold of skin growing backwards from the hyoid arch, i.e., the anterior boundary of the first cleft and supported by flat scale bones. This operculum occurs in the Holocephali or Chimaeroids, in which the gill-septum is so far reduced that it does not extend beyond the branchial lamellae; but the latter are still attached along nearly their whole length.

In the sturgeon, the septum extends little more than half the length of the folds, which project freely for the rest of their length, while in Teleostei as has been said, the reduction of the septum reaches its extreme and it forms merely an axis with a double series of gill-filaments extending outwards from it. In the lung-fishes (*Ceratodus*) the interbranchial septum is well developed, and the gill-lamellae are attached to it along nearly their whole length. In several other cases the adaptation for breathing air consists in a special modification of the gill-chamber enclosed by the operculum, as increase of surface being produced by complicated foldings of the walls, usually supported by bony plates. The organ so produced is known as a labyrinthi form organ. These organs occur in tropical fishes.

Under tropical conditions both temperature and decomposition are unfavourable to the supply of dissolved oxygen in water, and these fishes have the power and the habit of living for considerable period of time out of water and carrying on aerial respiration. One of the best known of such cases is that of the climbing perch, *Anabas scandens*. In *Anabas* and other allied fishes the organ consists of three or more concentrically arranged bony plates with frilled margins, attached to a bony base which in turn is attached to the upper end of the fourth gill-arch, and enclosed in a dorsal enlargement of the gill-cavity. The vascular membrane which cover the organ is supplied with venous blood by a branch of the fourth afferent branchial artery, and the efferent vessel from the organ joins the dosal aorta.

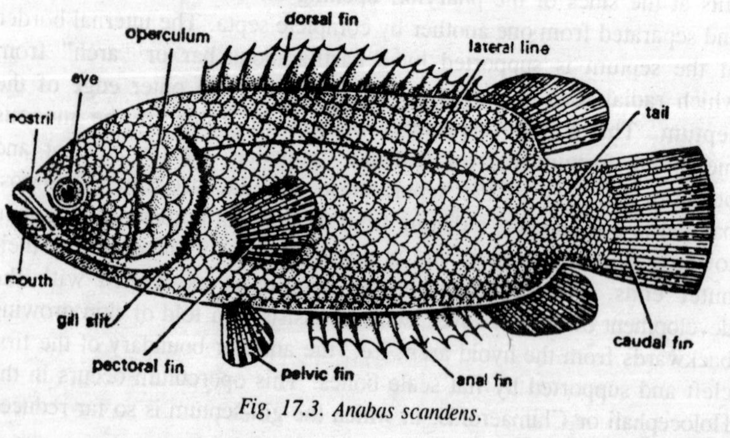

Fig. 17.3. Anabas scandens.

The possession of this organ enables the fish to breathe air for a long time, and although its climbing powers may have been exaggerated, there is no doubt that it has been observed frequently among vegetation on land. The Ophiocephalidae or serpent-heads are related to the Anabantidae, and have a similar distribution, but more restricted in Africa, where they occur only in the west central region. They also possess an air-breathing organ, but of a much simpler structure. It is in fact merely a cavity on each side in front of the gills and communicating with the gill-cavity: the membrane lining the cavity is full of blood-vessels and is thickened and folded, but there are no bony plates connected with the gill-arches.

There are only two genera, *Ophiocephalus* and *Chauna*, the latter without pelvic fins. They are found both in rivers, "tanks", and swamps, and are able to exist for considerable periods out of water and to travel over land: they also remain torpid or "aestivate" in the mud of dried-up ponds in the hot season. They are said by the natives to descen with down-pours of ratio, which is not impossible if they are taken up with the dried mud by violent whirlwinds. When living in muddy water they rise to the surface at intervals to obtain air. In the cat-fishes, *Clarias* and *Heterobranchus*, the air-breathing organ has the form of much-branched highly vascular out-growths of the dorsal ends of one or two branchial arches, enclosed within a posterior and dorsal chamber opening out of the gill-cavity.

Clarias and *Heterobrachus* are again fishes of the Asiatic and African region. *Clarias* has an elongated eel-like shape, and in Senegambia "aestivates" in burrows from which it emerges at night to crawl about for food. *Saccobranchus* is another cat-fish of India and eastern Asia, which has a special air-breathing organ. In this case the organ is a long sac extending along the side of the body from the first gill-cleft to the tail. The sac is situated in a deep internal position close to the back-bone. The right sac is supplied by a branch of the first afferent branchial artery, the left by a branch of the fourth, and

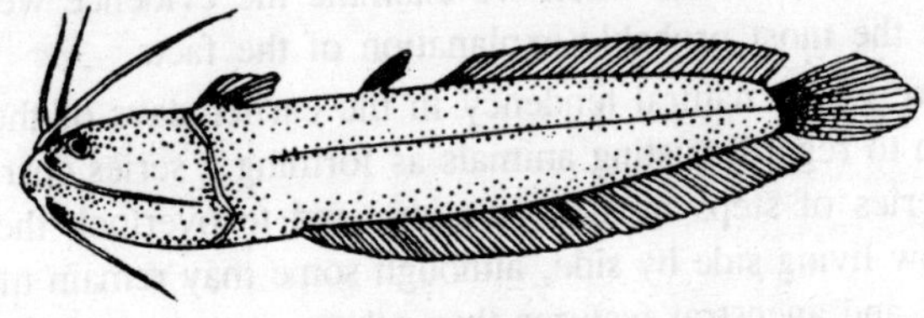

Fig. 17.4. Clarias.

the wall of the sac is highly vascular. These sacs therefore resemble lungs and have a similar function, but open into the gill-cavity not into the gullet. The fish can live for days out of water.

Amphipnous is a fish allied to the eels and of similar elongated shape, without either pectoral or pelvic fins, and even the vertical fins are rudimentary. The air-breathing organs consist of a pair of bladder-like sacs opening from the branchial cavity above the first gill-cleft. The true gills are in this fish much reduced, the first arch having no gill-flaments, all its blood going to the air-sacs. There are only three branchial arches altogether, and the second alone bears gill-filaments. The fish, known familiarly as the cuchia, is, like most of these already mentioned, an inhabitant of the fresh and brackish waters of India and Burmah. It grows to two feet in length. It rises to the surface of the water to inspire air, and spends much of its time out of water in the grass on the banks of ponds, like a snake. The functions of the air-bladder are considered in another place: here we have to discuss certain peculiar adaptations which enable fishes to breath air.

In some cases air is swallowed into the intestine, and the oxygen is absorbed by the blood-vessels of the mucous membrane. It was formerly a very general idea among zoologists that the lungs of higher vertebrates were evolved from the air-blader of fishes. There is very good reason for regarding the two structures from the point of view of comparative anatomy as homologous, which means that they correspond to each other when we compare the general plan of structure of a fish with that of an animal breathing air by means of lungs. The explanation given by the doctrine of evolution of homology in such a case is that one of the two structures compared has been evolved from the other, or that both have been evolved from the same structure in some common ancestral form which lived at a former period. The progress of knowledge, however, has led to the conclusion that the air-bladder was evolved from lungs rather than *vice versa*. This may at first seem a paradox, but it does not mean that fishes are descended from terrestrial animals, and when we examine the evidence we shall see that it is the most probable explanation of the facts.

There was a natural tendency in the earlier days of the study of evolution to regard existing animals as forming a series corresponding to the series of steps in the evolution, and to overlook the fact that forms now living side by side, although some may remain more of the primitive and ancestral features than others, cannot actually be in the relation of ancestors and descendants. Thus although terrestrial vertebrates are doubtless descended from ancestral fishes, we cannot conclude that a mammal such as the dog is descended from a fish like the gold-fish, or that the lungs of the dog are a modification of the air-bladder of the gold-fish, which is not a respiratory organs, but an organ for decreasing the specific gravity of the fish. Lungs are paired structures opening out of the ventral side of the oesophagus: the air bladder in the majority of fishes is a single structure opening, when it is not closed completely, from the dorsal side of the oesophagus. But we must endeavour to trace the origin and history of the air-bladder in fishes themselves by the kinds of evidence usually employed by the evolutionist, by palaeontology, or the evidence of fossils, by comparative anatony, and by embryology or the evidence of development.

It is known that the fishes with closed air-bladders, classed together by *Dr. Gunther* as Physoclisti, and comprising in Bouleger's classification the Spinyfinned Fishes (Acanthopterygii) and a few other Sub-Orders, are the latest fishes to appear in the palaeontological record, and in all these the air-bladder arises in development as an outgrowth of the gullet, the communication with the latter being afterwards lost. The Soft-finned Fishes (Malacopterygii) and a few other Sub-Orders, forming the old group Physostomi which retain, e.g., the salmon and carp, the communication of the air-bladder with the gullet throughout life, are therefore more primitive, and their fossil representatives are found in older geological strata.

Lepidosteus and *Amia* are confined to fresh water, and in both of them the air bladder is single and opens into the gullet dorsally: at the communication there is a special chamber called the vestibule which resembles a larynx, the organ containing the opening to the lungs in higher Vertebrates, and communicates with the gullet by a narrow slit, the glottis. Moreover, in these two fishes but especially

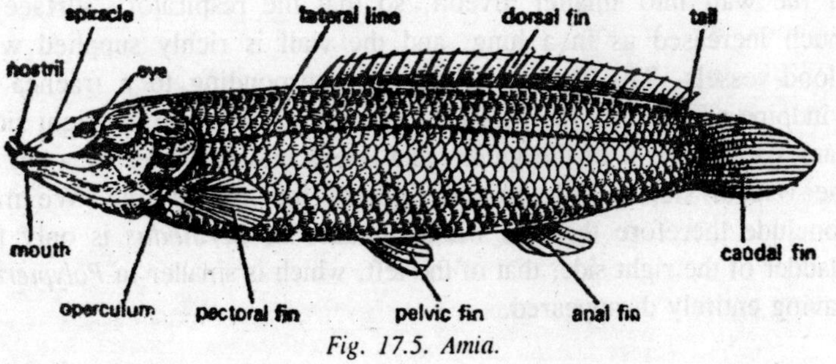

Fig. 17.5. Amia.

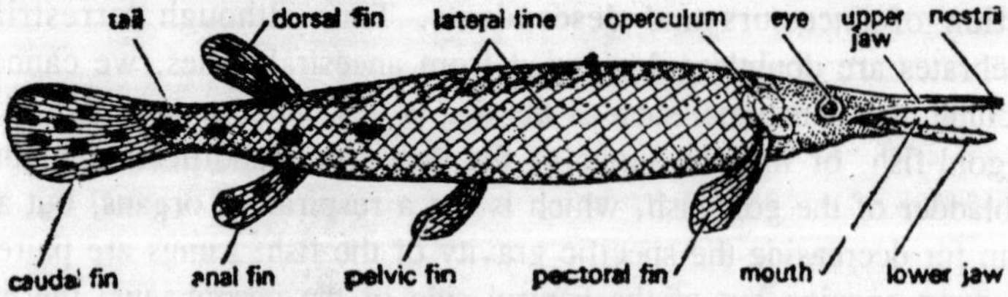

Fig. 17.6. Lepidosteus.

in *Legidosteus*, the inner surface of the bladder is much increased by being produced into lateral recesses called alveoli, and each recess is again subdivided or sacculated much in the same way as it a true long. In both of these fishes it has been shown by the American naturalist Wilder that the bladder is actually used for respiratory purposes.

In the Chondrostei or Strugeons, the respiratory character of the bladder is not so apparent, the lining membrane is smooth, and the bladder opens without a vestibule into the dorsal wall of the pharynx. The fossil forms related to them are very old, Palaeoniscidae being represented in the Devonian, but most abundant in the Carboniferous and Permian strata. Of the living forms, the spoon-bill sturgeons are confined to large rivers, but the common sturgeons, though they spawn in fresh water, descend to the sea between the breeding seasons. In the living lung-fishes the respiratory function of the air-bladder is of course quite obvious, and in the respect they are intermediate between fishes and Amphibia.

In *Ceratodus*, the Australian lung-fish only one sac is present which lies above the intestines and extends from one end to the other of the body-cavity. The sac is large and wide, and along the median line dorsally and ventrally are two fibrous bands projecting into the cavity. Transverse bands passing between these on either side form the borders of two series of lateral chambers, which are again divided by projections of the wall into smaller alveoli, so that the respiratory surface is much increased as in a lung, and the wall is richly supplied with blood-vessels. The pneumatic duct, corresponding to a trachea or windpipe arises from the anterior end of the bladder on the right side, passes to the right of the gullet, and opens into the latter by a slit on the *ventral* side a little to the right of the middle line. We may conclude therefore that the whole bladder in *Ceratodus* is only the bladder of the right side, that of the left, which is smaller in *Polypterus*, having entirely disappeared.

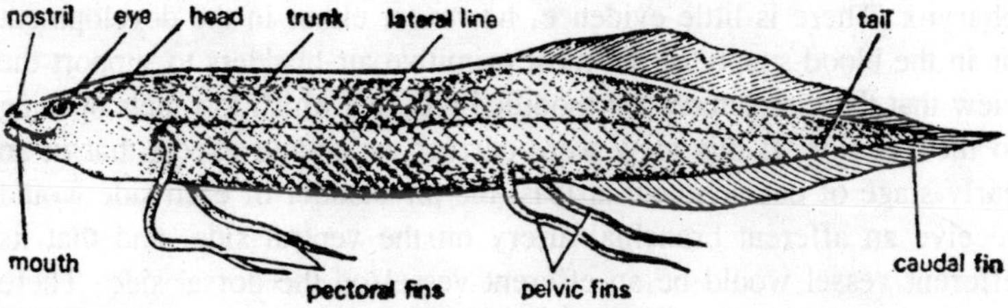

Fig. 17.7. Protopterus.

In the African lung fish *Protopterus*, and the South American *Lepidosiren*, the air-bladder although single anteriorly is completely divided posteriorly for the greater part of its length but even in these the gullet does not pass between the right and left sacs, but the air-duct passes to the right of the oesophagus and opens into the ventral side of the latter. The structure, however, is a symmetrical that it is scarcely possible to regard it as formed by the division of a single sac like that of *Ceratodus* belonging to the right side: it appears rather as a median divided outgrowth of the gullet which has passed to the side of the oesophagus and become dorsal in position. Although, however, it is probable that the air-bladder was primitively respiratory, it is by no means easy to explain its mode of origin. This question is intimately connected with the relations of the blood-vessels, arteries and veins, of the air-bladder to the heart, and these will therefore be described in discussing the question.

The mode of origin which naturally suggests itself first is that the paired air-bladders were modifications of a pair of posterior gill-sacs of the kind seen in Elasmobranchs. This view was proposed in connection with the evolution of Amphibia as long ago as 1875; it assumes of course that the gill-sacs are lateral outgrowths of the pharynx, which open by a superficial opening to the exterior. *Goette*, however, who proposed this theory, maintains that such pharyngeal gill-sacs, lined by hypoblast, i.e., by a continuation of the cells lining the pharynx, occur only in Cyclostomes, e.g., the lamprey, and in Amphibia, while the gill-sacs of Elasmorbranchs are lined by a continuation of the layer of cells covering the external skin.

In Teleostomes the gill-septa being reduced to narrow bars, there are no pharyngeal gill-sacs. *Goette*, therefore, supposes that the paired air-bladder arose in primitive fishes and that the Elasmobranchs were derived from these before this structure had been evolved. This is a view which other zoologists do not accept, for there seems no reason to doubt that the gill-sacs of Elasmobranchs are outgrowth of the pharynx. There is little evidence, however, either in the development or in the blood-supply of the most primitive air-bladders to support the view that these structures are modified gill-sacs which lost their opening to the exterior. If this view were correct we should expect that in an early stage of development at least the air-bladder of each side would receive an afferent branchial artery on the ventral side, and that its efferent vessel would be an efferent vessel on the dorsal side. There is no such arrangement known.

In the Dipnoi, in *Polypterus*, and in *Amia* the air-bladder receives on each side an artery which is a branch from the fourth efferent branchial artery (epibranchial) as it does in the development of the lungs in terrestrial vertebrata. In Dipnoi, the veins from the bladder open into the left auricle of the heart which is almost completely separated from the right, as in Amphibia: the connection is, therefore, behind the ventricle, not, as in gill-sacs, in front of it. In *Polypterus* the veins from the air-bladder join the hepatic veins, that is between the liver and the heart; in all other bony fishes they open for the most part into the hepatic portal vein, behind the liver, while some of them join the posterior cardinal veins; the arteries are derived from the dorsal aorta. Among Teleostei with open air-bladders there are some cases in which the respiratory function of the organ is retained, and in fact the majority of the fresh-water forms at least seem to have the power of respiring air to some extent, as we see in the common goldfish.

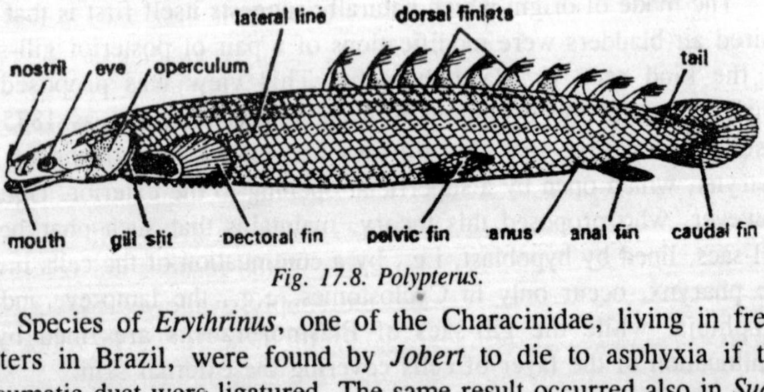

Fig. 17.8. Polypterus.

Species of *Erythrinus*, one of the Characinidae, living in fresh waters in Brazil, were found by *Jobert* to die to asphyxia if the pneumatic duct were ligatured. The same result occurred also in *Sudis gigas*, one of the Scopelidae. The hydrostatic function of the air-bladder, that is to say its effect on the buoyancy of the fish, especially in the sea where there is a great range of depth, invovles some interesting physical principles which are not generally appreciated. Every one knows that when fishes are caught and brought to the surface even from depths of only a few fathoms, the air in the bladder expands to such a degree that the stomach of the fish is often pushed out through the mouth. When the pressure on a gas is diminished by a half, the gas expands till its volume is doubled: at a depth of only five fathoms in the sea, the pressure on a fish is double the atmospheric pressure on the surface, and increases by one atmosphere for every five fathoms of depth. It is easy, therefore, to understand the great force of expansion exerted by the gas enclosed in the air-bladder when a fish is suddenly brought to the surface.

The effect of the air-bladder at any depth when the fish is alive and is normal condition is to render the weight of the fish equal to that of the water which it displaces: in these circumstances the fish has no tendency either to rise or sink, and can therefore keep at the same level in the water without any muscular exertion. A shark or dog-fish, on the other hand, having no bladder is heavier than the water and always, has a tendency to sink; it can only keep up by the action of its tail and fins. If, however, the fish with an air-bladder swims to a higher level, the gas in the bladder expands, the fish as a whole becomes larger and therefore lighter than the water it displaces, and has a tendency to rise more and more rapidly till it reaches the surface.

Conversely, if a fish with an air-bladder swims to a slightly greater depth the bladder is compressed, the fish as a whole becomes smaller and heavier than the water displaced, and tends to sink with increasing velocity to the bottom. The gas contained in the air-bladder has been analyzed, and found to consist of oxygen and nitrogen with a trace of carbon dioxide. These are the same gases as exist in the atmosphere, but the proportion of oxygen in the air-bladder is much greater than in atmospheric air, and is greater in marine fishes than in fresh-water forms, and greatest of all in certain deep-sea fishes. It has been maintained by *Dr. Thilo* that the gases in the bladder are originally derived from the atmosphere, and are simply swallowed by the fish, but this could not apply to the adults of those forms in which the bladder is closed, although it would be possible in their young stages, before the closing takes place.

There are, however, certain structures known as red bodies and red glands in the wall of the air-bladder which contain a net-work of blood-vessels, and thee is good evidence that these secrete gases and exude them into the bladder, while another structure known as the oval is believed to absorb some of the gas when necessary. The red bodies are modifications of the primitive general vascular supply in the walls of the bladder, the blood vessels having become concentrated in limited areas, instead of being generally distributed. The red bodies are situated on the ventral surface of the bladder, and receive their blood supply from a branch of the coeliac artery, the corresponding vein joining the portal veins so that the returning blood passes, with the blood from the digestive organs, to the liver. In fishes with closed bladders, for example the maigre, *Scioena aquila*, the red body consists of a dense collection of capillaries, above which is a thick mass of epithelial glandular cells: the vascular layer belongs to the inner of the three layers which form the wall of the bladder, the origin of the glandular mass is not known. Above the organ extends the layer of flat cells which everywhere lines the bladder.

Adaptations for Protection and Defence in Fishes

Armour is, by definition, defensive. The tough leathery skin of the Sun-fish is evidently a most effective armour. The spines of the Puffer-fish and the Porcupine-fish, particularly those of the latter, are, at least when stiffly erect, sufficiently formidable to deter all but the most resolute of predators. But to be fully effective, spines must not only be stiffly erect, they must be capable of holding that position. The Pike, for example, almost always swallows its prey head first: which means that if the spines of its prey are not locked in the erect position they are pressed backwards, and so rendered harmless. Spines which are self-locking in the erect position are rare in fishes. Our own Sticklebacks is one of the few to possess them. Puffer-fishes and Porcupine-fishes do not: and so they, and particularly the Puffer-fish (hence the name), attempt to make assurance doubly sure by inflating their bodies. They do this by swallowing water, and are capable of inflating their bodies to twice or even three times the normal size, thereby offering a very formidable array of erect spines to any would-be predator.

The Snapper (the name of appropriate) is an especially successful predator of Puffer-fishes, taking large numbers before they have had time to inflate. A number of fishes have overcome this inherent weakness in the spine as a means of defence by having it attached to a poison gland. Most of these venomous fishes are sluggish, bottom-dwelling, creatures. A number—among them, the Greater Weever, the Lesser Weever, the Father-lasher, and the Sting-ray—occur in British waters. The Greater Weever and the Father-lashes are more or less confined to our western shores. The Sting-Ray is an autumn visitor to southern coasts. The Lesser Weever (commonly known as the Sting-fish) is common enough on sandy shores everywhere. The tail of the Indian Sting-Ray is armed with a long serrated spine. The point of this spine is very sharp and the wound is widened by the toothed edge of the spine. The poison is powerful enough to kill a man by direct action on the nervous system. However, one is very unlikely to come across a Sting-Ray.

The Lesser Weever is altogether different from others. This fish has a habit of lying just, burried in the sand, only the upper part of the head being visible and then only to very sharp eyes. The poison is contained in the spines of the first dorsal fin, which is situated just behind the head. The danger, of course, lies in treading on one with bare feet. This is only a little fish, 10-12 cm long, but the wound it inflicts can be quite excruciating. Such wounds are best treated by rubbing in crystals of potassium permanganate. Some of the deep sea fishes—for example, the Eagle-Rays and the Rabbit-fishes—also possess poisonous defence mechanism.

In general, poisonous derives in fishes are always defensive and are not used in the capture of prey. A possible exception, since in this case, the poison is injected with the bite, may be the Moray of the West Indies. Another method of defence developed by some fishes is the use of electricity, produced by special organs which represent, with one notable exception, a modified muscular tissue. All muscle fibres, of course, have the capacity of current production as part of their mode of working, but the currents and voltage produced are normally very small. And this is true of the majority of fishes producing electricity. In many of the Skates, for example, the electric organs are formed from parts of the muscles in the whip-like tails, are very weak, producing low voltage at low current. These Skates seem to use their electricity very rarely and it is impossible to believe that it could, in any circumstances, be of any real value as a means of defence. But it must, presumably, have some value for some purpose.

The Thornback Ray, a common fish around British coasts, is capable of producing a discharge of 4 voltes, though it will do so only after considerable provocation. So low a voltage can only be of value in defence. But many fishes are capable of producing discharge of high voltage, and in these cases the electricity is used not only in defence but also, and more usually, for offence. The best known example, because a great many experiments have been conducted on it, is the Electric Eel. The Electric Eel, which is not a true Eel (though it looks very much like one), is an inhabitant of South American rivers. It grows to a length of 240 cm. or more and attains weights in excess of 50 kg. There is a huge electric organ, composed of columns of cells arranged longitudinally, on each side of the body; the front of the organ being positive and the rear negative. The more electric organs, each arranged in the same manner, are situated at the base of the anal fin. Both the lateral and the anal organs are connected to the spinal column by a vast array of nerves. This apparatus is capable of producing a shock of more than 500 volts. The discharge is not involuntary, but is controlled by the fish. Electric Eels use their batteries to stun other fishes, which can then be eaten at leisure. How the Electric Eel finds its prey, and at what range it can stun, is not known. The Electric Ray, which is by no means rare in British waters, also stuns its prey, producing a shock of 220 voltes.

The Electric Ray is a sluggish, bottom-dwelling fish, but specimens have been taken containing Salmon. The Salmon is a swift swimmer.

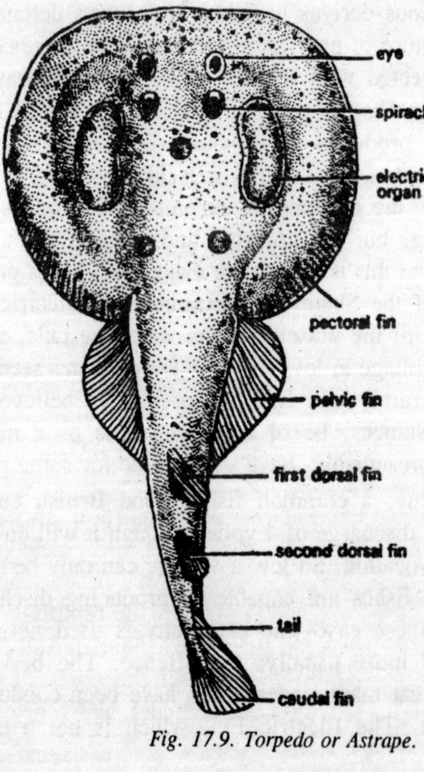

Fig. 17.9. Torpedo or Astrape.

Since the Electric Ray also controls its discharge, this suggests that its aim must be very good. Again, how it knows where and when to aim, and at what range it can stun, is not known? This applies equally to the Electric Cat-fish of tropical African rivers. In this fish the electric organs are not formed from parts of the muscles, but from a mass of gelatinous material lying between the skin and the muscles and enveloping virtually the whole of the body. And there is another important difference: in the Electric Cat-fish it is the rear end of the organ that is positive, the front negative. Again, the discharge which is very powerful, 350 volts, is controlled by the fish. The flesh of a number of fishes is poisonous to human beings. That of the Muki-Muki of Hawaii, for example, has often been used by those wishing to common suicide. Presumably, the flesh of some fishes will be poisonous to other fishes. But if this is so, it can hardly be regarded as a useful means of defence, since the owner of the poisonous flesh must first die for it to be effective. Nevertheless, it may be useful in an indirect way: predators learning by experience, or perhaps knowing by inherited 'race memory' that such fishes are best avoided.

Adaptations in the Sensory Organs in Fishes

According to the American zoologist *G.H. Parker*, there is no conclusive evidence that the sense of hearing exists at all in the majority of invertebrate animals, such as coral polyps, jelly-fishes, worms, star-fishes, crabs, oysters, snails, etc. These animals possess organs which resemble auditory organs, and were formerly so described, but recent experimental researches have shown that their function is the sense of equilibrium, a sense which is associated with the auditory organ in the vertebrates also. The only animals in which the sense of hearing is known with certainty to be present are the higher arthropods, especially the insects, and the vertebrates.

The fact that the sense is certainly most developed in the animals which live in air suggests the question whether the sense really exists in aquatic animals, and in the case of vertebrates this question must be tested by the investigation of the sense in fishes. *Bateson*, from his observations at Plymouth, found that the report or shock of the explosions made in blasting operations of rocks in the neighbourhood were followed by sudden movements in conger, flat-fishes, and pouting; while other fishes seemed to take no notice of the reports. When the side of the tank was struck by a heavy stick similar movements were produced. Pollack, however made no response to the vibrations produced by striking a piece of glass with a stone under water provided the objects used were not visible to the fish. Bateson concluded that fishes perceive the *sound* of sudden shocks and concessions when sufficiently severe, but do not hear the sounds of bodies struck in the water but not perceived by them. In several cases it has been stated that gold-fishes or other fresh-water fishes kept in ponds have been in the habit of assembling for food at the sound of a bell, which would be sufficient to prove not only that they possessed a sense of hearing but that they could hear sounds produced in the air and transmitted to the water.

The Viennese physiologist *Kreidl* concluded from his experiments on the gold-fish that it made no response to sounds produced either in the air or in the water but only reacted, as *Bateson* found, to the shock of a blow given to the sides of top of the aquarium. He further removed the auditory nerves and the attached ear-sacs from several specimens and found that they reacted to shocks in the same way as uninjured fish. He concluded therefore that the fish does not perceive vibrations by the ear at all, but only perceives strong vibrations by the skin; the ear, according to this result, would only be concerned with the sense of equilibrium and direction which is one of its functions in ourselves.

Kreidl investigated a special case of the assembling of fishes at the sound of a bell, namely the trout of a particular basin at the Benedictine Monastery of Krems in Austria. He found that the trout assembled equally at the sight of a person when no bell was rung, and if the bell was rung by a person whom they could not see the trout took no notice. If, however, a pebble or a piece of bread were thrown into the water the fish swam to the spot where the water was disturbed. *Kreidl*, therefore, maintained that the fishes were only affected by the sense of sight and the sense-organs of the skin. The American physiologist Lee made similar experiments and obtained results entirely in agreement with those of *Kreidl*. Comparisons were made between normal fishes, others in which the auditory nerves had been divided, and other in which these nerves were intact and the skin had been rendered insensitive by division of the spinal cord. When the string was set in vibration the fishes responded by movements of the pectoral fins and by other movements.

It was found that entire fishes regularly responded to the sound, and also those in which the sensibility of the skin was destroyed but the auditory nerves were intact, while those in which the auditory nerves had been destroyed gave no response. On the other hand, *Parker* was unable to obtain any evidence of the perception of true sounds in the smooth dog-fish, *Mustelus canis*. *Zenneck* in Germany found that three fresh-water species of fish, namely, the roach, the dace, and the bleak, showed distinct evidence of the perception of sound vibrations. At Parker's suggestion *Bigelow* repeated the experiments on the gold-fish, on which species *Kreidl* had obtained only negative results.

Bigelow used as source of sound a tuning-folk vibrating one hundred times per second and used great precautions to eliminate all other shocks or disturbances in the experiments. He found that the fish responded to the sound by characteristic movements, and gave the same response even after the auditory never of one side had been cut, but when both nerves were cut the responses disappeared. There are three different modes in which the air-bladder is connected with the auditory organs. In many marine fishes the bony capsule surrounding the auditory organ on each side has an aperture closed by a membrane, and a tubular outgrowth from the air-bladder comes into contact with this membrane on the outer side, which on its inner side is the liquid surrounding the membranous labyrinth of the ear, the liquid called the perilympyh. This arrangement occurs in certain species of the Gadidae or cod family, and in the Serranidae. Berycidae, Sparidae or sea-bream family, and the Notopteridae.

All these families belong to the spiny-finned fishes or Acanthopterygii, except the last, which is Malacopterygian, or soft-finned. In the second mode the apertutes in the bony capsule of the ear are open, and the ends of the tubular extensions of the air-bladder are in direct contact with outgrowths of the membranous auditory vesicle. This occurs in the herring and pilchard and other species of the same family. In the third method the air-bladder is not directly in contact with the bony capsule or the auditory vesicle but is connected with the ear by a series or chain of small ossicles. These bones, known from their discoverer as the Weberian ossicles, are derived from the first four vertebrae, of which they are separated and modified portions. The ossicles are named claustrum, scaphium, intercalarium and tripus, on each side. The scaphium is inserted into the wall of the auditory capsule, and the tripus into the dorsal wall of the air-bladder as its anterior end. The possession of the latter complicated apparatus is a common character of the Cyprinidae or carp family, the Siluridae, the Gymnotidae, formerly known as the electric eels, and certain other families, which are consequently united by *Mr. Boulenger* in the sub-order Ostariophysi, for, as he points out.

It has long been agreed that the lateral line organs are sense-organs, but it has been always very difficult to discover to what particular stimulations they were sensitive. There is no evidence that they are affected by light, heat or electricity. Experiements have been made to decide whether the organs responded to chemical stimulation, such as oxygen, the salinity of the water, the dissolved substances of food, etc., but the results were negative, *Mr. G.H. Parker* has recently carried out a systematic series of experiments on these organs at the Government Biological Laboratory at Woods Hole, Massachusetts. The fish which he principally used was *Fundulus heteroclitus*, which is known in America as the kilifish or mummichog, but experiments were also made on seven other species, including the smooth dog-fish, *Mustelus canis*, a skate, *Raja erinacea*, and a flat-fish, *Pseudopleuronectes americanus*.

The method of experimenting was to cut the nerves supplying the lateral line and also the fifth and seventh nerves supplying the similar organs on the head, and then to compare the behaviour of fishes so treated with uninjured specimens. Only one kind of stimuli were found to produce definite effects which were wanting in the fishes in which the nerves had been cut as described; these stimuli were vibrations of low frequency produced by pulling the aquarium slightly to one side and then letting it go. The rate of vibration was about six per second. Normal specimens of *Fundulus* when put into the aquarium swim first to the bottom, and after a time when they have become accustomed to their new surroundings they ascend to the surface. Any slight disturbance such as a quick movement of the observer or a slight jar to the aquarium causes them to descend to the bottom. The slow vibration above described has always the same effect.

Specimens in which the nerves to the lateral line organs have been cut, on the other hand, take no notice of such vibrations, although they were sensitive to ripples on the actual surface. Similar reactions to the same kind of vibrations were observed in the other species of fishes tested. On the other hand, it was proved that the lateral line organs were not sensitive to sound vibrations although the ears were. It has often been suggested that the lateral line organs are stimulated by currents in the water. All the fishes above mentioned in the normal condition, were found to turn their heads always towards the direction of the current, but they did this with equal certainty after the nerves to the lateral line organs had been cut, so that apparently the reaction to the stimulus of a current of water depends on the ordinary tactile sensibility of the skin.

The most important question arising from these experiments is: what are the stimuli which affect the organs in the natural life of the fish, and this question is not very satisfactorily answered by *Parker*. He found that when he produced strong ripples on the surface of the water by blowing upon it specimens of *Fundulus* invariably went to the bottom, while those whose nerves were cut only sank below the superficial water. The lateral line organs are originally developed on the surface of the skin, the tubes are formed by the closing in of a groove, and communicate with the surface by a series of pores. The tubes of the head, both the continuous ones of Teleostei and the separate ones of Elasmobranchs, are developed in the same way.

In many fishes groups of sensory cells like the sense-organs of the lateral line occur on the free surface of the skin, for instance in the cod; and in addition to this the skin has doubtless a general tactile sensibility due to nerve endings not connected with special sense-organs. Experiment has shown that the skin of fishes is sensitive not merely to the contact of solid bodies, but also to the motion of the water in surface waves and currents. A normal fish usually places itself with its head towards a current and swims against it; it does this instinctively or automatically, and it can be easily understood that it must habitually do so in order not to be at the mercy of every current it meets. *Parker* found that a fish in which the lateral line nerves had been cut swam against a current in the same way as a normal fish and also swam downwards out of the reach of surface waves; it was evident therefore that the waves and the currents were perceived by the general skin without the aid of the lateral line organs.

One of the most extraordinary adaptations of the eyes of fishes is that of *Anableps tetrophthalmus*, which are divided into two parts, the upper adapted for vision in the air the lower for vision in water. The eyes are large, and projecting across the surface of the transparent front part called the cornea is a horizontal pigmented band, the pupil is divided by a corresponding partition formed by anterior and posterior projections of the iris, and it is said that the curvature of the lower half to the lens is more convex than that of the upper part. This horizontal division of the eyes is exactly in line with the dorsal surface of the fish which is perfectly straight, and is exactly level with the surface of the water as the fish swims. In several species of fish the eyes are absent, or so rudimentary that vision is impossible.

As in other animals this degeneration of the organs of sight occurs in fishes which live in darkness, and the case has attracted much interest among biologists in connection with the question of the evolution of adaptations: by one school of evolutionists the fact are considered to afford support to the belief in the direct action of conditions on the organism, while those who deny the possibility of the inheritance of acquired characters maintain that the loss of the eyes can be explained by variations or mutations which were independent of the external conditions. The most celebrated blind fish is *Amblyopsis spidea* which lives in the stream flowing through the Mammoth Cave of Kentucky. This fish is not only eyeless but colourless, both conditions being either directly or indirectly the result of the absence of light.

Amblyopsis is closely allied to the Cyprinodontidae, and was formerly placed in that family. Several other species are, however, now known which closely resemble *Amblyopsis*, and they are separated from the Cyprinodontidae as a distinct family, Amblyopsidae, differing from the former chiefly in two characters, namely that the mouth is scarcely at all protractile and that the anus is placed far forward close to the gills. The pelvic fins are rudimentary or absent, but this is also true of some Cyprinodonts. With regard to the evolution of the blind cave-fishes, the development of the blindness in the individual must be admitted to have an important bearing on the question of its origin in the species.

The Amblyopsidae, like the Cyprinodontidae, are viviparous, and the development of the eyes in the embryo has been carefully investigated by *Eigenmann*. The young are born at a length of about half an inch, or 12 to 13 mm. The eye develops as an outgrowth from the brain at the same stage of growth as in normal fishes, when the embryo is 1.5 mm long, but grows little after its appearance. The lens develops when the fish is 2.5 mm long, but never loses its embryonic character, it soon degenerates and disappear altogether when the fish is 10 mm in length, that is before birth. The retina and other parts of the eye have entirely degenerated by the time the fish is 13 cm or about 4 inch and its highest stage is reached when the fish is 10 mm long. This history of the eye is, at least in the opinion of the present writer, more in harmony with the view that the blindness is due to the gradual effects of the absence of light, than with the view that it arose from spontaneous variations; certainly the gradual degeneration in the embryo is in no way similar to the sudden variations which are called mutations, which develop directly.

The only other fish, besides the cave-fishes and deep-sea fishes mentioned, which is totally blind, is a species named *Typhlogobius californiensis*; this is a small goby from 2 to 3 inch in length, said to be most nearly allied to *Crystallogobius nillsonii*, a transparent small fish which is common at depths of about thirty fathoms off the British and Irish coasts. *Typhlogobius* is found only in holes in the mud under stones at Point Loma, San Diego, California: the holes are excavated by a burrowing shrimp which lives in them in company with the fish. Both the animals are of a similar pink colour but the shrimp is not blind. It is a curious fact that according to *Professor Eigenmann* the same shrimp is found all over the Bay of San Diego and is accompanied by other species of goby such as *Clevelandia* and *Gillichthys*; but these fishes live outside the holes and only retreat into them when frightened, while the blind species is found only at Point Loma and never leaves the burrows of the shrimp. In this case therefore the relations of *Typhlogobius* and the other gobies of the neighbourhood are similar to those of *Typhlichthys* and *Chologaster* among the freshwater fishes of the Mississippi region, and it seems more probable that *Typhlogobius* is allied to *Gillichthys* or some other species of the same locality than to *Crystallogobius*.

In fact, it is difficult to avoid the conclusion that the blind goby is simply one of the ordinary species of the district which has lost its eyes and pigment in consequence of its subterranean habits. A very extraordianry condition of the eyes occurs in *Stylophthalmus paradoxus*, a deep-sea fish of the Indian Ocean: the organs are at the ends of long slender stalks projecting from the sides of the head. The stalks become shorter in the adult fish.

ADAPTATIONS FOR SELF-LUMINOUS IN FISHES

Many deep-sea fishes are self-luminous, producing their own light. It is not always easy to understand why they should do so. But in the Lantern-fishes, the significance would seem to be sexual; for the light of the female shines downwards, that of the male upwards. In the deep-sea Angler-fishes, very slow swimmers living in perpetual darkness, there can be no sexual significance, for the male, no more than 10 cm long, is attached to the female, 360-480 cm long, throughout life and so completely that it is impossible to say where male begins and female ends. In this case the light is possessed only by the female and is situated at the end of a movable rod which is attached to the front of the head. There can be no doubt that the light is used as a lure for prey.

ADAPTATIONS FOR FLIGHT IN FISHES

It is commonly thought that Flying-fish are the only fishes which fly. As a matter of fact, the Flying fish does not fly: it glides. It does so by holding the pectoral fins (in some cases also the pelvic fins) stiffly out. Before it can take to the air it has to attain a high speed in the water and it is this speed that maintains the glide, which only lasts for a few seconds. Nevertheless, quite long distances may be covered and heights of some 350 cm attained. But there are fishes which actually fly; which use their pectoral fins as 'wings' in true flapping flight after the manner of a bird. The Butterfly-fish of West Africa has evolved this method of escaping its enemies. It cannot fly very far, no more than 3 metres, not get very high, no more than 30 cm. But this is no mean achievement for a fish 5 cm long. We have considered in the above pages some of the most striking facts concerning the life and structure of fishes and may add a few general reflections suggested by these facts.

It will be seen that while the truth of the general principle of evolution, the more proposition that fishes like other animals have descended from remote ancestors through innumerable generations with various changes of structure, and wide divergences from the original type, is supported by the evidence of fossil remains. With regard to adaptations the question is different; here, admitting the advantage, we may doubt whether the variations were spontaneous, that is to say, independent of the conditions of life. If, for example the bladder arose as a spontaneous variation we may ask how it is that it arose only when the fish was compelled in the shark type in the sea. It seems most reasonable at present to conclude that the structure and evolution of fishes are due to the constant interaction of two general causes or processes, spontaneous variation or the occurrence of "mutations," on the one hand and the influence of the conditions of life on the other. The struggle for existence drives fishes, as other animals, to adopt various habits in order to obtain food, to escape enemies, and to produce offspring; the actions involved in these habits by "functional stimulation" produce change of structure and the results of these changes are adaptations, while at the same time the direct action of the conditions may produce results such as blindness in cave-fishes, and absence of pigment on the lower sides of flat fishes.

18

LONG DISPLACEMENT

Among the various interesting peculiarities of fishes, migration is perhaps the most remarkable and fascinating. The word "Migration" has been derived from the Latin word (*Migrare*, meaning to go from one place to another) and is used to represent long journeys undertaken particularly by animals such as fishes and birds. The term "migration" is frequently employed to express three different kinds of movements, In the early stages of life the eggs and fry may be carried by the currents some distance from the place where they were born. This helpless drift is at first involuntary, but later the young fry are quite able to make their way in and out of currents or remain stationary. Then the young fishes and the adults spend most of their time seeking food; they scatter about within their area; but good feeding grounds or incoming currents rich in plankton food bring many together in a concentration or shoal which persists for some time. These may be called voluntary movements on the part of the fish. Not all fishes are migratory.

In fact, a majority of them restrict their movements to small territorial limits and do not go out of these during any part of their lives. Few species, however, travel over long distances for feeding and spawning either in the same type of water to which they belong or in search of water naturally different than their own, e.g., from freshwater to sea water or *vice versa*. Rightly considered, the true migration is something very different from the drift and wanderings. Here the fish is urged to leave its ordinary hunting grounds and even to stop feeding, in order to undertake sometimes long and perilous journeys. The call of the blood, a deeper necessity common to all living things, takes possession of the fish, and the thousands and millions in like case for gather year by year almost at the same place and time—a circumstance of great importance both in the past and in the present. Most of the great fisheries are based on the spawning migrations of fishes, and at the present day we can observe how fishing populations move from place to place and even from one country to another in pursuit of their calling. History tells us also, how the New Englanders would hardly have surmounted the hardships they encountered in the first years of their endeavours to settle in America but for the rich fisheries of the coast; and again, how the Dutch fishing population spread from one country to another in the seventeenth century, and how England and Holland went to war on a small matter of the fishing industry.

In ancient times also the Greek and Phoenician colonists chose their settlements where the fishing was of importance, and the nature of the fishing is sufficiently indicated by the representation of the Tunny on the old medals of Carthage and Cadiz. These under-currents in the life of nations are not apparent in the surface histories, yet *Edward Forbes* was justified in his quaint comment, "the existence of Alexander may have been determined by the migration of a shell-fish." It is a question whether the leading or the feeding of peoples is of the greater importance. *Heape* (1931) has defined migration as "a class of movement which impels migrants to return to the region from which they have migrated."

Migration can be said to typed as follows:

1. *Alimental Migration*. For the need of food and water.
2. *Gametic Migration*. For reproduction and development.
3. *Climatic Migration*. For securing suitable climatic condition.
4. *Osmoregulatory Migration*. For better watery conditions to live as far osmoregulation is concerned.

A fish can make migratory movements by several methods.

By Drifting

The fishes are carried passively by water currents. This may result in 'directional movement' is the overall water movement is in one direction.

Random Locomotory Movements

Random movements in any direction may lead to distribution or aggregation. If fishes move uniformly from one point to all direction, it is called *dispersal*, which results in distribution.

Oriented Swimming Movements

The fishes swim in a particular direction (i) either towards or away from the source of stimulation or (ii) at some angle to an imaginary line between them and point of stimulation. The migration pattern is believed to be related to the currents. The young ones drift with the water current to the nursery grounds. The spawning migration is against with the current. Two terms are used to describe the movements of fishes in relation to water currents. Swimming or migration alongwith the direction of current is called *denatant* and swimming or migrating against the water current, and the migration of the adult fishes towards the spawning area is called *contranatant*.

Types of Migratory Fishes

Anadromous

Those fishes which spend a major part of their lives in the sea but migrate to freshwater during breeding season for spawning. *Salmon* and *Hilsa* have been found to travel several thousand miles in the sea and then several hundred miles inland to reach the spawing grounds.

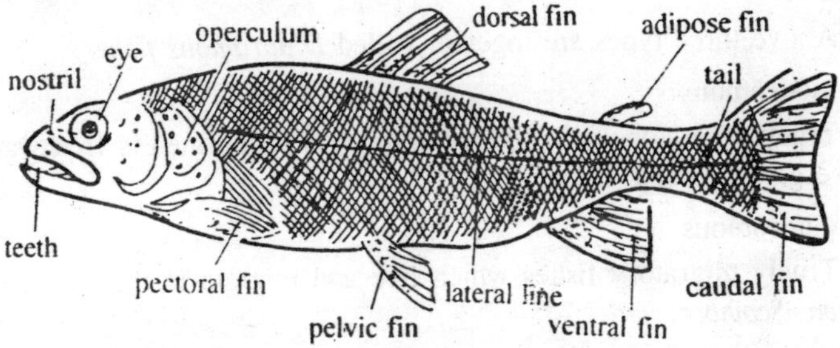

Fig. 18.1. Salmo.

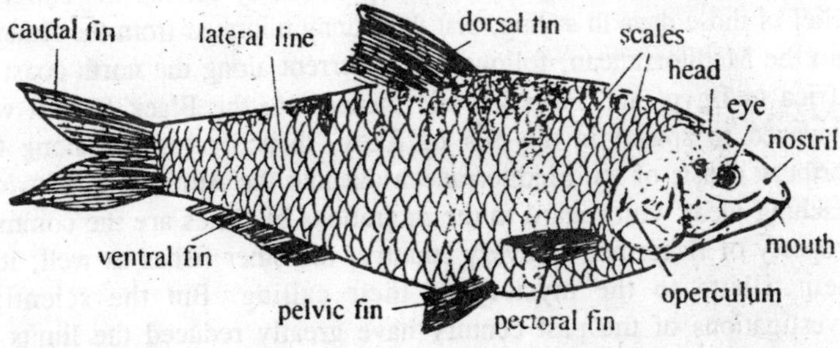

Fig. 18.2. Hilsa.

Catadromous

Those fishes which spend a major part of their lives in freshwater but migrate to the sea for breeding purposes. *Angular* travels several thousand miles starting from the rivers and reaching the spawning grounds in sea.

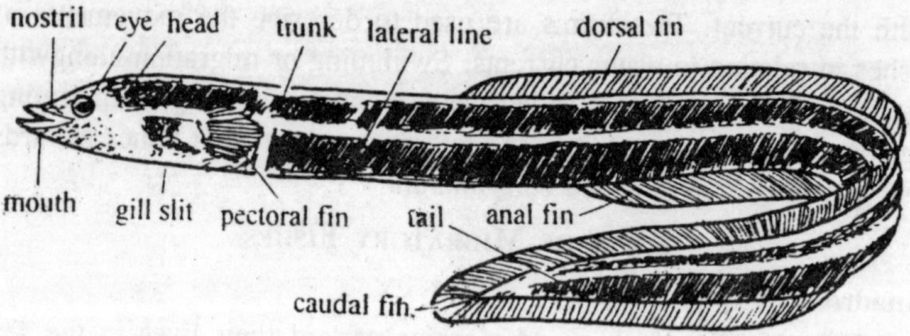

Fig. 18.3. Anguilla.

Amphidromous

Those fishes which migrate from freshwater to the sea, or *vice versa*, not for breeding but regularly at some other stage of the life cycle.

Above three types are together called *Diadromous fishes*.

Potamodromous

Truely migratory fishes whose migration is confined to freshwater only e.g. Carps and Trout.

Oceanodromous

Truely migratory fishes which live and migrate in the sea, e.g., *Clupea*, *Scomber*.

Migration of Different Fishes

Aristotle and the ancient writers presumably echoed the common belief of those days in saying, that the Tunny migrated from the Atlantic into the Mediterranean, following the current along the north coast of Africa to Egypt and Palestine and thence into the Black Sea. It was supposed to spawn in the Sea of Azov, thence returning along the northern coasts of the Mediterranean out into the Atlantic. Such wide-reaching ideas with regard to the migrations of fishes are the common property of fishermen of every country, for other fishes as well; it is their tribute to the mystery of their calling. But the scientific investigations of the past century have greatly reduced the limits of the migrations, except in one remarkable instance to be noted later. The presence of the young in all stages and the appearance of the adults in all parts of the Mediterranean at practically the same time (summer), are among the phenomena which led *Pavesi* to believe that the Tunny lives for most of the year in the deep water off the coasts. As the spawning time approaches, the fish first of all make a vertical migration, from the depths to the surface.

It is said that new arrivals in the surface waters can be distinguished from the earlier by their darker colour. Then they move towards the shallower waters to spawn. It is the latter, the horizontal movement, which has given rise to the idea of immigration from the Atlantic and which accounts for the elaborate preparations made by the fishermen for the reception of the Tunny. These are not spawners; they are the larger fish which, after spawning in the summer on the Spanish or Portuguese coasts, follow the Sardine and Mackerel shoals northwards, and later feed on the autumn Herring of the North Sea.

When the trawl with its load of Herring is hauled up, the Tunny follows and snaps at the protruding heads and tails. Possibly it was just as common in earlier years; there is a record by *Schonevelde* that it was fairly common near Eckernforde in the Western Baltic as far back as the beginning of the seventeenth century, also in connection with the presence of Mackerel. The Porpoise may often have been blamed for its depredations, but until the trawl was used for the capture of Herrings in large quantities, its occurrence was regarded as a rarity. A curious thing about the Herring, not known in any other species, is its ability to spawn at all seasons of the year and in fresh as well as salt water, thus under the most diverse conditions as to temperature and salinity. Its variability in this respect suggests that from the beginning it has been a shallow-water or brackish-water form; it can readily adapt itself to any conditions from 0 down to 100 fathoms.

There are winter Herrings which spawn in the Firth of Forth and in the west; spring Herring in Norway, in the Schley, and north of Scotland; summer Herring off the north-east coast; autumn Herring spawning in the southern North Sea, and so on. But the Herring keeps to the temperate zone; in the tropical and subtropical regions the variability has changed it into other species. Another interesting problem in distribution is furnished by its relative, the Anchovy. This is a southern, subtropical form, with its main spawning places in the Mediterranean and adjacent waters. Yet a northern branch appears regularly year by year, and that in great quantities, in the Zuidersee and neighbouring brackish waters. In southern regions the species spawns in water of comparatively high temperature and salinity; here the water is brackish and cold.

There is a wide gap between Spain and the Zuidersee; few specimens have been taken in the Channel or Bay of Biscay. Are we to suppose that it migrates this distance yearly, or that it remains somewhere in the Channel, where it eludes capture? In either case the origin of this branch remains a mystery; possibly it represents the original form of Anchovy, as we can more readily understand a migration to the pleasanter conditions of southern waters than the reverse. On the other hand, this particular branch may have suffered greatly from its enemies, Mackerel, Tunny, and others, and found a safe retreat from their pursuit in the Zuidersee. But no definite solution can be given to such a problem; of greater interest would be the investigation of the structural and developmental differences between forms of the same species growing up under such wide differences. The principal thing about the species so far mentioned is, that the migrations are from deeper to shallower water in the sea.

Some Clupeids, like the Shads (*C. finta*, *alosa*, *sapidissima*, etc.), go further and ascend river to spawn in fresh water. Some are permanent inhabitants of the inland seas (Caspian, North Italian Lakes, and Great Lakes of America). The same phenomena are found among the nearly allied Salmonoids. Whilst some forms, like Argentina, spawn in deep water in the open sea, and others, 'like Mallotus' (the Capelan), come close inshore to spawn, thus constituting a most important source of food, for example, for the Eskimos of Greenland, others like the Smelt enter brackish and fresh water, and the principal species of the group ascend rivers right to their sources, whilst whole families belong only to the fresh water.

The *Salmon*, it is now believed, does not move far from the mouth of the river to which it belongs; yet, like the Anchovy, it is rarely found in the sea except where it assembles to feed, as in the straits of the Pacific and in the Southern or South-eastern Baltic. The latter case is of near interest; the *Salmon* are taken on hooks, but some escape and are caught later, with the hooks still attached in the lakes of Finland. The *Salmon* is thus able to spend its whole life in fresh or nearly fresh water.

In the Rhine again, when it was a clean river, the winter *Salmon* used to remain nearly a whole year in the middle reaches below Basel before proceeding to spawn. It sounds almost incredible, therefore, that the *Salmon* does not feed after it has entered fresh water. Yet the evidence from all countries supports this view, and further, the digestive organs undergo a certain amount of degeneration. How then can we explain that the *Salmon* rises to the fly, and has even been taken with a Trout in its stomach? For one thing, it is certain that the *Salmon* of different regions, even different rivers, vary greatly in their powers and habits; for another, we have only to take a wider perspective to understand what happens.

In the larger Norwegian rivers the *Salmon* rise to the fly in the lower reaches, but provide no sport in the upper. In the still longer American waters the *Salmon* provide no sport at all. They feed, therefore, or endeavour to feed only when fresh from the sea. All the time they are in fresh water, and it may be a whole year, they live on the materials stored up in their body. This has been considered a wonderful provision or adaptation of nature. Regarded from the Salmon's point of view, it seems more like a tragedy; the *Salmon* wishes to feed but cannot! Certainly, its body must be an accumulator of energy of very good quality and enormous power. The fish ascends rivers against strongly running waters, jumps numerous falls (probably not more than 10 ft. at a single leap however); if a male it has to fight other males and keep off enemies, then help to prepare a bed in which the female lays its eggs.

It is not to be wondered at, when the spawning is over, that the fish is completely exhausted and allows itself to drift tail-first down the stream. The Atlantic salmons spawn mainly during November and December. They ascend the rivers as the breeding period approaches, travelling several thousand miles in sea and then inland. On entering freshwater they stop feeding and lose weight. Their bright silvery colour changes to a dull reddish-brown shade and the skin becomes thick and spongy. The body of the male salmon becomes spotted with red, orange and large black spots. After selecting suitable spawning grounds the fish segregate into pairs. Shallow saucer-like depressions are prepared in the river bed by the females where spawning takes place. After spawning, the return journey to the sea begins but few males survive to breed a second time. However, many females are able to reach the sea and start feeding. They soon recover their normal condition and silvery colour.

The *salmon* does not usually spawn more than three times in its life-span of eight or nine years. From the time it reaches freshwater till spawning, the salmon lives on its reserve food accumulated in its beop in the form of fat. The king *salmon* of the Pacific coast of North America travels about 2250 miles from the sea in order to reach the rivers to spawn. However, all of them die after spawning. No male or female survives to undertake the return journey. Studies have shown that young salmon hatched in freshwater cannot return to the sea until their salt-secreting cells have developed.

Studies indicate that migratory behaviour in both adults and the juveniles is not due to any special hormone regulation other than that which influences their general metabolism. Changes in their body states and behaviour may expose them to different environmental conditions to which they react in specific ways, such as travelling up freshwater streams by adults and downstream by juveniles. Several theories have been proposed to account for the unique way salmons are able to return to the very places where they were hatched. Some authorities believe that the fish have an extremely sensitive chemical sense that enable them to perform this feat. Another theory links the accelerated metabolism at spawning time with the need for more oxygen which increases as they ascend the hard waters of a stream.

Still another idea stresses the importance of a carbon dioxide gradient in the water as being the determining factor. Recent researches strongly favour the theory that sense of smell in fishes is the determining factor in locating home grounds, for different streams have different odours.

The *Hilsa* of the Indian Ocean also migrates to rivers for spawning. This species is known to ascend up the Ganges and go upto Varanasi and Allahabad. During the monsoon months *Hilsa* also migrates to many rivers of southern India. However, the details of its migratory habit are not known. On the whole, it may be said that the run of Salmon up the rivers is connected with spawning, yet some important exceptions have to be made. Many specimens will go some distance up a river only to turn back (and this may happen several times), or even remain months in definitely fresh water before proceeding to spawn. There may be a simple explanation of these eccentricities: it must be remembered that the Salmon is not far from being a freshwater fish.

According to Paton and Newbigin (1900) the factor determining migration from sea to river is not the "*nisus generativus*, but the state of nutrition." The spawning comes later. With the Salmon we may finish this account of the one stream of migratory fishes, those which journey from deeper to shallower and even fresh water in order to fulfil their life purpose. This is by far the most common form of migration; a very large number of marine fishes, even the Sharks, Skates, and Rays, follow in the main the same course. As suggested by Meek, the term "anadromous," formerly used only for those entering fresh water, may well be applied to all those fishes which follow the same general rule in their migration, namely, an ascent from deeper to shallower waters.

To a very large extent, this at once indicates the nature of the problem to be solved and even suggests the solution. But there are two groups of apparent exceptions. In the one the fishes exhibit but little tendency to undertake any spawning migrations. They spawn, as it were, where they find themselves. This applies naturally to the stationary fishes, though even among these a certain amount of migration can be noted. Thus, among the Blennies the Butter-fish or Gunnel has a short, anadromous migration of a few miles, whilst the Angler or Frog-fish shows a longer migration in the reverse direction. Among the Gadoids, again, most species prefer the deep water for spawning; but the Cod, if it has any preference, comes from the deep water to shallower banks. These groups, in which the migratory movements are of little account or in which different tendencies are shown by different species, may be called neutral for the movement.

The probability is, indeed, that they are not so much affected by the spawning process. The seeming migrations may be either a return to the natural homes places, following upon the drift of the young away from these, or they may be mixed up with the movements in search of food. The Cod, for example, is a rover with a circumpolar distribution like the Herring, and it shows a decided preference for the lattar as food. It pursues the latter at all seasons of the year and at the proper time it is able to spawn whenever it goes—within certain limits, of course, but they are very wide and correspond closely to those of the Herring. From the biological standpoint these neutral groups are of great importance in any theory regarding the causes of spawning migrations. We may take it, that the species composing them are so well-adapted to their surroundings that they are not urged to make any great change when the need of spawn comes upon them. Viewing their life-history as a whole, they show a better balance between the inner and outer conditions.

The second exception is represented by the Eel group. With the Common Flounder the Eel constitutes the old group of Katadromous fishes, those which migrate from the fresh water to spawn in the sea. The Flounder is certainly Katadromous like the Plaice seeks water of greater density than its ordinary feeding grounds for the purpose of spawning; but other Flat-fishes like the Soles show the reverse procedure, and it seems better to regard to group as neutral. There is in any case no comparison with the life-history and migrations of the Eel. The Eel seems to belong to quite different world from other fishes; it would be difficult to imagine a greater contrast.

Whilst other fishes are content with a reasonable area of distribution, thus admitting the limits of the piscine nature, the Eel ranges between the most extreme conditions on the globe, from mountain stream to the depths of the Atlantic. It passes through two transformations in the course of its life, when it enters and when it leaves fresh water, other fishes have only one or none at all. The Eel takes some ten to fourteen years (in fresh water, plus three more in salt water) to become mature, other fishes everywhere only one to three. Other Eels, Conger, Muraena, etc, do not show such a great contrast, so far as we yet know their life-histories. All seem to be Katadromous fishes, migrating out from shallow to deeper water, but only the *Anguilla* genus enters fresh water. Eels have two distinct phases in their life-history, each morphologically different from the other. Yellow eels represent the feeding and spawning phase while silver eels represent breeding phase.

Yellow eels are found both in salt and freshwater and vary in length from a few inches to five or six feet. During autumn a number of yellow eels become silvery and prepare to leave for spawning grounds. They stop feeding, their eyes become large, lips thinner, and the snouts sharper. Their colour becomes silvery along the sides with a blackish back. The reproductive organs develop while the alimentary canal shrinks. The eels migrate down the rivers to reach the sea. They travel about three to four thousand miles in the Atlantic to reach their breeding grounds in the western Atlantic, south of Bermuda. It is believed that the eels spawn at a depth of about 400 metres below the surface and the parents die soon after. The young larvae that hatch out are called leptocephali. They are flat, leaf like, tiny creatures provided with long needle-shaped teeth for feeding. They begin their homeward journey in the easterly direction. They grow rapidly during the first few months and when reach the coast of western Europe they are about three inches long and two years old. At this point they undergo metamorphosis and stop feeding. The needle-like teeth are lost and the transparent body assumes a cylindrical shape. They are now called elvers, or glass eels.

The elvers, when about three years old, start ascending the rivers in large numbers. They are able to cross all obstacles and reach their suitable resting places where they feed and grow for some years to develop into yellow eels. Finally on attaining maturity, they change into silver eels and start moving towards their breeding site in the sea like their parents. The American species, *Anguilla rostrata* also has similar breeding habits. In the western Atlantic larvae of both species are found living together. The interesting and mysterious aspects of this strange migratory phenomenon are, how the tiny elvers without any guidance are able to find their way across the sea, and why the two species which practically hatch in the same region do not get mixed.

The explanation lies in the fact that the American eel grows more rapidly (only in one year as against three years in the case of the European species). Thus if the larva of the European eel travels westward instead of an easterly direction, it will reach the coast of America long before it is ready to change into an elver. And conversely, if the larva of the American eel migrates in an easterly direction it will undergo metamorphosis in the middle of the Atlantic. In other words, the larval life in each case is geared to the distance it has to travel. A recent theory, however, suggests that both American and European eels belong to the same species and that the adult European forms die before reaching the spawning grounds. All eggs are laid by American eels but, when hatched, the larvae in the northern part of the spawning grounds are carried away by currents to Europe. Those in the southern part are carried towards the American coast. By a series of brilliant observations and deductions *Johs. Schmidt* has at length traced the Eel to its spawning places (1921), in an indirect yet quite convincing manner.

The post-larval Eels (Leptocephali) were found in ever increasing quantities the further he proceeded westwards, but they became gradually smaller. Successive lines or zones could thus be marked off, according to the size, until at length he found the smallest larva with the yolk-sac still attached and even the eggs out over the deepest parts of the Atlantic, near the West Indies and nearer the American than the European coasts. These were taken in the middle and upper layers, and there is therefore some doubt whether the adults actually spawn at the bottom. The depths on the "spawning grounds" were about 3000 fathoms. Two more of Schmidt's discoveries are of interest. At the same place where he found the larvae of the European Eel he also took those of the American Eel.

The spawning grounds of two forms, living thousands of miles apart, are practically the same, and the young of both species were occasionally taken in the same haul of the net. The second discovery is even more extraordinary. The two streams of post larvae spread out and move northwards towards the Bermudas; there they separate, one going east to Europe, the other west to America. The latter form spends only one year in the Atlantic and is then ready to enter fresh water; the European form takes three years to cross the Atlantic before it is ready to change its constitution and enter the rivers. How these fishes can settle which direction to take, even allowing for a difference in constitution, is a profound mystery. In illustration of the immense quantities of these young Leptocephali it may be mentioned that *Johs. Schmidt* took as many as eight hundred specimens in one haul out in mid-Atlantic.

Factors Affecting Migration

Many efforts have been made to connect the migrations of fishes with the external conditions, temperature, salinity, light, etc. and it is probable that each has some influence, but the problem has another side. In the first place, it is essential to distinguish the food concentrations and wanderings from the spawing migrations, a matter not so easy as at first appeared. It would seem, indeed, that a feeding concentration, in many cases if not at all, precedes the actual spawning, and it may be that the storing of food-stuffs leads directly to the internal changes in metabolism which give the stimulus to change of grounds and conditions (*Paton*, 1990). Physiologically the feeding and the migratory impulses are very different. In the one the fish is seeking to obtain something, in the other to get rid of something.

The latter may be called a morbid, even pathological condition, affecting every part and sufficiently powerful to make the fish ignore terrible injuries and even to overcome the impulse to feed. The digestive organs partially or wholly degenerate, and the fish is them urged to seek relief by unusual or extraordinary means, that is, to find if possible external conditions in better agreement with its changing or changed constitution. One cannot say definitely what happens in any particular case, owing mainly to the protein nature of fishes.

A Dog-fish, for example, is more in harmony with the surrounding medium than a Salmon, and a Stickleback will survive under conditions that kill most fishes. These phenomena depend apparently on their ability to alter the blood-pressure, or concentration of salts, as they pass from one set of conditions to another. And this ability, again, depends upon how far the gill-membrane and the skin permit of the passage of water outwards and inwards, a question that has not yet been sufficiently investigated. It is possible, therefore, that the internal stimulus to migration comes from some internal change in the blood; probably in blood-pressure. This is always much less than the pressure of the surrounding water, but it varies.

In the Salmon, *Greene* found that the concentration of salt in the blood was far below that of the salt water in which the fishes were feeding voraciously, and gradually altered as the fish entered fresh water. It may be that a slight reduction in the pressure within the blood is a source of discomfort to the fish and urges it to enter the rivers. In the case of Eel, on the other hand, there would seen to be a great increase of the internal pressure accompanied by an increase in the amount of carbonic acid at the time when the fishes are beginning to migrate from fresh to salt water. If they are detained on their journey in any way, the body thickness and swells and even increases greatly in weight, though the fish are apparently not feeding. *Petersen* has shown that the digestive tract is in process of degeneration at this time and even the air-bladder decreases in size.

The average Silver Eel in its breeding dress is about 70 cm. (27 to 28 ins.) in length and weighs about 1½ lb. The large specimens referred to were from 110 cm. and more, and weighed from 4 to over 8 lb. Most remarkable of all is the great increase in the size of the eyes. These become more than four times their former size and weigh twice as much. Without being able to say positively that such is the case, one may conclude that these phenomena point to a great increase in the internal concentration of salts and absorption of water. And this increased pressure with seems to be the cause of the Eel leaving the rivers to seek the greater pressures of the salt water. When the impulse to migrate comes upon them, the Eels are restless and ill at ease, moving up and down for a time in a manner quite unlike their usual sluggishness, then definitely turning down the rivers.

At every stage of their descent they come under increasing pressures without. And it may be recalled here, that the Eels are well-known to be extremely sensitive to the differences of pressure produced by thundery weather. In the case of marine fishes like the Plaice or Cod, we know that when the spawning fish are brought up in the trawl, the spawn continues to run even when the fishes are still alive. This indicates that the internal pressure persists for some time at the level of the deep water whence the fishes came. Since the low pressure inshore would help in getting rid of the spawn, the migration of the Plaice from shallow to deep water would be unnecessary for that purpose. Hence we may conclude, that an increase of pressure within has occurred with the maturation of the sexual organs, and the resulting discomfort has urged the fish to seek deeper water and greater pressures. Similarly in the case of other fishes. Each species is urged to seek the external conditions, whether of lower or greater pressure, which corresponds best to the change in its own constitution.

MIGRATION IN FISHES AND BIRDS

Though the migrations of fishes may appear the same superficially, they are widely different in important matters. They are not seasonal migrations in the same sense and the young take no part in them. Fishes have nothing to fear from the winter, not those we are dealing with at any rate; both the Salmon and the Herring spawn during the winter months. Further, the migration brings an advantage or benefit to the birds. On the one hand, they are urged on by the constitutional fever within; on the other, they are entitled forwards by what lies at the end of the journey. This is wanting in the fishes; the urge is there but not the inducement or reward.

If an Eel could foresee what lay before it in the depths of the Atlantic, would it undertake the journey. We may say, then, that the migrations of fishes cannot be regarded as adaptations, as we apply that term to birds. They are on a lower level, more primitive and crude. In all probability they are an inheritance from remote ancestors, which the fishes from lack of sufficient intelligence perhaps have not been able to overcome. Again, it should be noted that the birds do not show contradictory phenomena; the migrations are always to avoid the severity of the external conditions.

We might say this of the summer spawners among fishes; but all sorts and conditions are to be found, and the extreme cases, Salmon and Eel, seem willfully to be seeking the most dangerous and most deadly condition to be found on the globe. How far the fish in its migrations is guidéd in any way by its own sense organs, we cannot say. The external conditions—temperature, salinity, dissolved gases, etc.—all make for differences in pressure in diveree ways. Possibly, as Hofer has suggested, the extreme sensibility of the lateral line system to slight variations in the strength of currents, may enable the Salmon to detect the entrance of a river or side-stream. But on the whole, one may doubt whether the senses play much part in the directing of the spawning migrations.

INDEX